Practical Process Design for Chemical Engineers

Practical Process Design for Chemical Engineers

Keith Marchildon and David Mody

WILEY

Published by John Wiley & Sons, Inc., Hoboken, New Jersey.
Published simultaneously in Canada.

Library of Congress Cataloging-in-Publication Data is Applied for

Hardback ISBN: 9781394203840

Cover Design: Wiley
Cover Image: Courtesy of David Mody

Set in 9.5/12.5pt STIXTwoText by Straive, Chennai, India

SKY10093703_121224

Contents

Preface

Our book is intended to be significantly different from other process design books. The expected readership will include recent graduates from chemical engineering schools as well as engineers re-assigned to process design. The book will be shorter relative to some previous books. We believe this – along with the content – will recommend itself to selective and even complete reading. With regard to recent graduates, this book may serve as a "finishing school" to fill in concepts that the curriculum omitted or covered inadequately.

It is divided into two main sections: the first covering topics such as synthesis and process modelling, one of two essential prototyping (process verification) methods discussed in this book. Sage advice on prototyping using pilot plants is also provided. The process verification steps – from paper design to pilot plant, to semi-works, to commercial facility – to minimize the chance of project problems are discussed. In the chapter "Operating Problems: Solution by Design," ten common problems are identified that can crop up in an operating plant, and some means are proposed to circumvent them.

In the second section (Chapters 11 through 20), the selection of technologies for a design is approached by explaining the technology (e.g. heat exchangers) but by focusing on the process needs and relating those to technologies that can be utilized. An effort has been made to ferret out the whole range of technologies, some obscure, that are possibilities for each need. Furthermore, the book utilizes chemical engineering principles to help a designer prioritize the design needs.

In addition to the usual suspects of technologies used in the design of chemical processes – such as reactions, separations, heat transfer, mass transfer, and so forth – a variety of process needs not covered in many undergraduate programs or process design books are discussed, such as:

- Formation, Modification, and Transportation of Solids
- Mixing, Blending, and Agitation
- Integrated Process Operations.

Patents are rarely discussed, but they should be. This book discusses intellectual property for the process designer.

Keith Marchildon and David Mody

To Keith: Dedication and Foreword

Keith Marchildon didn't live long enough to see this joint book with his friend and colleague David Mody published, but it is the culmination of his life-long dream. From as far back as we, Keith's five children, can remember, writing a book had always been on Keith's list of things to do. However, this list was long, encompassing Keith's many responsibilities, ambitions, and interests. Keith placed a high value on family, community, and service to others, which had to be balanced with his professional responsibilities and academic pursuits.

In addition to raising the five of us with his wife Joan, Keith led boy scout troops, was an active member of his church, organized university reunions, taught engineering courses at Queen's University and elsewhere, created and presented courses on nylon polymerization fundamentals to his DuPont colleagues around the world, wrote and acted in skits for the annual Dupont Christmas party, and hosted grandchildren, cousins, DuPont and extended family at his cottage, all while becoming one of the foremost technical authorities at Dupont Canada and being promoted to DuPont fellow in 2000. From 2010, Keith was the primary caregiver for Joan on her journey with Alzheimer's, noting that for the first 40 years of their marriage she had cared for all of us, and now it was his turn to take care of her.

True to his engineering training, Keith believed that any problem could be resolved with a sufficient application of thought and analysis. Moreover, the solution to any problem starts with a consideration of first principles. He mulled over and researched issue after issue, including getting up to speed as much as necessary on new technologies. Nevertheless, he drew the line when it came to computer programming, insisting until the end on doing all his programming in Fortran. Getting the right answer was always more important to Keith than a deadline.

It was only after Keith was diagnosed with cancer in February 2019 at the age of 82 that the finiteness of his life – and the need to finish this book – came into stark view. Always deliberative, Keith joked that his terminal cancer diagnosis was the first deadline that he truly couldn't ignore.

Keith threw himself into the final stages of his book-writing the way he tackled most challenges in life, with an intensity and energy that doesn't come easy to many of us. Blessed with a steel trap memory and a boundless love for learning and teaching, Keith spent many hours bent over in complete stillness reading papers he had written or collected over the span of decades. Other times we would find Keith lying in bed, hands behind his head as he worked through mathematical problems in his mind, a look of pure contentment on his face.

Thankfully, Keith had more than two years after his diagnosis to collaborate with his friend and colleague, David Mody, on this book. David completed the remaining work after Keith died in July 2021. We are exceptionally grateful to David for bringing their joint project to fruition and fulfilling Keith's dream.

Keith's efforts in this book reflect the high priority he put on teaching and the transfer of knowledge to his kids, his students, and his colleagues. We hope that this book contributes to the practice of chemical engineering process design in a way that Keith had hoped.

April 2024 *Lori, Lee, Lynn, Bernie, and Marc*

Acknowledgments

Keith and I (Dave) have a long list of people whom we thank for contributing to the book. The book's origin dates back to professional development courses that Keith and other DuPont Canada engineers developed and taught. I think we owe a thank you to those individuals for the beginning of this endeavor. The notes for those courses turned into professional development courses that Keith and I taught in various locations across Canada. Each course we taught brought unique and interesting perspectives and questions from engineers across Canada, and we thank those participants for that. Keith taught many courses at DuPont, Invista, and Queen's and I know he would want to thank those individuals. David's work at Fluor Canada with the engineers, designers, and managers made possible his lifelong enjoyment of learning from others and teaching engineering ways of thinking and the skills needed to do design. Those technical and management leaders helped me to build a skill set and an enthusiasm for the creativity that underlies all of engineering. Several of those individuals have advised and contributed directly to this book, and I'm deeply grateful for their time and commitment to helping myself and the next generation of designers. I echo Keith's sentiments of never letting an opportunity to learn pass you by, and to that, we owe a lot to the many operations and maintenance people we worked with over the years.

In my second career, while working at Queen's University, I worked with so many bright, aspiring engineering students who, in many ways, contributed to this book through their enthusiasm, their questions, and their monumental efforts to become engineers. I never really considered myself to be teaching, as much as working with the students. In addition, a host of engineers and professors helped with the courses I mentored, and they refined my understanding and teaching of design, project management, and chemical engineering. I was blessed with the opportunity to see and understand engineering from a remarkable variety of industries. The names are numerous, but I would like to say thank you to all of them and to the ones who have answered my questions and provided feedback and suggestions, an even larger thank you! Your contributions are truly appreciated.

Keith was a devoted family person. We both owe a huge thank you to our wives and families for their unwavering support and love.

1

A Plan for Process Design

This book is about process design, the kind of design in which the knowledge and talents of a chemical engineer are significantly required. Design may include selection, modification, and even invention of equipment, as well as specification of operational sequences and conditions. Design always involves choices.

Engineers perform many functions, but for most engineers, the chance to be a designer is at the top of their ambition. Design is a chance to be creative, to integrate existing knowledge, and possibly the opportunity to be famous – doubly so if the design does not work!

A design assignment may be small – a modification to an existing process – or it may be a whole new plant. It may be a close duplicate of something else or a radical departure. However, the operative word is always "choice".

At the very outset, the activity of *process design* must be distinguished from that of *project implementation* or *project management*. Process design is part, but only a part, of a project. The project only ends when everything is up and running. The project leader – usually an engineer – has responsibility to see that all required engineering components are attended to, not just the chemical ones, that regulatory permits are obtained, that funding is obtained, and that the system is built and operational. The process designer is an integral part of the project implementation team. If the project is small, the process designer may be the leader of the implementation team. In any case, the designer needs to take this role as a member of the team seriously because there are no accolades if the project fails, even if the design was great.

To be specific then, this book is about process design. One chapter is included to better explain how projects are managed and how process design and project management mesh with each other.

The plan comprises the following three sections:

Principles of Project Design
Operations and Equipment
Appendices

We review each section here.

1.1 Principles of Process Design

In the 10 chapters of this section, the considerations that accompany all designs are presented.

Documents and Communications are given pride of place because the designer must get his or her plans into the hands of others who will take the next step, which is in the hands of the rest of the project team. There are formal ways of doing this, and they are explained.

Practical Process Design for Chemical Engineers, First Edition. Keith Marchildon and David Mody.

The mention of "choice" above implies that the designer must *Synthetize* the process. This is a creative operation with no step-by-step procedure. However, there are configurations and guidelines that chemical engineers know and which can help.

To back up the synthesis, the process designer engages in *Experimentation* and also in *Modeling*. The gold standard for experimentation is the pilot plant, which, if designed thoughtfully and the results interpreted properly, provides a huge help to the success of the commercial unit. But pilot plants are expensive, all-inclusive, and have issues with scale-up. Modeling of key elements can be cheaper, whether it is physical modeling, computational fluid modeling, or mathematical modeling of the transport and chemical phenomenon. Sometimes a simple homemade model suffices: Math modeling should be in every engineer's bag of skills.

Solution by Design addresses some of the untoward problems that often arise in the operation of an eventual process, such things as corrosion, interaction between vessels, etc. Ten items are listed and probably many more could be. Better to deal with them at the design phase rather than be surprised at the operational stage.

Process Monitoring and Control are integral parts of the process design. Control is obvious, although subject to various strategies to make it sharper and to various equipment choices. Monitoring is not so obvious, especially to management who are trying to cut costs. However, the conscientious process designer insists on heavy monitoring to assist the personnel who will subsequently operate and troubleshoot and modify the system. Who knows, it might be the designer themselves.

Safety and Health are paramount concerns these days for all industries, in spite of past transgressions over earlier decades and centuries. The chemical industries are highly regarded in their efforts to keep their workers safe, to make safe products, and to protect the integrity of their worksites and the people and land around them. There are formal procedures for making this happen.

Protecting the Environment is part of safety and health, namely the well-being of the world around the plant or process. It has risen over the years to be a major part of new industrial installations. It is highly regulated and rightly so.

Project Economics, Capital Cost Estimates, and Operating Cost estimates may seem like more of a task for the project manager or for management in general. However, there is no use building a wonderful process design if it is obvious that it will be a technical success but an economic failure. The designer may want to keep an eye on what others are doing with the cost figures to make sure that items are evaluated fairly and that costs for extraneous items are not being added on.

And finally, *Project Implementation* is addressed since the process designer is an integral part of the project team and may often find her or himself lying on the critical scheduling path.

Thus, ends the general considerations that all process designs must adhere to, to a greater or lesser extent.

We turn now to resources for designing specific processes.

1.2 Operations and Equipment

The approach here is to list according to process needs, rather than according to technologies. For instance, a design might require a miscible solution of two liquids to be separated. The obvious choice might be distillation, but there are other methods, and these will be described. In some cases, the technologies are described more than once because they serve more than one design need. Hopefully, a full palette of technologies is described, but of course, the genius of chemical engineering from time to time results in new and improved methods.

The sequence was meant to start at the plant door and to proceed stepwise through the plant. But plants do not operate in a linear fashion. However, the first of the eleven parts of this section does start at the beginning.

Bulk Shipping and Storage of Gases, Vapors, Liquids, and Solids are common to many operations, with a variety of modes depending on the phase and nature of the material.

In-Plant Transport of Liquids, Liquid–Solid Mixtures (Slurries), and Liquid–Gas Mixtures, with the combinations quite common but generally not taught in initial chemical engineering education.

In-Plant Transport and Compression of Gases, which because of variable density, are more complicated than other phases.

In-Plant Transport of Solids: Solids being very common in the chemical industry, yet their transport is not much touched on academically.

Formation and Modification of Solids are again not much studied in initial chemical engineering education but important because solids are more susceptible to modification than liquids or gases. Solids may be comminuted or built up, including by 3-D printing these days. Solids may be melted and recast, and they may be made into film or laminates, or spun into fibers, or converted into extremely tiny nanoparticles.

Energy Transfer is a very frequent accompaniment of other operations, i.e., in heating and cooling and changing phase of materials. This includes sensible effects as well as vaporization and condensation. Liquids, gases, and solids as well as combinations require various treatments.

Mixing, Blending, and Agitation are ubiquitous in chemical plants. Similar phases and mixed phases generally require different equipment. Material properties – e.g., the viscosity of a liquid – may call for radically different agitators.

Separation of materials is also found generally somewhere along a process. *Mechanical Separations* remove separate immiscible phases from one another using a surprising number of techniques suited to specific occasions.

Molecular Separations, by contrast, are used to break apart a single phase into more than one phase, with the object of having the different phases richer or poorer in various components.

Not every process design includes chemical reaction, but, of course, chemical engineers require the ability to design reactors. Many general *Chemical Reactors* are described.

For biochemical applications, there are additional restrictions, so a small treatment is provided of *Biochemical Reactors*.

One of the talked about topics of the day is process integration to reduce costs, increase efficiency, and reduce "footprint." We finish this section with a part on *Integrated Operations*.

2

Documentation and Communication

All communication is essentially good. For chemical and chemical engineering projects, a set of formal documents has evolved over the years, which is almost universally used. The production of some of these documents is the sole prerogative of the process designer or process design team. Others have shared responsibility with other members of the project team, including management, to a greater or lesser extent.

Here is a summary of the possible documents that can accompany a project. The list is arranged chronologically. The names may differ from organization to organization, but the content and sequence are universal.

1. Basic data – responsibility of management sponsor, laying out the goals and constraints of the projects.
2. Process flow diagram (PFD) and the heat and material balance – graphically displaying the result of process synthesis, showing the flow and properties of internal and external streams. An accompanying process description document may also be provided for the purpose of clarity.
3. Equipment list – evolving as equipment becomes better defined, providing key parameters that can be used for an initial and ongoing estimate of capital cost, and providing key information for other disciplines such as weights, size, and so on for preliminary design and estimating.
4. Piping and instrumentation diagram (P&ID) – showing interconnections and interactions between vessels and the required instrumentation hardware and controllers.
5. Equipment data sheets – showing details, as they are calculated, about each major piece of equipment; added by other specialists and vendors so that quotes can be provided.
6. Instrument data sheets.
7. Functional specifications for distributed control system (DCS) to augment the P&IDs.
8. Scope of work and cost estimation.
9. Notes from process hazard review.
10. Input to applications for environmental permits.
11. After the project has been approved and before plant start-up, training manual and operating instructions.
12. Maintenance instructions.
13. Record of design calculations for future troubleshooting and modification of the system.

This is a formidable list and does not apply in full to every project. For a large project, several people may be involved in creating these documents. Of course, if management eventually decides

Practical Process Design for Chemical Engineers, First Edition. Keith Marchildon and David Mody.

against proceeding with the project, then some documents (e.g. operating instructions) would be curtailed or never be written.

Here is an overview of the general content of each document.

2.1 Basic Data

Projects always start with a sponsor, i.e., someone who can authorize the money. That person or persons may be wishing to make a new product and will convey in the basic data document the production rate and characteristics (e.g. purity) of that product. The sponsor may be a government agency and perhaps wish to build a facility to deal with a waste stream. Again, the rate is given, and, in this case, the purity of the stream leaving the facility is specified. In particular cases, there may be additional requirements and constraints. The key is to ensure that the process designer and the project team deliver what the sponsor asks for. The document also serves as a reference point for any subsequent negotiation between the sponsor and the team as the details of the project start to emerge.

Sponsors and designers will engage in producing a workable document and may wish to refer to writings on this subject, such as Read (2000), Manganaro (2002), Ainsworth and Brocklebank (2003), Pavone (2006), Buckbee (2010), Lagace (2011), Ogle and Carpenter (2014), and Toghraei (2015).

2.2 Process Flow Diagram (PFD)

The PFD can be drawn for any process but is most commonly illustrated for continuous processes. It consists of a sheet on which schematics of the major vessels are located. The interconnections between the vessels are shown with arrows. Every flow stream is indicated with a number. The streams are listed in a table outside the diagram, showing their properties, e.g., phase, flow rate of each of the components in the stream ($kg\,s^{-1}$), temperature, pressure, and physical properties. In some jurisdictions, this information is listed on the diagram itself, but this can introduce clutter and make it difficult to list extra information about the streams, so it is simplest to have it separated in a table. The details of the flows, in the initial version of the PFD, may be simply what the process designer wishes them to be, with no reference to the actual equipment that will produce these flows. Design of the equipment can come later. Nasby (2012) saw PFDs as communication tools.

The example shown in Figure 2.1 is a chemical reactor feeding into the side of a distillation column, with recycle from the top of the distillation column back to near the entrance point of the reactor. Near the top of the column, there is a waste or purge stream. At the bottom of the column, almost pure product B emerges ready for sale. This is a very common scenario in chemical process systems, which deals with the situation where one possibility is that the reaction of A to B approaches equilibrium. Another is that B has a tendency to degrade into C and B is removed as quickly as possible. The seven streams are numbered as follows:

1. raw material feed stream,
2. the exit from the reactor,
3. the recycle return flow to the reactor,
4. the pure B product stream from the bottom of the distillation column, and
5. 6 and 7 – the streams around the C-purge unit.

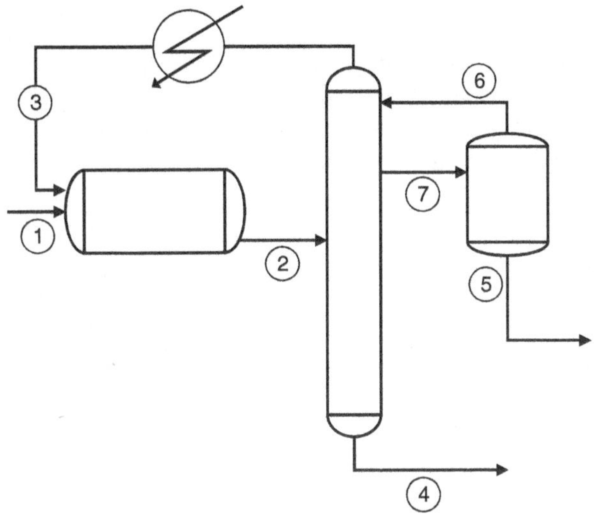

Stream	1	2	3	4	5	6	7
Flow	100.0	250.0	150.0	96.0	4.0	60.0	64.0
A	100.0	150.0	149.0	0.0	1.0	59.0	60.0
B	0.0	96.0	0.0	96.0	0.0	0.0	0.0
C	0.0	4.0	1.0	0.0	3.00	1.0	4.0

Figure 2.1 Reactor with recycle.

This is the way the process is intended to operate and for which the designer has specified the size, geometry, and operating conditions. Subsequently, the details of the vessels will be worked out and will appear in the equipment data sheets. A process description similar to the description above is a useful document for communication to other members of the team.

For simplicity in this example (Figure 2.1), all components A, B, and C have the same molecular weights. The order of volatilities is "A" (most volatile), "C", and "B".

The flow data are useful to other members of the team, particularly the pipe designers who need to know the flow and properties of the stream and any corrosion potential. The principal vessels on the chart may have names attached to them, but in many cases, this is not done to avoid giving a competitor information in case the sheet falls into the wrong hands. PFDs are generally quite guarded by the enterprise and released to other employees only with a need to know. For aesthetic purposes, the vessels are usually shown with rounded corners rather than squared corners.

The choice of flows for this process is dictated by the amount of product that the manufacturer wants this process to yield. Some of the vessels, when eventually chosen, would be quite capable of outperforming their rate in this particular flow sheet. All of the flows have to be consistent with one another and one way would be to count the atomic species in the streams in the PFD. The numbers in the flow table should be checked for consistency with the reaction kinetics or conversions and the vessel sizes.

The flow table can and should display more data than shown, especially the phases of the stream. In this case, only stream **6** and stream **3** before the condenser are vapor; all the rest is liquid.

2.2.1 PFD for Batch Processes

The continuous version of the PFD does not lend itself to batch processes because the flow streams are not constant. There is, in fact, no hard and fast rule that diagrams have to appear on a flow sheet; however, they do assist engineering personnel in interpreting the sheet for any process. In batch processing, there may be a single vessel which undergoes various cycles or there could be two or more vessels all of which need to be described.

Consider the example of a single vessel undergoing four cycles of operation from start to finish (Figure 2.2). In the first cycle, the pressure rises autogenously with temperature. In the second cycle, the pressure is kept constant by virtue of the release of volatile material from the vessel. In the third cycle, the pressure is reduced. In the fourth cycle, the material is expelled from the vessel. A single image of the vessel would suffice, but an alternative is to show an image for each cycle since the valves or pumps can be shown as on or off for each cycle. In this case, four copies of the vessel could be printed on the process flow sheet. The figure shows two of the cycles, with the time-wise progression of the variables during the initial pressurization and then during the emptying.

Sometimes a batch process uses an extra feed somewhere along its process path. This is known as a *batch-fed* process. A particular cycle in which this feed is introduced is separated into two cycles. The additional feed also has a time line for the time it takes to be introduced: its time, pressure, temperature, flow, and composition are listed out in the same manner as the actual emissions of the main process vessel. Note that this is an addition rather than an emission. Once again, the full rate out of the vessel is useful to the piping designer. In these situations, it is useful to document the instantaneous flow rate for sizing pipes, etc., and the time average flow rate for material consumption and yearly production.

In some cases, one batch vessel feeds to another. In that case, the above treatment is required, again with data along horizontal time lines. The time lines have to be coordinated between the two vessels. In multiproduct batch plants, vessels may have to be shown many times because their cycles are specific to given products. In such an arrangement, if common pipes are employed for

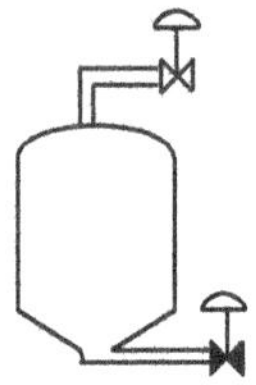

Time	min.	0	5	10	15	20	25
Pres	kPa	6000	6500	8500	11 000	14 000	17 000
Temp	°C	150	155	160	170	185	210
Topflow	kg min^{-1}	5	5	10	10	15	20
Botflo	kg min^{-1}	0	0	0	0	0	0

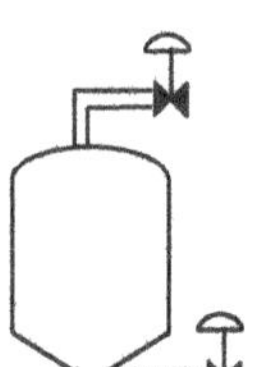

Time	min.	0	2.5	5.0	7.5	10	
Pres	kPa	10 000	9000	8000	7000	6000	
Temp	°C	270	270	270	270	270	
Topflow	kg min^{-1}	0	0	0	0	0	
Botflo	kg min^{-1}	120	100	100	80	0	
Batchwt	kg	1000	700	450	200	0	

Figure 2.2 Batch process material balance.

different flows, then the diagram requires more imagination. Creating a PFD for a batch process is discussed by Cooper and Moore (2013).

* * * * * *

For both continuous and batch processes, the PFD is a living document. It begins with sketching the proposed equipment items but only indicating their function and their interconnectivity. The flows and properties of the streams are then added, but these are what is intended by the process designer based on the chemistry and phase behavior of the materials. The design of equipment follows, in order to effectuate the desired material transformations. The process flow sheet will live on to assist future workers in understanding the process. The calculations which led to the PFD and its development need to be available to future workers. We have not discussed simulators and optimization at this point, and, in some cases, the final PFD is not complete until those tools and processes have been utilized.

2.3 Equipment List

Although the equipment has not been fully designed at this point, the nature and size of the vessels and major auxiliary equipment may be approximately determined. Assembling the equipment list and applying well-known cost factors allow an initial estimate of capital cost. This is an early step in the ongoing communication among the project team and the sponsor. Toghraei (2014A) acknowledged that equipment is designed with margins, but these margins must be thought through and managed.

2.4 Piping and Instrumentation Diagram (P&ID)

The P&ID is actually more recognized among project managers and engineers than the PFD. This is because, when fully completed, it shows essentially every piece of equipment in the system including pipe sizes, reducers, expanders, valves, strainers, sample points, points of measurement, and points of control action along with all of the main vessels in the form of simple blocks. Like the PFD, the P&ID is a living document. It begins with the process engineer adding essential controls, and, eventually, the diameters, wall thickness, length, and code of every pipe are included. In a large corporation, the piping would be handled by a piping specialist. In a smaller operation, it may be handled by the process designer. This is the diagram to which the operators and maintenance personnel refer. In many cases, the P&ID is taped to the wall in the control room for ready reference. For a P&ID on multiple sheets, it is highly convenient to match up the outflows from one P&ID to the inflows of the subsequent P&ID (Figure 2.3).

At some point, the process designer must specify the process monitoring and control equipment and add these details into the P&ID. Essentially, this means, for *monitoring*, points of measurement of temperature, pressure, level, flow, and possibly concentration. It also includes sample points. For *control*, some of these points of measurement are included as parts of feedback loops that activate final control elements such as valve position, pump speed, and energy flow to heat-transfer devices. The designation of these loops forms part of the P&ID. As part of the initial calculation of cost, there is usually an average cost-per-loop which can be used for initial estimates. Rinker (2017) provided a valuable compendium of symbols for piping and instrumentation and also for valves. The Instrument Society of America (ISA 5.1) provides a standard of instrument symbols for P&IDs.

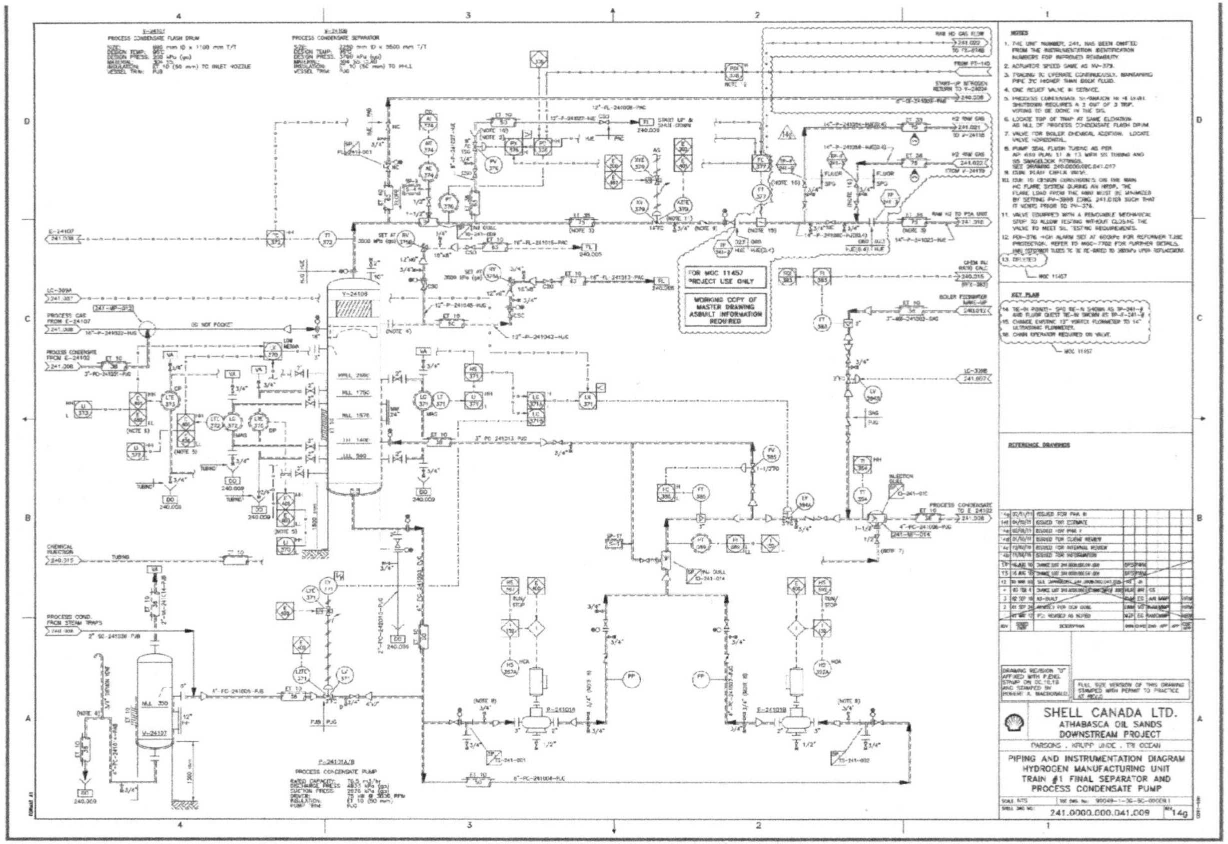

Figure 2.3 P&ID ready for construction. Source: Used under permission by the Alberta Government.

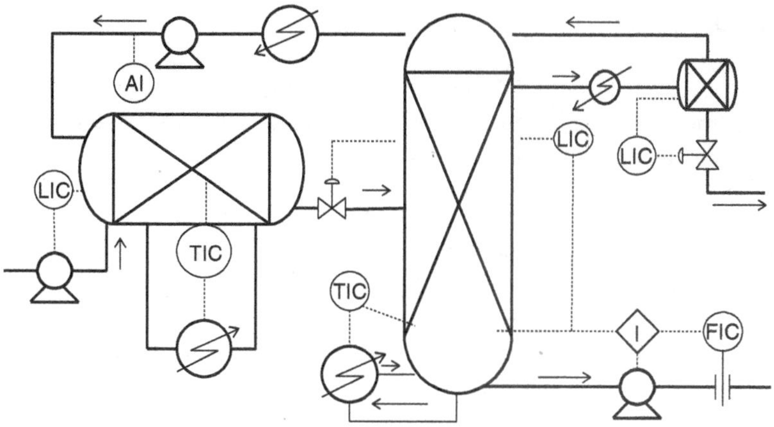

Figure 2.4 First stage piping and instrumentation diagram for the recycle reactor.

Of course, an enterprise often has its own symbols, which should be used. Shah (2020A, 2020B) spoke about P&IDs. Toghraei (2014B) provided principles of P&ID development.

A PFD with controls is the first step in the transition from PFD to P&ID. Some aspects of P&IDs are illustrated in Figure 2.4 for the reactor-with-recycle example for which the PFD is shown in Figure 2.1. Starting in the lower right-hand corner, it is seen that the product is desired to be delivered at a constant rate, with the flow indicator and controller (FIC) measuring the flow, comparing it with the set point, and manipulating the pump to deliver the desired product rate. Heat is provided at the bottom of the distillation/separation column, and the heating rate is manipulated to maintain the desired bottom temperature – that temperature indicates the composition of the bottom product. This is the function of the temperature indicator and controller (TIC). The bottom liquid level is also measured (level indicating controller (LIC)) and is controlled by manipulating the flow from the reactor into the column.

If some upstream malfunction were to cause the liquid level at the bottom of the column to fall below its set point and possibly to zero, the outlet pump would run dry, with damaging effects to itself. For this reason, the level control measurement connects to an interlock ("I"), which shuts off the pump regardless of the signal from the FIC (which would have speeded up the pump if it detected reduced flow).

The purge of component "**C**" near the top of the column is a small vapor stream which gets partially condensed with the liquid running off to waste. The level in the separation tank is controlled by manipulating the outlet valve.

The recycle line contains vapor from the purge separator which is condensed and fed to the reactor. An analytical detector (Analyzer Indicator (AI)) for component "C" could be installed in the recycle line since it is desired to keep the impurity from building up in the system. This detection could consist of just a sample point, where occasional samples are taken and transmitted to an online analyzer or to the lab for analysis. The fresh feed to the reactor is manipulated in order to control level. The temperature is maintained by external heating. The reader is invited to consider what equipment or controls would be needed to provide liquid reflux for the column, and what would be required to handle an impurity with a lower volatility than component A.

Usually, there are more than one P&ID. The very practical custom is to make the piping on one drawing line up with the same piping on the next drawing. The P&IDs are often posted, side by side around the walls of the control room.

2.5 Equipment Data Sheets

The process designer begins the specification of the equipment which is going to be used to carry out the process. Referring to the example in Figure 2.1, the major items will include (i) the size and internals of the reactor, (ii) the dimensions and tray schedule of the distillation column, (iii) the dimensions and cooling requirements of the "C" separator, (iv) the duty and type of the heat exchanger for condensing the recycle flow, and (v) the characteristics of the two pumps. Other specialists may examine the vessels and auxiliaries and decide on changes. Mechanical specialists will pronounce on the wall thicknesses and the material of construction, and the company that supplies the distillation column will likely make some changes. One of the key considerations is the actual size of the two main vessels. Because it is a split process, the duty could be varied between the two vessels within certain limits. Hence, if the shop in which, say, the reactor is being built has space only up to a given dimension, then there is no use specifying a reactor that will exceed that space. It can be made somewhat smaller and a greater load shifted over to the distillation column. If, as part of this project, a mathematical model has been built, then it could be used to optimize the whole process, relative to temperature, pressure, and reactor-column split. At this stage, it is appropriate to revisit the PFD and modify it in accordance with the performance of the actual equipment. As stated previously, the PFD, the P&ID, and the equipment data sheets are "living documents". Shelley (2008) advised how to streamline data sheets. Toghraei (2013A) stressed data sheets as the key document.

Ultimately, the equipment data sheets have been so augmented and put into the final comprehensive form that a vendor or manufacturer can quote a price.

2.6 Monitoring and Control Data Sheets

The process designer will have added the process monitoring and control equipment into the P&ID. Most of the control loops are SISO – single input single output. Some control schemes are more complicated. Every controlled variable and every manipulated variable has an expected range. Values outside this range must trigger a visual and/or audio indication and, at excessive deviation, an interlock on some process variable – typically a pump or valve. The PMC data sheets capture all of these details.

2.7 Functional Specification for Distributed Control System (DCS)

In large plants, the monitoring and control are generally part of a much larger entity called a DCS. The process designer has a significant responsibility for writing specifications to communicate with vendors of DCSs. These systems not only handle the final control loops in the process but also provide internal calculations, graphic displays, historians, and recipe handling. The details of the functional specification are the shared responsibility of the designer and manufacturing. This is a task of considerable effort. As early as possible, the vendor needs to provide a quote. Deitz et al. (2003) offered help in this area.

2.8 Scope of Work

This document is the culmination of the foregoing efforts. Although not all of the above work has been done, there is enough for an estimate of the time, effort, and money to bring the plant or process to physical completion. The process designer is expected to contribute, but usually engineering planners and estimators take the lead. Management expects the cost to be estimated to within some limit – even as small as 10%. Maas (2005) described a very full content of the scope of work. Uppal (2002) urged as much upstream work to have been done as possible, in order for a scope of work to yield an accurate capital cost. Planners and estimators will object if they perceive uncertainty in the specifications of the process designer. Their reputation depends on how well their report agrees with eventual reality. Rosentrator (2001) divided the scope of work into "preliminary" and "final", which gives some space for reconciliation.

2.9 Notes from Process Hazard Reviews

As the design becomes more fixed, and particularly as the PFD and P&ID become more complete and comprehensive, a review team is set up to seek out possible malfunctions and safety hazards. The team meets for a few hours, possibly more than once, and notes are taken. The process designer will obviously be part of this team and may likely be the notetaker. He or she then has to follow up on the team recommendations: either to satisfy that some point that was raised is not in fact a hazard or that a point requires action because it was overlooked in the design so far. All of this needs to be documented and followed up on, for the following reasons:

- so that no one gets hurt,
- so that the plant does not malfunction, and
- so that there is evidence of due diligence if something untoward still occurs.

2.10 Input to Applications for Environmental Approval

All plants create some change to the surrounding land, water, and air. Governments have strict rules on the form and extent of these disturbances. In the initial design, these restrictions will have been honored. However, whatever emissions still remain must be approved by the regulatory authorities. Quantitative data must accompany applications, and the process designer is in the best position to provide this information. Deakins (2010) provided advice on generating regulatory documents.

2.11 Operating Instructions

Providing detailed instructions for starting up, operating, and shutting down the plant is a major task. A great deal of thought is required because there may be more than one possible procedure in some cases. Usually, more than one person handles the task because there is often debate. This document is often conjoined with the process hazard review. By this time, the elements (instrument numbers, equipment numbers, and pipe or 'line' numbers) in the P&ID need to have numbers assigned to them, in order to tie in with the instructions. At some point, there should be training sessions for the operators and first-line supervision. Gentilcore (2001) assisted here.

2.12 Maintenance Instructions

Long-term viability requires inspection and repairs over time. The process designer can help by, at least, advising on setting up a maintenance program to anticipate and ward off process and equipment problems.

2.13 Record of Design Calculations

At some time in the future, the plant will likely be required to increase production or move to an altered product. The original process designer may still be available to lead or assist the changes, but more likely this will not be the case. Then, it will be important for his or her successor to understand how the existing plant was conceived. Design calculations done by hand should be scanned into the computer, and any mathematical model input and output should be preserved – along with the model itself. These records show the existing constraints and show where selected changes to the equipment or flow path can bring about the desired upgrades. The extent and quality of the records depend, of course, on the conscientious of the original process designer or design team, on the willingness of management to devote time for this post-project step, and on the records-management system in use.

One other communication is the *Cost Sheet*, which sums up the cost per unit mass of the product, including entries for labor, supervision, materials, energy, and depreciation. This document is the summation of the total plant design and of its fairly close cost estimates, and it indicates what the product will have to sell for in order to make a profit. The process designer will have started the progress toward this ultimate document. This document is described in Chapter 9, Economics.

2.14 People Communications

Every project in which an engineer or other designer engages produces an impression in the mind of management. It may be a feeling of a job well done or it may be a feeling of incompetence. Futures depend on it. Here are some references on how to make the most of one's efforts. Often just the quality of writing and presentations is enough to turn heads.

Anthony J. (2001) on calculational style; Halle (2002) technical style; Bly (2003A) on presentations; Bly (2003B) on reading speed; Chopey (2003) twelve tips for clearer writing; Roberts (2006) optimizing presentation of numbers; Greer (2008) carrying lessons learned from one project to the next ones; Driskill (2010) optimizing the impact of a poster at a conference; Civan (2013) units of measurement; Toghraei (2013B) calculations in process engineering; Sofronas (2014) writing concisely; Conley (2015) talking to and convincing management (i.e. making a pitch); Kasprzak (2019) speak so your boss will listen. The article on flowcharts in Wikipedia is also helpful in process design.

References

Ainsworth, D. and Brocklebank, M. (2003 Jul). Multiproduct plant design. *Chem. Eng.* 110 (7): 42–49.

Anthony, J. (2001 Nov). Elements of calculational style. *Chem. Eng. Prog.* 98 (11): 50–55.

Bly, R.W. (2003 Jan). Give memorable presentations. *Chem. Eng. Prog.* 99 (1): 84–87.

Buckbee, G. (2010 May). Closing the gap between engineers and management. *Chem. Eng. Prog.* 106 (5): 38–41.

Chopey, N.P. (2003 Jul). Twelve tips for clearer writing. *Chem. Eng.* 110 (7): 73–75.

Civan, F. (2013 Feb). Avoid problems with units of measurement. *Chem. Eng. Prog.* 109 (2): 43–49.

Conley, R. (2015 Jan). DCS migration: lessons learned. *Hyd. Proc* 94 (1): 75–80.

Cooper, T. and Moore, C. (2013 Aug). Use process modeling to improve quality. *Chem. Eng. Prog.* 109 (8): 49–52.

Deakins, J. (2010 Aug). Foolproofing regulatory documentation. *Chem. Eng.* 117 (8): 40–41.
Deitz, C., Ham, T., and Murray, S. (2003 Aug). Writing a functional specification for an S88 batch project. *Chem. Eng.* 110 (8): 46–52.
Driskill, L.P. (2010 Aug). Optimize your conference poster's impact. *Chem. Eng. Prog.* 106 (8): 43–48.
Gentilcore, M.J. (2001 Feb). Take the tedium out of writing process descriptions. *Chem. Eng. Prog.* 97 (2): 31–36.
Greer, S. (2008 Aug). A lessons-learned knowledge management system for engineers. *Chem. Eng.* 115 (8): 50–52.
Halle, J.M. (2002 Jun). Technical style. *Chem. Eng. Prog.* 98 (6): 82. Macatea Productions, Central South Carolina.
Kasprzak, L. (2019 Jun). Speak so your boss will listen. *Chem. Eng. Prog.* 115 (6): 35.
Lagace, J. (2011 Mar). Clarifying the role of the lead process engineer. *Chem. Eng.* 118 (3): 44–50.
Maas, S.W. (2005 Nov). Prepare a top-notch scope of work. *Chem. Eng. Prog.* 101 (11): 29–32.
Manganaro, J.L. (2002 Aug). Estimate the capacity of simple batch processes. *Chem. Eng. Prog.* 98 (8): 70–75.
Nasby, G. (2012 Oct). Using process flowsheets as communication tools. *Chem.Eng. Prog.* 108 (10): 36–44.
Ogle, R.E. and Carpenter, A.R. (2014 Aug). Calculating the capacity of chemical plants. *Chem. Eng. Prog.* 110 (8): 59–63.
Pavone, A. (2006 Jun). How to prepare a process design basis. *Chem. Eng.* 113 (6): 48–53.
Read, C. (2000 Jan). Start to improve batch operations. *Chem. Eng. Prog.* 96 (1): 69–74.
Rinker, J. (2017 Dec). P&IDs (piping & instrumentation diagrams) and P&ID valve symbol library. *Assured Automation.com* 21: 1–12.
Roberts, B.E. (2006 Oct). Optimize your presentation of numbers. *Chem. Eng.* 113 (10): 59–61.
Rosentrator, G. (2001 Dec). Preliminary and final engineering scopes of work. *Chem. Eng. Prog.* 97105 (12): 74–77.
Shah, B. (2020A Apr). Piping and instrumentation diagrams (P&IDs): part 1 – origin and evolution. *Hyd. Proc.* 99 (4): 65–69.
Shah, B. (2020B Jun). Piping and instrumentation diagrams (P&IDs): part 2 – causes and management of change. *Hyd. Proc.* 99 (6): 53–56.
Shelley, S. (2008 Jan). Online configurations: streamlining equipment specification. *Chem. Eng. Prog.* 104 (1): 6–9.
Silbert M. D., Tips to improve your PM (project management), l'Actuality Chimique Canadienne 2005 Avril, page 8, 2005
Sofronas, A. (2014 Apr). Case 78: writing concisely is a definite work skill. *Hyd. Proc.* 93 (4): 39.
The Royal Bank of Canada, The Royal Bank of Canada Newsletter, 1975 Aug. https://www.rbc.com/en/wp-content/uploads/sites/4/2022/07/august1975.pdf.
Toghraei, M. (2013A Jul). Getting the most out of data sheets. *Chem. Eng.* 120 (7): 44–46.
Toghraei, M. (2013B Dec). Calculations in process engineering. *Chem. Eng.* 10 (12): 51–53.
Toghraei, M. (2014A Jan). Wide design margins do not improve engineering. *Hyd. Proc.* 93 (1): 69–71.
Toghraei, M. (2014B Apr). Principles of P&ID development. *Chem. Eng.* 121 (4): 62–71.
Toghraei, M. (2015 Sep). Improving the operability of process plants. *Chem. Eng.* 123 (9): 72–78.
Uppal, K.B. (2002 Sep). Project cost estimation: scope of work is vital. *Chem. Eng.* 109 (10): 72–76.
Wikipedia, Flowchart, 2024. Retrieved from the web at https://en.wikipedia.org/wiki/Flowchart.

3

Introduction to Synthesis

Synthesis, as defined in the dictionary, is a combination of ideas to form a system. For a chemical engineering process designer, that system is the process and those ideas are the thinking which is put into the design.

The process can have a range of scales. At maximum, it may be for a totally new product of which there is no precedent for its manufacture. At minimum, it may be a minor improvement in part of an existing facility. In all cases, the design of the process is a synthesis.

The idea of carrying out a synthesis in a systematic manner has been pursued by many workers with the objective of putting synthesis onto the same step-by-step basis as the design of unit operations or standard chemical reactors. These efforts have been very fruitful in the area of heat integration, but, in general, they are successful only for design problems that are much better posed or defined than is commonly the case in practice. If process synthesis could truly be done in a direct linear fashion, then there would not be the many patents that are taken out every year on "novel" chemical processes. Process synthesis is more akin to the work of an artist who, while drawing on common principles of technique and using tools that are available to all, uses his or her experience and inner imagination to create an original work. The objective of the present chapter is to present to the "artist" who creates processes, some insights that augment experience and fire imagination.

A caution to the designer with an *oversupply* of "artistic" ambitions: There are often existing examples of successful designs for facilities somewhat similar to the present objective. They may be found within the company or in the literature, especially in patents. There may also be licensing opportunities. The designer will be criticized for failing to make a survey before launching into his or her creative phase. There will still be opportunity within the design for ingenuity and the choosing of options.

The present book, and especially Part II, Chapters 10–20, is about synthesis. Each of these chapters takes up a specific task in chemical processes. The 11 chapters are structured to provide options for meeting the physical and chemical requirements that arise as the designer puts together the concept for the process. For instance, at some point in the process, it may be necessary to separate the components of a fluid in which the components are intermingled at the molecular level: Chapter 17, Molecular Separations, describes possible methods including, hopefully, the method most appropriate for the particular fluid and components.

In this chapter, a higher-level, more general examination is made of the synthesis task. This introduction to synthesis generalizes about process configurations, e.g., batch and continuous, plug flow and mixed flow, and countercurrent and cocurrent – all of which are central to the way in which chemical engineers set up material transfer and material transformation.

Practical Process Design for Chemical Engineers, First Edition. Keith Marchildon and David Mody.

3.1 Economic Basis of Synthesis

The goal of a process is usually but not necessarily to make money for a company: Sometimes it is a public work like a waste-treatment system. However, if the objective of the design is, as it usually is, to maximize profit for the company, then it turns on the twin desires to minimize capital cost and operating cost per unit of production. Using standard methods of accounting, these two costs are brought into a single quantity (i.e. a single "objective function") such as net present value, return on investment, or cost of manufacture. In the latter case, the capital cost is converted to a component of the operating cost by assigning a portion of it (e.g. 10%) as an annual depreciation or write-off. Then, it can be combined with and compared with the other operating costs, which are mainly

- raw materials,
- energy and other services,
- human resources,
- maintenance, and
- waste disposal,

in order to arrive at the total cost of manufacture. With this scheme in place, quantitative judgments can be made on such trade-offs or choices as

- adding equipment (capital cost) to capture process heat and reduce energy consumption (operating cost),
- using energy to power purification columns that increase yield from raw materials – i.e., increasing one operating cost to reduce another capital or operating cost,
- automating to reduce the number of operating personnel, and
- increasing the vessel size and hold-up time to allow a decrease in the reactor temperature that lessens waste production, or to reduce operating cost in a separation and recycle system.

Implicit in these considerations is that the facility will operate at a high percentage of the time, typically at more than 8000 hours out of the annual 8766. If there is significant "down time", then the plant will not return the expected profit. Since many costs such as depreciation, heating and ventilating, and operating salaries are at least partially independent of the production rate, the cost of manufacture per unit of product goes up. To help ensure that the process will stay largely "on stream", the process designer needs to champion the following inclusions:

1. adequate process monitoring and sampling, for early detection and diagnosis of problems,
2. storage capacity for raw materials, product, and intermediate streams, in order to buy time and keep the plant operating if there is a difficulty at one point,
3. redundancy of ancillary equipment such as pumps, and
4. ability to handle a range of throughputs, below and above the flow sheet values.

The boundaries that are set around the synthesis assignment are typically

- required production rate,
- required product quality,
- unit cost for raw materials and for services,
- raw material characteristics, and
- environmental regulations.

The process designer has little or no control over these parameters but should be aware that some of them may change in the future. Most commonly, if the process and product are successful,

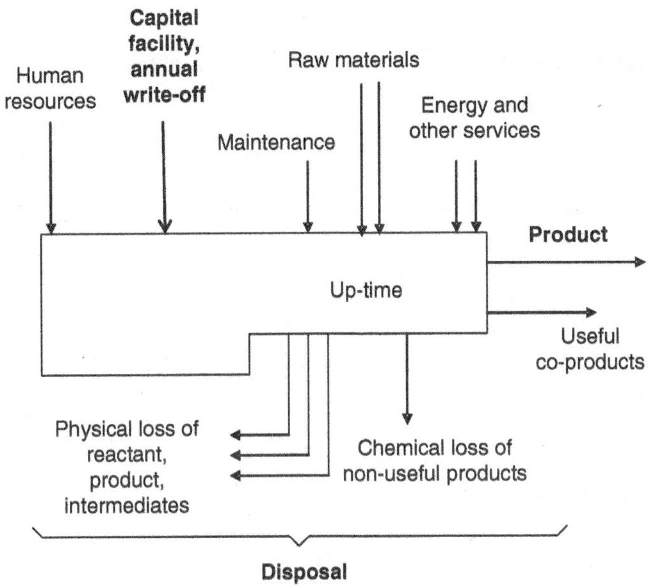

Figure 3.1 Cash-carrying streams in a chemical process.

then the business may begin requesting higher rates of production. It would be unfortunate if the plant immediately had to have an expensive replacement of equipment because everything had been designed at its very limit. Pumps and piping can usually be considerably oversized, and operating vessels can be sized so that the initial throughput falls in the middle of their operating range. The extra money which these provisions cost is quantifiable, and the extra expenditure should be strongly proposed to the project team and to business management.

Figure 3.1 shows the cost-bearing and revenue-making streams in a chemical operation.

3.2 The Rate Concept

It can be said that the *motivation* for process synthesis is “all about money”, and it can equally well be said that the *execution* of process synthesis is “all about rates”. Taking first a “macroscopic” view of the process, it is seen from Figure 3.1 that there are good rates and bad rates: Good are the rates of production of product and useful coproducts; bad are all the other rates, which are associated with costs. The goal is to maximize the good and minimize the bad in such a way as to maximize the return on investment or minimize the cost of manufacture.

Achieving this happy state requires a “microscopic” view of the process, coming to grips with all the interior rates. For the most part, they are the chemical engineering rates of momentum transfer, heat transfer, mass transfer, chemical reaction, and biological transformation. There are others: rates of corrosion, rate of supplier delivery, fluctuating rate of sales, etc., which will be examined in other chapters. The present discussion focuses on the three transfer phenomena (momentum, heat, and mass) and on chemical reactions. For all of these happenings, a very general but very instructive equation can be written:

$$\text{Rate} = \text{Rate coefficient} \times \text{zone of action} \times \text{driving force} \tag{3.1}$$

In the case of convective heat transfer, for instance, Eq. (3.1) becomes

$$\text{Rate of heat transfer} = \text{Heat transfer coefficient} \times \text{area normal to the flow of heat} \times \text{temperature difference} \qquad (3.2)$$

Equation (3.1) has two key characteristics:

(1) If any one of the three terms on the right side is increased, the whole rate is increased proportionately, and
(2) if any one of the three terms goes to zero, the rate goes to zero.

Thus, the process designer is presented with both an opportunity and a warning.

We desire to increase the rates of favorable phenomena, e.g., chemical production of useful product, necessary heating and cooling of streams, and necessary separation of components. We try to suppress the rates of unfavorable phenomena, e.g., chemical production of unwanted products, heat loss to atmosphere, evaporation of components, and frictional loss in piping. Collectively, all these interior rates determine the exterior cash-carrying rates of Figure 3.1. The discussions of this chapter are aimed at how the chemical engineer can manipulate the interior rates in order to bring about the desired exterior rates that the business is interested in.

In many cases, an overall outcome depends on the occurrence of two sub events, each with its own rate constant and its own zone of action. An example is the conduction of heat through a composite sheet of two layers of unlike materials. Each layer has its own thermal conductivity, its own thickness, and its own area normal to the flow of heat. An overall driving force exists across the sheet, but it is quite possible that one layer imposes a significantly greater resistance to the flow of heat than does the other and, as a consequence, uses up a greater share of the driving force. In terms of Eq. (3.1), that layer has a lesser value of the product "Rate Constant × Zone of Action", a quantity which we may call the *Rate Product*. In this simple case, the relative resistances of the two layers are easily calculated, but in many other cases of this common resistances-in-series situation, the relative resistances are not so obvious. If the designer is to improve an overall rate, then he or she must make a judgement on which individual resistance requires attention. The following are four examples:

1. Heating of pellets with a hot gas: Two heat transfers are going on simultaneously: from gas to the pellet surface and from the pellet surface to pellet interior. If the former has the smaller rate product (Case 1 in Figure 3.2), then it may be advantageous to improve that heat transfer coefficient by operating as a fluidized bed with lots of gas–particle surface agitation and turbulence. If the latter has the smaller rate product (Case 2), then consideration should be given to making the pellets smaller.
2. Drying of pellets with a hot gas: Two mass transfers are going on simultaneously: moisture diffusion within the pellets and moisture vaporization from pellet surface into the hot gas. If the former has the smaller rate product, then it would be beneficial to make the pellets smaller. If the latter has the smaller rate product, then fluidize the bed of particles.

   In a particular case where these questions arose, it was calculated that, for heat transfer, the external rate was limiting, but for mass transfer, the internal rate was limiting. By making these quantitative determinations, it was possible to take the appropriate steps, leading to the desired improvement in overall rates.
3. Chemical reaction and heat transfer in a batch polymerization: A condensation polymerization is being carried on in a boiling reactor with concurrent boiling out of a reaction by-product as

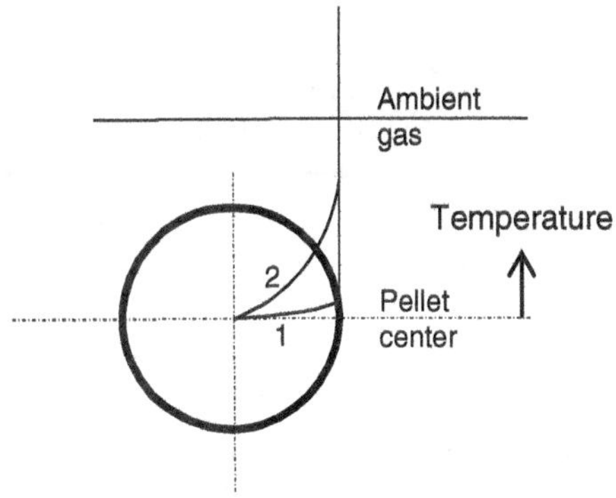

Figure 3.2 Pellet heating.

a vapor. If not continuously removed, the by-product causes reverse reaction. It is desired to reduce batch time, but it is not known whether the rate of chemical reaction or the rate of heat transfer is determining the overall speed of the process. If the former, then a catalyst should be sought. If the latter, then an extra heat transfer surface should be added. Both routes will be expensive and a hunch-based approach may give the wrong route. Some possible aids in making the determination:

- an experiment in which extra – electrical – heat is supplied,
- lab measurement of the polymerization kinetics, and
- engineering calculations.

In this particular case, the rate of heat transfer turned out to be limiting, which led to an increase in the production rate by installing more heat transfer surface.

4. Chemical reaction in a sparged reactor: A reactant "**C**" is sometimes introduced as a vapor into a liquid-phase reaction mixture. This "sparged" reactant must
   (a) first be absorbed into the liquid and then
   (b) enter into the reaction.

Both actions depend on the dissolved concentration of "C":

> The rate of absorption is at least approximately proportional to the difference between $[\mathbf{C}]_{\mathbf{vle}}$ the liquid-phase concentration that would be in equilibrium with the concentration in the vapor and $[\mathbf{C}]$ the actual dissolved concentration; the reaction rate depends in a positive way on the dissolved concentration.

At steady state, the two actions – absorption and reaction – must take place at the same rate, but if one has a significantly lower rate product than the other, then there is potential for improving it to increase the overall rate. Figure 3.3 shows three cases: (i) The absorption rate limits the process – more agitation would help, (ii) the chemical reaction rate limits the process – a catalyst would help, and (iii) both actions are significant and either could be profitably improved to make a limited increase in the overall rate.

These examples are given to show the importance of applying chemical engineering know-how in order to determine where to concentrate effort. The literature is replete with correlations for making most of these predictions. Sometimes a targeted experiment must be done to isolate and measure a rate. Chapter 4 presents some predictive methods and also discusses the use of pilot plants for obtaining key data.

Returning to the general concept of *rates* and looking again at Eq. (3.1), it is logical to examine separately the three independent factors – rate coefficient, zone of action, and driving force – and

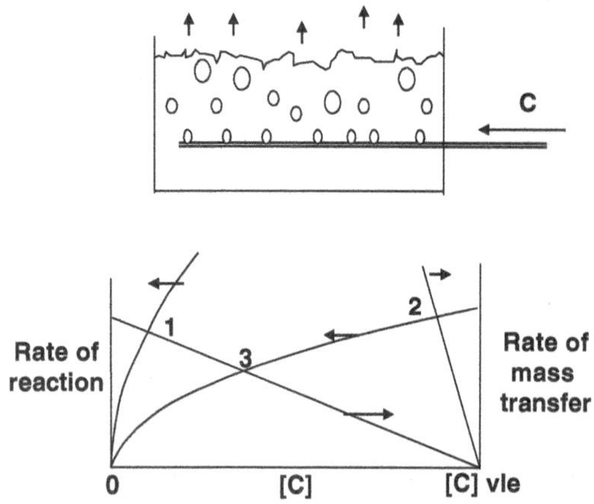

Figure 3.3 Reaction and mass transfer in a bubbling reactor.

to explore ways in which to increase each. It turns out that the creation and maintenance of *driving force* largely dictate the arrangement of chemical processes, and, therefore, this factor is studied first.

3.3 Achieving Driving Force: Some Patterns in Single-Stream Processes

The observer of chemical processes soon notes that there are certain standard configurations that occur regularly. For single-stream processes, they are

- batch and continuous
- plug and back-mixed
- multistage back-mixed, the stages being similar or stages being dissimilar, and
- separation and recycle.

These configurations are now examined for their influence on driving force and ultimately on process efficiency and economics. The general objective is to maximize driving force in order to minimize equipment size and capital cost.

Single-stream systems are those in which changes in a mass or stream of process material, although affected by other systems, can be considered independently of its own effect on those other systems. An example is the flow of fluid through a pipe: The frictional pressure drop caused by the force of the pipe wall on the fluid is calculated without reference to the fact that the fluid is also exerting a force on the pipe (a fact that the pipe designer must recognize in specifying the piping supports). Another example is heat transfer to the same fluid from a fixed-pressure condensing medium (e.g. steam): The rise in fluid temperature is calculated without reference to the fact that steam must be continuously supplied in order to maintain its pressure. Another example is the chemical reaction in a vessel or pipeline where all reactants begin together and flow concurrently.

Consider first the *Batch – Continuous* option. Many processes begin life as batch operations because batch is the way they were developed in the laboratory and batch is considered the simplest route to commercialization. By batch, it is meant that discrete quantities of raw material are fed to the system at some times and not at others and that discrete quantities of product are issued from the system at some times and not at others. A batch operation may comprise more than one vessel in series, with an upstream vessel discharging intermediate material from time to time to the next vessel. Batch versus continuous is a comparison that is often made in a general manner, with many good arguments on both sides, but for a specific process, the choice is sometimes obvious. Some advantages of batch processing are as follows:

- It is generally simpler, with less vessels.
- Process development tends to be done by changing operating conditions rather than the design of vessels.
- There is relatively easy transition between successive product types.
- Incremental expansion can be low cost: just add duplicate vessels.
- For some biological processes, batch is the only option due to the nature of the organisms being used.

The disadvantages of batch processing are usually stated to be greater human resources to operate, greater batch-to-batch product nonuniformity, and more challenging to recover and reuse energy. The first two conditions are being alleviated because

- modern-day systems of distributed control incorporate recipe handling and automated addition of raw materials and additives, which relieve many operator functions;
- advanced control schemes, particularly model-based control, can track batches and keep them all to an identical process path and/or detect any that stray and require segregation.

In the common view, batch operation is usually seen as less "advanced" than continuous, so that there is a progression from batch to continuous processing. In this case, the batch experience provides valuable information for the design of the continuous process. It should also be noted that there are hybrids, with

- a continuous process that has batch operation somewhere along its length, usually for raw material introduction or for product handling;
- a batch process that has a continuous feed of some component during all or part of its course.

Continuous processes normally operate at steady state, where conditions vary along the path but are constant at any point. They also operate at unsteady state during start-up, shut-down, transition periods, and unusual incidents: They are designed to handle all of these modes. One possible configuration of a continuous system is *plug flow*, where the process stream follows a path of "first in, first out". Typically, this condition is approximated in a pipeline with internal fittings that prevent or minimize axial mixing but that promote good lateral mixing. Plug flow makes the best use of vessel volume, as will be examined shortly. Plug flow and a well-mixed batch system are analogous because the batch can be considered to be "plug flow" with respect to time. In practice, perfect plug flow is difficult to achieve because of wall resistance to flow and because of turbulence.

The other extreme of continuous processing is the *fully back-mixed vessel*. The contents of the vessel are deliberately made as uniform as possible, and the material leaving the vessel has the same properties as the material within. As will be seen, this type of operation makes less efficient use of vessel volume and is therefore, at first examination, more expensive. However, it has the following two advantages:

- It is easier to achieve than plug flow.
- It is uniform and avoids local regions of adverse conditions such as hot spots.

A hybrid between plug flow and fully back-mixed flow is the *multistage back-mixed* configuration in which the fluid flows continuously from one fully back-mixed stage to the next.

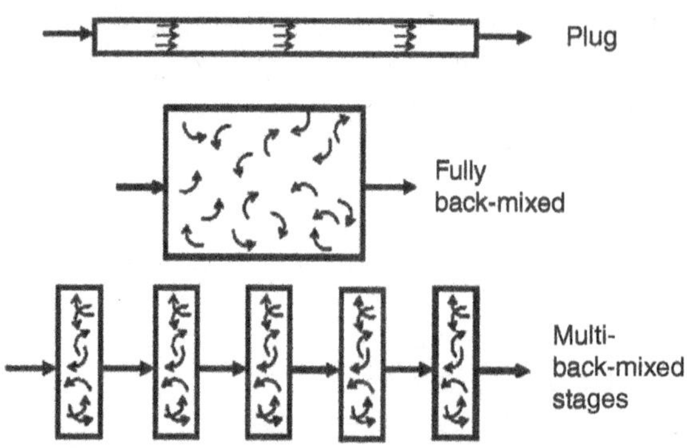

Figure 3.4 Single-stream flow configurations.

Figure 3.4 shows these three flow configurations.

The relative efficacies of these vessel types can be assessed by considering them as reactors for a first-order reaction in which an entering component "C" is consumed according to the equation,

$$\text{Rate of consumption} = \text{k} \times \text{liquid mass} \times [\text{C}] \tag{3.3}$$

The extent of reaction is defined as

$$\text{Extent} = ([\text{C}]_{\text{no reaction}} - [\text{C}])/[\text{C}]_{\text{no reaction}} \tag{3.4}$$

It can be shown that, starting with extent of zero and finishing with extent equal to $\text{Ext}_{\text{final}}$:

- For plug flow,

$$\text{Required hold-up time} = \text{HUT} = (1/\text{k}) \times \ln[1/(1-\text{Ext}_{\text{final}})] \tag{3.5}$$

- For a single-stage fully back mixed vessel,

$$\text{Required HUT} = (1/\text{k}) \times \text{Ext}_{\text{final}}/(1-\text{Ext}_{\text{final}}) \tag{3.6}$$

- For a series of m equal-hold-up-time back-mixed stages,

$$\text{Required overall hold-up time} = \text{HUT} = (m/\text{k}) \times \{[\text{Ext}_{\text{final}}/(1-\text{Ext}_{\text{final}})]\hat{\ }(1/m) - 1\} \tag{3.7}$$

Table 3.1 shows the required hold-up times for various final extents of reaction in comparison with a plug-flow reactor at a final extent of 0.99.

Table 3.1 Relative sizes of various reactor types.

	Final extent of reaction, $\text{Ext}_{\text{final}}$				
	0.10	**0.20**	**0.50**	**0.90**	**0.99**
Plug Flow HUT	0.02	0.05	0.15	0.50	1.00
Back-mixed, one stage	0.02	0.05	0.22	1.95	21.50
2-stages – each fully back-mixed	0.02	0.05	0.18	0.94	3.91
3-stage reactor	0.02	0.05	0.17	0.75	2.37
10-stage reactor	0.02	0.05	0.16	0.56	1.27

The following observations are clear:

(1) Required hold-up time falls off greatly as the final extent of reaction drops.
(2) All configurations behave about the same at extents up to 0.5.
(3) At high (0.99) extent, the single well-mixed reactor requires very large hold-up time.
(4) A sequence of well-mixed stages is much more efficient than one stage and, with enough stages, can even approach the performance of plug flow.

These results all show the effect of driving forces and are most readily understood in terms of the concentration [C], which is the driving force for the reaction. Take $[C]_{\text{no reaction}}$ as 100. In plug flow (or in a batch reactor), the concentration of the reactant is high (100) at the beginning and produces a rapid reaction. The concentration of reactant is everywhere as high as it can be, subject to the fact that it has been reduced by the reaction that has taken place up to that point. O the other hand, in a continuous fully back-mixed stage, the concentration is everywhere equal to its final (low) value, which limits the rate everywhere: At $\text{Ext}_{\text{final}}$ of 0.99, [C] is 1. In the compromise case, e.g., the two-stage system, if the stages are of equal size and if the second stage is producing a stream of 0.99 extent of reaction, then it is calculated that the first stage contains and produces 0.90 extent of reaction, i.e., from Eq. (3.4), its concentration [C] is 10. Thus, the first stage, with [C] of 10, achieves 90% of the reaction and the second stage, with [C] of 1, achieves 9%. Between them, they cut the overall required residence time to less than a quarter of the single-stage time. The benefit of multistaging is obvious and is greatly taken advantage of in chemical process design. Some typical configurations are shown in Figure 3.5.

In these examples, it is implied that the stages are similar. Sometimes it is advantageous to adopt what could be called a *Vari-Stage* approach in which the successive stages are customized for the nature of the reaction mass and for the task to be handled. For instance, in a polymerization process, the initial reaction mass may be of low viscosity, while the final mass may have a viscosity that

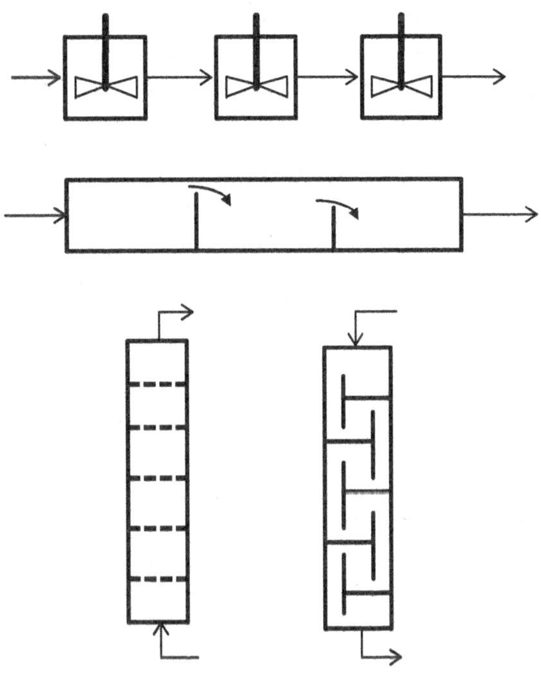

Figure 3.5 Some multi-well-mixed-stage configurations.

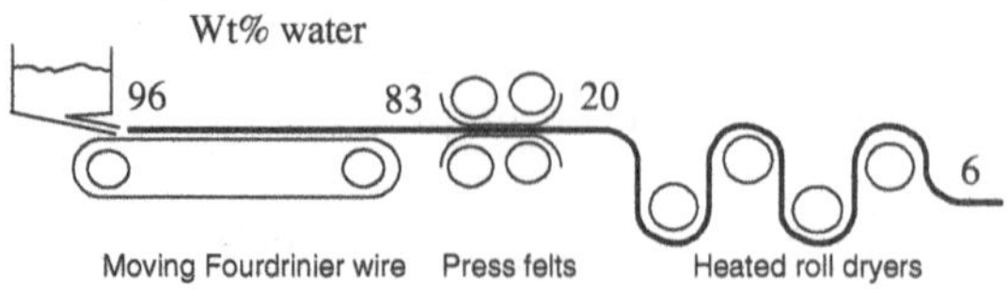

Figure 3.6 Paper making.

is greater by many orders of magnitude. The early stages may require no mechanical agitation, while the later stages must be heavily stirred. A classic example of vari-staging is the paper-making machine, where paper is produced starting with a slurry of wood pulp. The process is illustrated in Figure 3.6.

Vari-staging has its counterpart in batch processing where a sequence is used of two or more vessels of different types, each appropriately designed for its part of the process.

In another configuration for single-stream processing, use is made of *Separation and Recycle.* In the case of chemical reaction, the inspection of Table 3.1 shows that the reactor volume for low final extent of reaction is very much smaller than for high (e.g. 0.99) final extent. An option, then, is to take the reaction to only a low extent, accept the fact that the outlet stream contains a large amount of unreacted reactant, separate this unreacted material, and return it to the reactor. This separation is typically (but not necessarily) done by distillation. The other stream from the separation is rich in the desired product and proceeds to final processing. Figure 3.7 shows the principle.

The advantage here is the smaller reactor size. The disadvantages are the capital cost and the energy requirement of the separation device and also the possibility of the time-wise buildup in the recycle stream of unwanted reaction by-products. If saving of the reactor size is the only benefit, then separation-and-recycle may not be a good choice. However, including this case, there are four situations where the approach should be considered:

1. The process must be taken to a high final extent of reaction, either for reasons of product purity or because of high cost of the raw material.
2. There is a significant reverse reaction which slows the process and limits the achievable extent.
3. The product is susceptible to a further undesired reaction if it remains at reactor conditions.
4. The product has a poisoning effect on a catalyst.

The latter three conditions may make separation-and-recycle mandatory.

This technique has some applications other than in chemical reaction. Figure 3.8 illustrates the comminution of solids – e.g., in rock crushing – the product from the crusher may move to a screen

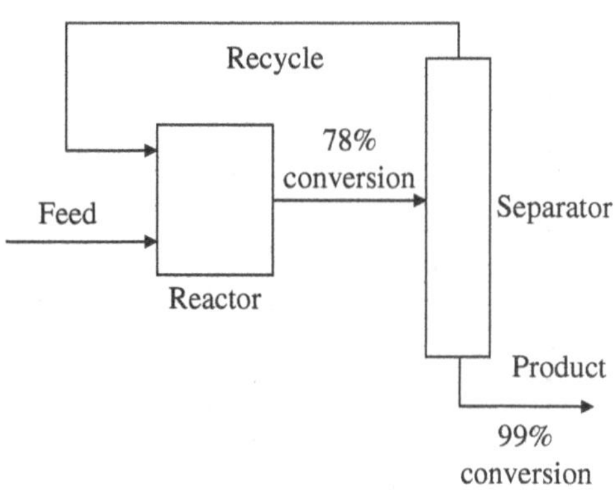

Figure 3.7 Recycle reactor.

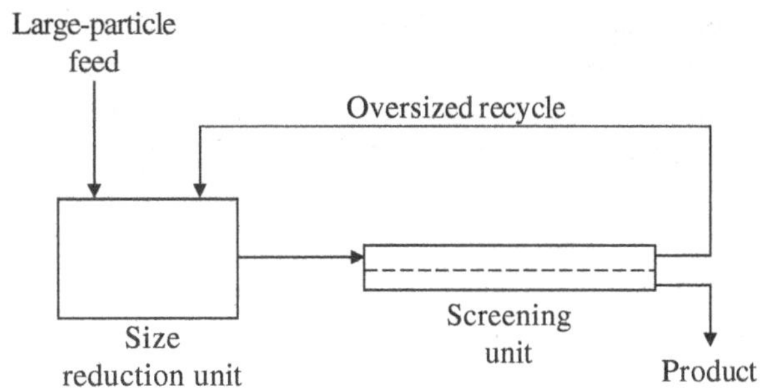

Figure 3.8 Comminution with recycle.

(the separation device), where the acceptably small pieces pass through and the oversize pieces return to the crusher.

In using a centrifugal pump to generate a specified flow, the pump may be set up to pump a flow greater than desired and the excess is split off and returned to the supply tank. In feeding a settleable slurry to an intermittent user, a "ring" feed pipe may be used, with continuous recycle to the source and withdrawal at the user as needed.

3.4 Achieving Driving Force: Some Patterns in Two-Stream Processes

Two-stream processes are defined as those in which it is necessary to consider the interaction between two process streams or, in the case of a batch process, between two masses of material. There are processes where the interaction among more than two streams must be considered, but, for the present discussion, two streams will be enough to deal with.

It will be recalled that, for single-stream systems, the purpose of maximizing the driving force is to minimize the equipment size and capital cost. This objective also applies to two-stream systems and a second objective now comes into play: to minimize such operating costs as

- Generating utility streams: a utility stream could be a liquid (e.g. a hot oil) used for supplying heat to the process. Designing to efficiently maintain driving force (temperature difference) allows both the flow of oil and also the temperature at which it must be generated to be minimized.
- Recovering secondary process streams: a secondary stream could be a gas used to strip a component from a liquid. Efficiently maintaining driving force (concentration difference) allows less of the gas to be used and minimizes the effort required to subsequently separate the component from the gas.
- Treating waste streams: a waste stream could be a discardable flushing liquid. Efficient usage minimizes the amount required and the amount sent to waste treatment.

Many standard operations of chemical engineering involve two interacting streams. Examples are as follows:

- two-liquid heat exchangers, where the temperature of each liquid affects and is affected by the temperature of the other,
- gas absorbers, where the composition of gas and of absorbing liquid are interrelated,
- liquid strippers, where the composition of liquid and of the stripping gas are interrelated,
- liquid–liquid extractors, where the compositions of the two liquid streams interact,

- pneumatic conveying, where momentum is continuously exchanged between particles and conveying gas, and
- drying, where the concentration of vapor in a gas stream is determined by and helps determine the rate of transfer from the material being dried.

Some of these processes are shown schematically in Figures 3.9 and 3.10.

As with single-stream operations, those with two streams display some commonly recurring patterns:

- batch and continuous,
- plug and back-mixed, and
- multistage back-mixed.

Additionally, two-stream operations are distinguished as

- cocurrent, crosscurrent, and countercurrent.

This new distinction is illustrated, for a continuous process, by the following example of drying, as shown in Figure 3.11.

In this operation, hot air is being passed over the surface of a moving bed of particles with the objective of heating them and extracting moisture from them. The air can be oriented to move in the same direction as the particles, at right angles, or in the opposite direction. The rate of transfer of

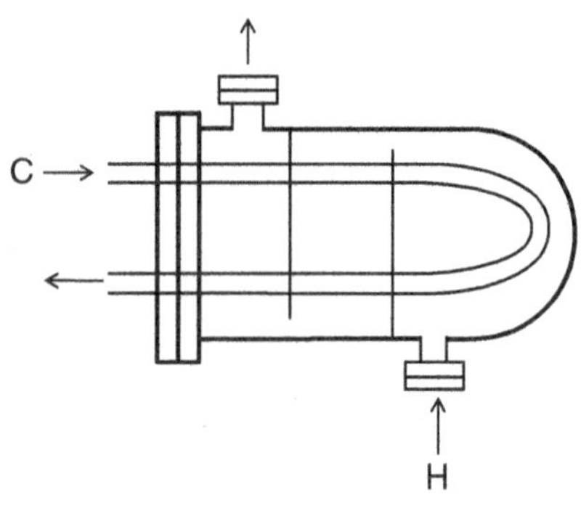

Figure 3.9 Two-liquid heat exchange.

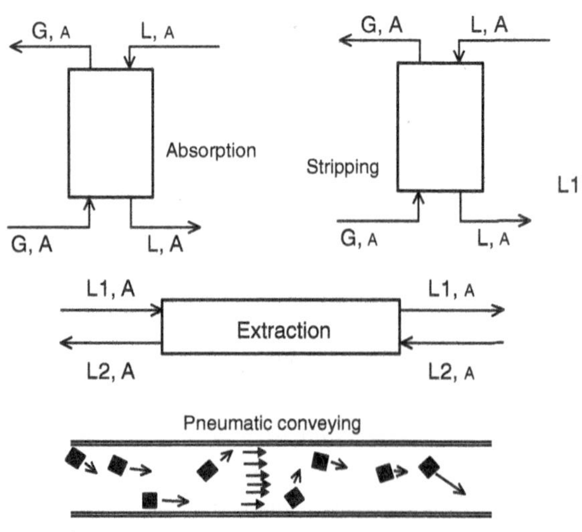

Figure 3.10 Other two-stream operations.

Figure 3.11 Hot-air drying of solids.

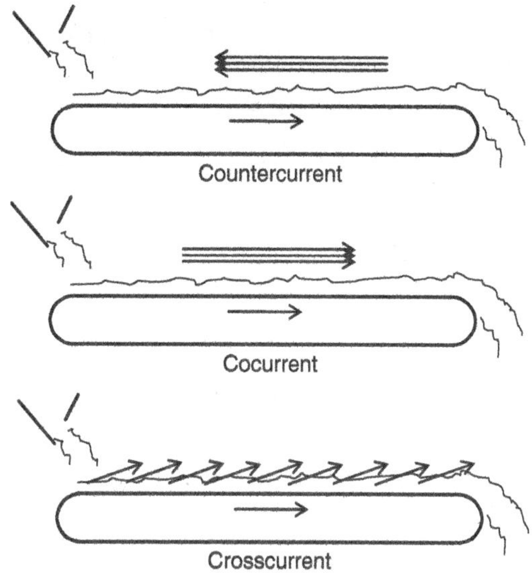

moisture is governed by the rate of heat transfer from air to solid, which is governed by the temperature difference between air and solid. For a specific case, the distance-wise profiles of temperature are shown in Figure 3.12. The following conditions applied:

- Ratio of (solid's flow × specific heat) to (air flow × specific heat) was 0.7.
- Parameter (heat-transfer-coefficient × area/(solid's flow × specific heat) was 2.
- Initial temperature of solids was 20 °C.
- Initial temperature of air was 200 °C.

Countercurrent operation is seen to be most efficient, raising the solids to a temperature of 152 °C. The cocurrent arrangement has the disadvantage that the air cools as the solid's heat up, reducing the driving force and making the final section of the dryer relatively ineffective. The solid's reach a temperature of only 122 °C. The cross-flow arrangement ensures that all parts of the dryer see the full temperature of the hot air but is still not as effective as countercurrent, raising solid's temperature to 139 °C. For either the cocurrent or the crosscurrent configuration to match the 152 °C solid's outlet temperature, either the flow rate of the hot air would have to be increased, entailing more operating cost, or the supply temperature of the hot air would have to be increased, probably also entailing more operating cost and, if the solids are temperature sensitive, possibly degrading them. If the required increase in flow or temperature is made, then a calculation of entropy production in each of the three configurations shows the lowest amount for countercurrent and the highest for cocurrent. A general principle is thus illustrated of maximizing efficiency by minimizing entropy.

The three configurations of counter, cocurrent, and crosscurrent operation are found in batch processing as well. Consider a vessel which accumulates a small amount **D** of residual fouled material which must be flushed out from time to time with a solvent. Assume that whatever amount of solvent is used it completely dissolves the undesired residue, but, upon draining, 5 kg of solution remains behind. One procedure would be to charge a single amount – say 100 kg – of solvent, let it dissolve the fouled material, and then drain it, i.e., 95 kg that comes out. Left behind would be (5/100) or 5% of the original fouling. This procedure is shown in Figure 3.13.

More effective would be the crosscurrent approach, which is also shown in Figure 3.13. The solvent is divided into two 50-kg aliquots. The first is used where it leaves behind (5/50) or 10% of the

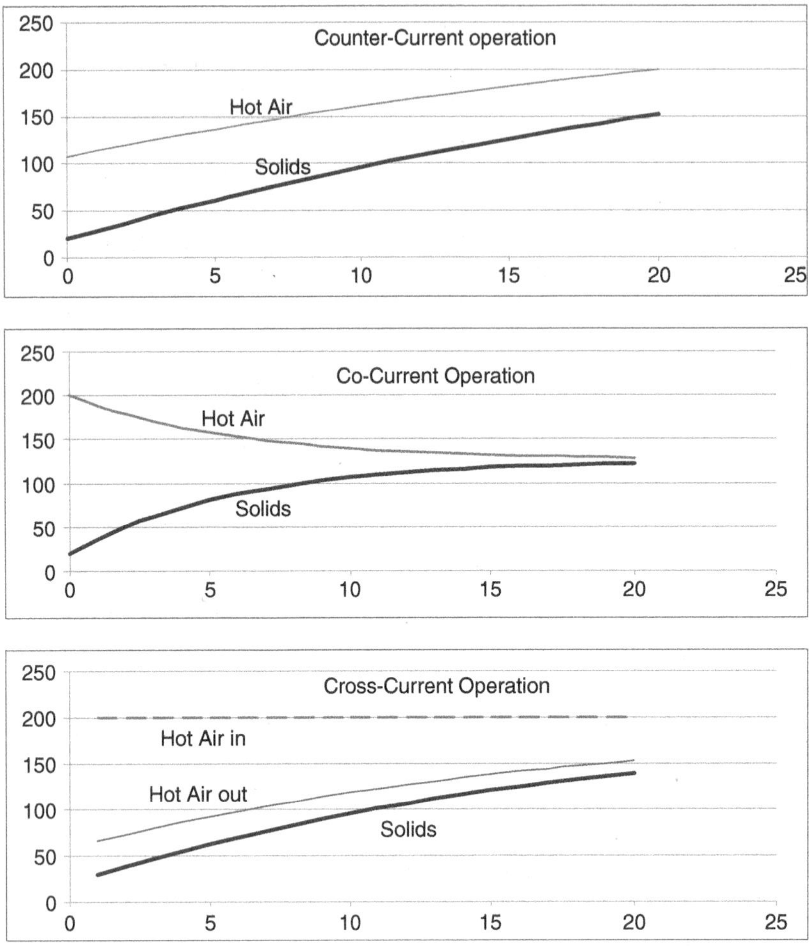

Figure 3.12 Solids hot-air drying performance curves.

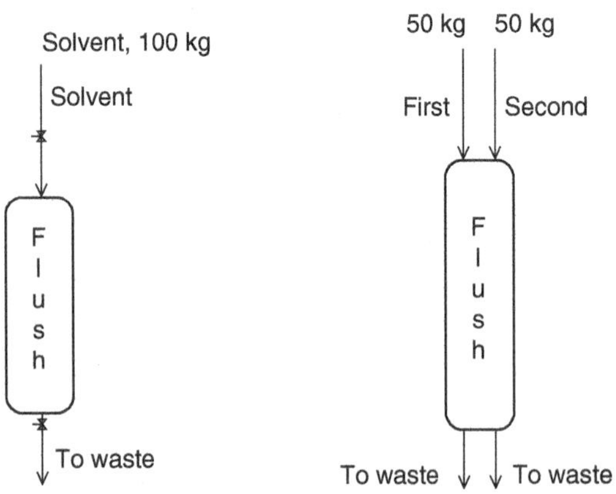

Figure 3.13 Single-pass flush and dual-pass flush ("crosscurrent").

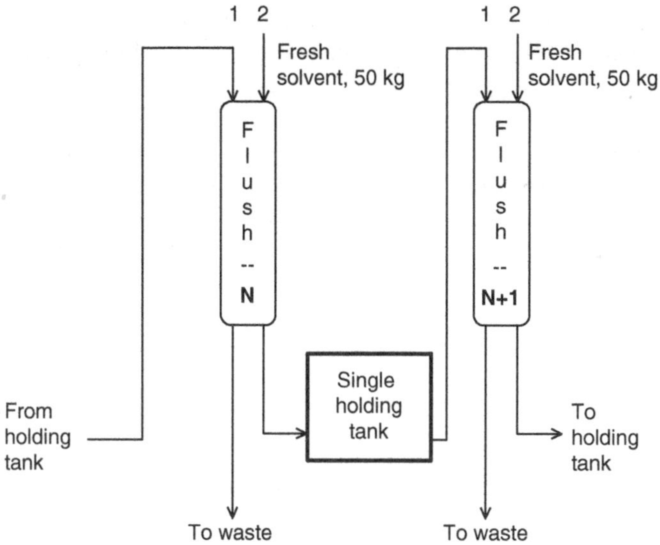

Figure 3.14 Countercurrent batch flushing.

fouled material. The second is immediately used where it leaves behind (5/55) × (5/50) or 0.9% of the fouled material.

The third approach is the countercurrent, as shown in Figure 3.14. It is also a two-pass procedure, but the liquid from the second pass, which is relatively clean, is saved in a storage tank and is used for the first pass the next time the vessel has to be cleaned.

The advantage of the countercurrent idea is that with only one charge of 50 kg of solvent, per flush, almost the same cleaning can be attained as with two 50-kg charges in crosscurrent operation. About 1% of the fouled material is left after the second pass. A domestic example of the principle is "suds saving" in a washing machine: The relatively clean rinse water from the second wash of one load of wash is saved in a sink and then used as the wash water for the first wash of the next load. The example has focused on application to a single vessel: It becomes more attractive when applied to a whole bank of vessels, where the intervals between cleaning are shorter. The suds-saving example shows a case where the "process material" rather than a vessel is being treated.

3.5 Summary of Synthesis

Time constraints make it necessary for the process designer to focus on aspects with greatest pay-back. In Figure 3.1, the cash-generating and the cash-consuming "streams" are shown. One or more might be identified as having potential to be optimized to give a significant increase in the *net* cash flow and a decrease in the unit product cost. These possibilities should be discussed by the project team early in the design.

The rate equation, Eq. (3.1), provides the framework for process improvement. When more than one action governs the rate, it is essential to focus on the one which is dominating.

A number of flow regimes may present themselves to the designer: batch and continuous, plug flow and back-mixed flow, countercurrent and cocurrent, single-stage and multistage, and single-mode and multi-mode. Some may not be appropriate to the situation but others may be options.

Sometimes the configuration that has the theoretically highest rate, e.g., plug flow, gets superseded by a less-efficient configuration, e.g., mixed flow, simply because true plug flow is difficult to achieve and therefore to characterize.

4

Experimentation and Modeling in Support of Design

The goal of chemical engineering process design is to create a facility that will transform materials. It begins with the designer's concept, aided by the principles of Chapter 3, of the steps in making that transformation. Unit operations and chemical reactors are chosen from Chapters 10–20. The properties of the materials along the process path are accessed from the literature or measured. The internals and the dimensions of the equipment are determined from chemical engineering principles. A test facility, of greater or less similarity to the commercial end product, is built to verify the design. These are the steps for which the present chapter provides guidance.

Some general recommendations can begin this guidance.

- For material properties, the task can comprise lab measurements, literature review, estimation methods, and access to property data banks or to a commercial math model that includes a property's database. A principle to keep in mind is that an hour in the library can save a month in the lab.
- If a chemical reaction is part of the design, then a separate set of experiments may be needed to establish rate, order, and selectivity: These experiments may well be done batchwise even though the ultimate design is for a continuous process.
- For design of the equipment, the task can start with knowledge from undergraduate studies, but most engineers safely resort to further study, advice from specialists, and recommendations from vendors. If a similar existing commercial unit exists, then useful data may be had, especially if the unit is allowed to vary in its settings.
- The verification task generally requires some approximation of a pilot plant. A comprehensive well-designed pilot plant is the gold standard for verification, but it does require time and expense, some of which can sometimes be judicially circumvented.

The chapter contains three parts. In the first part, a suggested three-pronged approach to making and checking designs. The second part discusses pilot plants and the associated problem of scaling up or down between units of small scale and commercial scale. The third part presents a very simplified example of how a mathematical model can support the process design.

4.1 A Systematic Review of Process Design

The designer must not neglect any details of her or his design, but, of course, it is very easy to miss something. Because every design is different, it is not possible in general to make specific recommendations for given cases. However, with application of some sort of organization, it is

Practical Process Design for Chemical Engineers, First Edition. Keith Marchildon and David Mody.

possible to split up the study and to lessen the omission of essential elements. The organization that is proposed is the threefold division of chemical engineering science, namely

- momentum transfer,
- heat transfer, and
- mass transfer.

These three phenomena are the core of chemical engineering science, and so a good place to start from.

Momentum is mass times velocity (units $kg\,m\,s^{-1}$). The transfer of momentum is described by choosing a quantity of material and setting the time-wise derivative of its momentum to the sum of all the factors that would influence the momentum. This is the momentum balance,

$$\frac{dM}{dt} = \sum \text{factors influencing momentum} \tag{4.1}$$

These factors are

- gravity,
- physical force (as in Newton's second law) (i.e. pressure and viscous forces), and
- change of density because of phase change.

The quantity on which we are doing the balance does not have to be homogeneous. A mass is static if all the forces balance out.

Momentum transfer is a consideration not only in chemical processes but in mechanical, civil, aerospace, and other disciplines.

Momentum balances can be used to analyze the following fluid and phase flow situations:

- Pipe flow: Analyzing fluid flow through pipes is a common application. Momentum balances can be used to predict pressure drops, flow rates, and velocities in different sections of a pipe network.
- Reactors and mixers: In chemical reactors and mixers, momentum balances help optimize mixing efficiency and avoid dead zones where fluid stagnation could lead to inefficiencies or safety issues.
- Fluidized beds: In processes involving fluidized beds, momentum balances are used to determine the minimum fluidization velocity and to understand particle behavior in fluidized systems.
- Distillation and separation: In distillation columns and separation units, momentum balances are important for understanding fluid distribution and ensuring even vapor–liquid contact.
- Pump and fan sizing: Momentum balances can help in the determination of the required pump or fan size to achieve a desired flow rate and pressure in a system.
- Environmental fluid dynamics: In environmental engineering, momentum balances play a role in understanding natural fluid systems like rivers, lakes, and oceans, as well as the dispersion of pollutants.

Heat, in the sense used here in "heat transfer", is partially a misnomer and is in fact the internal energy (U, joules) of the molecules, atoms, and partly of electrons of a chosen mass. It depends on the thermal properties, the temperature, and the phase(s) of the mass. The transfer is expressed by choosing a quantity of material and setting the time-wise derivative of its internal energy (or "heat") to the sum of all the factors that would influence the energy,

$$\frac{dU}{dt} = \sum \text{factors influencing heat transfer} \tag{4.2}$$

These factors include the following:

- heat entering or leaving by conduction, convection, and radiation,
- work, and
- internal energy consumed by or created by chemical reaction.

Like momentum transfer, heat transfer considerations are shared with other disciplines besides chemical.

Mass transfer is almost exclusively the domain of chemical engineering. The "mass" being referred to is some defined portion of a single chemical component: more specifically all of that component that resides in a particular single phase, by itself or in a molecular mixture with other components. The time-wise derivative of the mass is set equal to the sum of the factors that would augment or deplete it,

$$\frac{\mathrm{d}M}{\mathrm{d}t} = \sum \text{factors influencing mass transfer} \quad (4.3)$$

These factors are as follows:

- transfer of the mass out of its phase, one molecule at a time, into another phase and
- chemical conversion of the mass into another substance.

Many chemical engineers have forgotten (or never realized) that their specialty revolves around these three quantities and these equations. However, implicitly most of the chemical engineer's technical activities consist of identifying and quantifying the right-hand sides of these equations. Consequently, this structure helps to sort out the issues which a particular design must address.

4.1.1 Momentum Transfer Considerations (see also Chapter 12)

Momentum transfer for chemical processes is concerned mainly with the flow of substances – gases, liquids, solids, and various mixtures of two or three members of these three states of matter. One common assumption about liquids is that their rheology is *Newtonian,* that is, that shear stress is linear with the shear rate. See Figure 4.1. This is the basis for the usual equations relating frictional pressure drop to flow and geometry. In support of the design, it is necessary to check that this is true for all liquids that are part of the proposed process. Some "aberrant" behaviors that might complicate the predictions of flow and pressure drop are when the liquid is:

- *rheopectic*, with its apparent viscosity (stress/strain) increasing with time of strain,
- *thixotropic*, where the apparent viscosity decreases with time under strain,

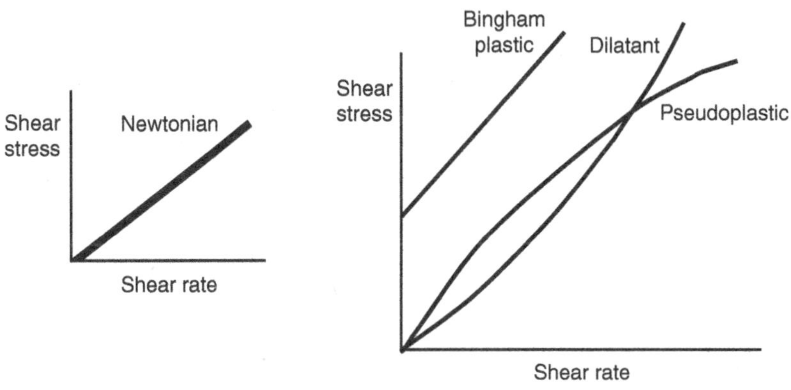

Figure 4.1 Rheology of non-Newtonian liquids.

- *a Bingham plastic*, where a critical stress is needed to move the liquid at all,
- *dilatant*, where the apparent viscosity increases with shear rate, and
- *pseudoplastic*, where the apparent viscosity decreases with shear rate.

Identifying a liquid that has one of these behaviors will allow appropriate modifications in piping and other pressure-drop devices. Some advice from a rheologist or fluid flow specialist may be necessary.

Another momentum-transfer problem is dealing with two-phase gas–liquid flow. There have been many studies over the years attempting to relate (i) flow friction factor and (ii) *in situ* composition to the flow and properties of the gas and liquid. One of the common and most quoted is the graphical procedure of Lockhart and Martinelli (Chem. Eng. Prog., vol 45, p. 39–48, 1949), which generally predicts pressure drop to within about 40%.

For horizontal flow, there are several types of flow – stratified, wave, annular, bubble or froth, spray or dispersed, plug, and slug – which are displayed on a map devised by Baker (1954). The axes of the plot are gas mass flow and the ratio of liquid flow to gas flow, each with some multiplicative modification based on fluid properties. Much to be avoided is slug flow, which causes masses of liquid at high velocity to be intermittently flung through piping and equipment, with possible mechanical detriment. The Baker plot is shown in Chapter 5, along with a similar plot (the Charles plot) for vertical flow.

Momentum transfer also takes place where one phase passes through another, either as immiscible droplets or as bubbles. Densities for liquid phases and the viscosity of a continuous liquid need to be confirmed.

4.1.2 Heat (or Energy) Transfer Considerations (see also Chapter 14)

Transfer of heat is almost always a consideration because (i) mass-transfer operations such as distillation require heat to power the generation of vapor and (ii) most chemical reactions require elevated temperatures to proceed at acceptably high rates. At the same time, heat must be extracted at some point in order to regain the liquid phase (if appropriate) and to get back to room temperature. The tasks are to

- estimate the required heating and cooling rates,
- choose heating and cooling media that are able to deliver the rates, and
- estimate overall heat transfer coefficients to allow the calculation of required transfer areas.

The simplest and most common heat transfer media are steam for heating and water for cooling. However, a process stage to be heated may be at too high a temperature for the usage of steam at a safe pressure or at any pressure at all. Analogous to steam, there are condensable heat transfer fluids which operate at significantly higher temperatures and at significantly lower pressures than steam systems. Commercially, there are several such products, among them a eutectic mixture of diphenyl and diphenyl oxide, e.g., Dowtherm A™. Comparing this substance with steam (Table 4.1)

Water is the first thought for cooling, but if the temperature of the hot source is too high or too low, again a commercial heat transfer liquid may be more appropriate.

Table 4.1 Dowtherm A and steam pressure.

Temperature (°C)	Dowtherm pressure (bar)	Steam pressure (bar)
255	0.97	43.2
355	5.8	175.6

Heat is lost from inter-vessel piping. It may be kept low enough by insulation, but other times an expensive jacket of steam or hot oil is needed at critical points.

The scale-up of heat transfer requires preserving the ratio of heating rate to mass of material, which becomes a challenge if a vessel is being heated though its walls. Internal heating coils or external pump-around coils may be required for large vessels. Heat transfer can be a problem with scale-up and will be discussed later at more length.

4.1.3 Mass Transfer Considerations (see also Chapters 17 and 20)

The transfer of mass takes on six variations:

1. gas/vapor phase + gas/vapor phase,
2. gas/vapor phase + liquid phase,
3. gas/vapor phase + solid phase,
4. liquid phase + liquid phase,
5. liquid phase + solid phase,
6. chemical reaction.

Here, we are including chemical reaction as an aspect of mass transfer inasmuch as it is one of the actions that can augment or deplete a "mass".

For interphase transfer, the key considerations are rate and equilibrium. For chemical reaction, the following characteristics are required to be determined:

- reaction order,
- equilibrium between reactants,
- rate of reaction, and
- selectivity for desired product.

For a given process design, all of the following processes that are relevant have to be recognized so that they can be studied and evaluated quantitatively. This activity may well carry over into the mandate of a pilot plant (Figure 4.2).

Case 1 requires the two phases to be separated by a semipermeable barrier through which the component under study may pass but not the entire phase. Such barriers, in the form of membranes, do exist for specific cases. Membrane-separation methods constitute a specific and relatively recent technology.

In **case 2**, if the gas is a permanent one which has a limited solubility in the liquid, Henry's law describes the equilibrium amount. Oxygen, for instance, has solubility in water at 25 °C and atmospheric pressure of 40 ppm. Its Henry's Law constant is approximately

$$\frac{\text{800 atmospheres pressure}}{\text{(oxygen concentration, moles/liter, in water)}} = \frac{1\text{ atm}}{\left(\frac{40\times10^{-10}\text{g O}_2}{\text{g water}}\right)\times\left(\frac{1000\text{ l water}}{10^6\text{g water}}\right)/\left(\frac{32\text{ g O}_2}{\text{mole O}_2}\right)} \tag{4.4}$$

Figure 4.2 Mass transfer combinations.

	Gas	Liquid	Solid
Gas or vapor	1	2	3
Liquid		4	5

For water in contact with air, the solubility is 0.2×40 or 8 ppm.

A component (i.e. the "mass" in which we are interested) in the gas may have a higher solubility than the carrier gas. Bubbling of gas through the liquid is typically the way to bring about the transfer of mass: *stripping* to remove component mass from the liquid and *absorption* to add it to the liquid.

In **case 2**, if the gaseous phase is a condensable vapor, then multistage distillation can bring about transfer of mass from one phase to the other. Vapor–liquid equilibrium data are needed and particularly identification of azeotropes. There are many compilations of such data for binary systems, e.g., Gmehling & Onken (2000). Predictive methods (e.g. UNIFAC/UNIQUAC) exist based on the molecular structure of the components. For a particular combination of components, it may be necessary to set up a lab-scale vapor–liquid measurement apparatus.

Case 3 is generally about transferring the gas or vapor (the "mass") by absorption into or adsorption onto solid particles, or about stripping it out of solid particles.

Case 4 is liquid–liquid extraction, which brings two nonmiscible or slightly miscible liquids into physical contact with each other. The component in question passes molecularly through the interface between the liquids, the interface being made as large as possible generally through agitation. The equilibrium is often measured by a simple process of shaking the liquids together vigorously in a flask, then letting them settle, and withdrawing from the two phases for analysis.

Case 5 may be about absorption or stripping of the "mass" from solid particles. It may be the dissolution of a solid into the liquid. It may be gradual accretion of the component of interest onto existing solid particles, as in crystallization.

Case 6: Rarely does chemistry have a simple "feed substance → desired product". Other products are produced to a lesser or greater extent, some of them costly waste and some useful and saleable. In many cases, some purification of the product stream is needed. Often the unwanted products are produced from the further reaction of the desired products, thus necessitating early withdrawal of the product stream, separation of the desired product, and recycle of the residue back to the reactor. This is shown in Figure 2.1. If it turns out that not enough is known about the reaction, then experiments may be necessary. One could follow the change in composition in a batch vessel at various lengths of time at reaction temperature.

One way to interpret the data is to simulate, on a spreadsheet, the reaction process, assuming guessed values for the reaction coefficients. Then, when data become available, adjust the coefficients to match the measured changes in composition. As an example, a simple first-order reaction scheme is shown here,

$$\text{Raw material A} \rightleftarrows \text{desired product B} \tag{4.5}$$

$$\text{B} \rightarrow \text{by-product C} \tag{4.6}$$

The initial estimate k for the forward rate constant (reciprocal minute) is 0.015 for reaction 1; equilibrium constant K is 200 for reaction 1; rate constant kc is 0.002 for reaction 2.

The results of the spreadsheet simulation are shown here (Table 4.2).

When the results of an experiment of 30 minutes or more become available, the three constants k, K, and kc can be adjusted to provide a match, bringing them closer to their true values. This procedure should be repeated for different times and, perhaps, for different compositions of starting materials in order to yield confidence in the estimates of the constants. The example is for first-order kinetics only so, in practice, a fourth parameter – order of reaction(s) would have to be included. However, once these constants are known with confidence, then they can be used in the chemical calculations for any geometry – batch, well-mixed, plug flow, and recycle flow.

Table 4.2 Simulation results of the batch reactor.

Time (min)	5	10	15	20	25	30
A ($mol\ l^{-1}$)	9.28	8.61	7.98	7.41	6.87	6.37
B	0.72	1.38	1.99	2.54	3.05	3.51
C	0.00	0.01	0.03	0.05	0.08	0.11

The reaction scheme is the same as that used in the example of Figure 2.1. With the above set of reaction constants, it is calculated that the reactor would require 0.95 m^3 of working volume (Example 2.1 conversion of 100 mol min^{-1} with a reactor residence time/conversion selected of 30 minutes).

The objective in this section has been to alert the designer to some of the considerations that may have to be addressed. Hopefully, many of the answers are in the literature or have been found in comprehensive bench-scale studies. Information from bench-scale should be maximized, often by negotiation with the chemists.

4.2 Pilot Plants and Scale-up

The degree of assurance that the process will function as intended rests with the owner of the facility. Least prudent would be to go from bench-scale, through process design, and then directly to commercial scale. In contrast with this minimalist approach, Figure 4.3 shows a stepwise verification process, with the size and comprehensiveness of the second stage – the pilot plant – undefined. In order to save money, it is often proposed to combine the second and third steps. The combined mandate necessitates a larger plant than the imperatives of a pilot plant by itself requires, but the imperatives of the pilot plant must still be observed. The present discussion concerns only the pilot plant.

A pilot plant must verify the following five aspects of the process design:

- The stability, once the steady state is achieved, of all recycle streams and purge points so that impurities do not build up with time.
- The absence or measurement of fouling, scaling, corrosion, and polymerized buildup with time.
- The conditions required for adequate mixing in vessels which are intended to be fully back mixed.
- The strategy and characteristics of controller feedback loops.
- Start-up and shut-down procedures.

The pilot plant provides the opportunity to correct deficiencies in these and any other aspects.

Berg (2011) urged the review of piloting plans early and often using a peer group that is external to the project team. They recommend any steps involving solids as important to be part of an integrated pilot plant. Working with a contractor who has experience with pilot plants can speed up the work and avoid mistakes.

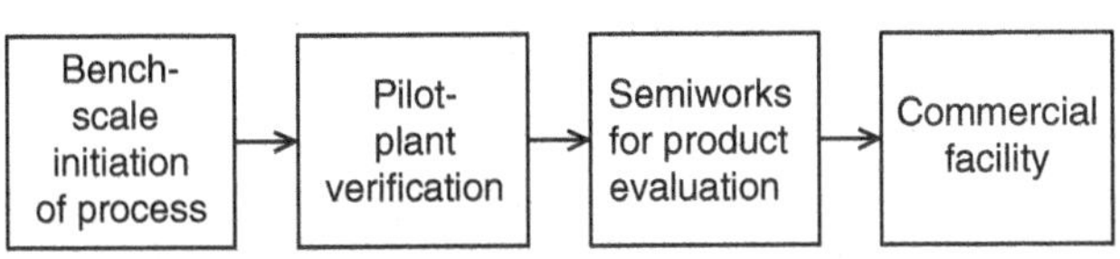

Figure 4.3 Full verification of process.

To what extent should the equipment of the pilot plant mimic the eventual commercial plant? If it is lumped in as part of a semiworks unit, then the resemblance will have to be more or less complete. However, by itself as a pilot plant, it requires only those parts that satisfy the above five aspects. Some other parts of the process may be handled by vendors in their labs.

4.2.1 Size and Cost of Pilot Plants

The cost of a pilot plant depends significantly on how big it is and how elaborate and comprehensive. The cost, if grossly beyond expectation, may even cause the project to terminate. Hence, a realistic estimate is called for and is a challenge to generate.

The first decision is always how big the pilot plant will be. Basically, it should be as small as possible. There are several factors to consider.

1. Although the unit is not intended to make products for customer testing, it will generally be tasked with producing small quantities for in-house evaluation. It should have a production rate equal to the needs for these tests.
2. Are raw materials plentiful or scarce?
3. Are there any "break points" in the handling of raw materials and/or product? At what rate does the handling require more labor or more or different equipment?
4. Are certain necessary pieces of small-scale equipment available?
5. For vertical columns, a rule of thumb is that an eight-inch internal diameter is the point below which wall effects have a significant influence. Wall effects would complicate scale-up. Hence, if a column is part of the process, then its size could dictate the scale of the whole pilot plant.

Estimating the cost of a pilot plant can be difficult because of the following:

1. The task may fall to the process designer, who does not have the costing skills of people working with commercial plants.
2. There are few cost comparisons because pilot plants are not plentiful within a company and details are not published by outside companies.

However, there are well-known types and methods used for estimating, at various accuracies, full-scale plant costs. They are the best one can use for estimating the capital cost of pilot plants. The well-known 0.6 power could also be applied:

$$\begin{aligned}\text{Pilot plant estimated cost} &= (\text{commercial plant estimated cost})\\ &\quad\times(\text{throughput of pilot plant/throughput of commercial plant})^{0.6}\end{aligned} \tag{4.7}$$

4.2.2 Equipment and Piping

The challenge is the small size of the facility. Two of the most common vessels are discussed here, namely, liquid-phase chemical reactors and separation columns, as presented in the example of Figure 2.1.

Many suppliers of laboratory equipment provide lab-scale jacketed batch reactors in the 25–250-liter size range, with temperature capability −60 to 200 °C. Glass reactors can take pressures from full vacuum to 0.5 or 1.0 bar above atmospheric: If this range is acceptable, then the visual capability allows the verification of flow patterns and makes possible the measurement of the liquid level through optical techniques. For higher pressures, many of the same suppliers

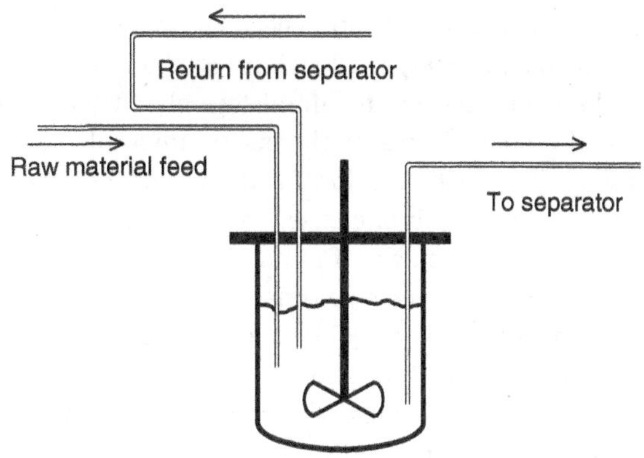

Figure 4.4 Continuous flow reactor vessel.

provide vessels of stainless steel or other metals that can work at several hundred bar. Suppliers provide not only the vessels themselves but also, with the customer, develop whole custom systems to support the operation.

Although the vessels are advertised as "batch," Figure 4.4 shows how they are easily converted to continuous for the example from Figure 2.1.

For the scale-up or scale-down of columns with trays, there are two somewhat conflicting considerations. One imperative is to match the mass transfer between columns of different sizes, which means including the same number of stages. The other is to approximate the hydraulic behavior: the tendency for the liquid phase to flood and the upward carry-up of liquid from tray to tray. A large part of that matching is achieved by retaining the ratio of cross-sectional area to column throughput. However, the inter-tray distance is also a factor and can only be kept constant by making all columns the same height – not very practical for a pilot plant. Figure 4.5 shows a commercial column and a scaled-down pilot column. The cross section corresponds to the throughput, but the height is made as high as practical to minimize the effect of inter-tray distance. Another approach is to test the removal of one or two trays from the pilot column to check if there is any effect on flooding and carryover. The Souders–Brown equation should be applied to the geometry to further check on hydraulics. For packed columns, there is no inter-tray

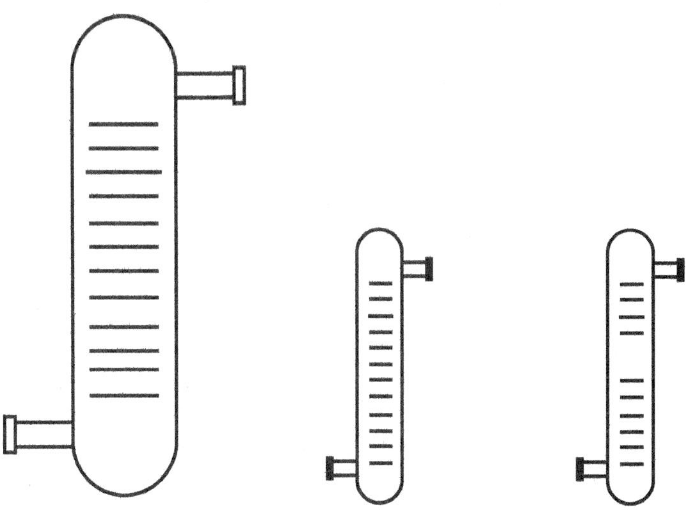

Figure 4.5 Scaling of distillation column.

issue, so height consideration becomes moot. However, packing efficiency may still depend on the column diameter, especially if it is less than 8 inches.

Pilot plants may use the multistage glass Oldershaw column, which is sold in various forms by several vendors. However, the maximum available internal diameter is 100 mm or about 4 inches (failing the 8-inch criterion for the absence of wall effects), so this device is more suitable for bench-scale use. It is often used in vapor–liquid studies.

The first requirement of piping is that it must not leak. One sure way is to use welded piping everywhere. However, pipe runs are never continuous, so piping must be joined at places by openable connections. These joints may be permanent, but others need to be openable unions. There is a choice between tubing and standard threaded piping. It is generally recommended that, for pilot plants, tubing and compressible fittings are preferred for resistance to leakage, for size ranges, and for flexibility. Tubing and fittings are generally made of stainless steel but are available in other materials needed to withstand corrosion.

In the valves of full-scale plants, the flow in valves is generally turbulent. This is the assumption in the commonly used equation for liquid flow and/or pressure drop,

$$C_v = Q\left\{[SG/(P_1 - P_2)]^{0.5}\right\} \tag{4.8}$$

- where C_v Is the valve flow coefficient when fully open,
- Q is the volumetric flow of liquid in US gallons per minute,
- SG is the specific gravity of the fluid relative to water, and
- P_1 and P_2 are the inlet and outlet pressures in pounds per square inch.

For small pilot plants, needle valves are almost exclusively used because of their ability to regulate low rates of flow. The flow is generally laminar or in laminar-transition. Equation (4.8), which is just a definition, holds, but the characteristic curve of pressure-drop versus flow for various valve openings becomes more complicated. The valve vendor should have an equation or a characteristic curve since this is required for sizing the valve and for including it as a final control element in a feedback loop. Whereas valves in a full-scale plant often include various refinements, it may be best to opt for simplicity for pilot plant valves.

4.2.3 Scale Considerations for Heat Transfer and for Flow Regime

The small vessels of a pilot plant have larger surface-to-volume ratios than the vessels of a full-scale plant. For heat transfer, this is an advantage and a disadvantage. When a vessel requires heat, it is easier to supply it from outside, i.e., from a jacket or by electrical heating. On the contrary, if heat is supplied at only one point – for instance, at the bottom of a distillation column – then it is easy for it to be lost from the rest of the vessel and completely disrupt the mass transfer function. Minimizing the loss requires insulation and/or jacketing. Some small distillation columns (e.g. Oldershaw) provide vacuum jacketing to impede heat loss.

The full-scale commercial vessels have lower surface-to-volume ratios and that too can be an advantage and a disadvantage. Heat losses might be controlled by insulation only, whereas the pilot plant required jacketing. However, if the vessel requires positive heat input, then heating through the walls may need to be supplemented by heating from internal coils or by an external heat exchanger through which contents of the vessel are continually passed. A poor solution is to raise the temperature of the heat transfer medium: This causes local excessive temperature in the process fluid next to the heating surface and potential degradation.

The designer plans for a given flow regime in a vessel, quite often fully back-mixed unless internal baffling is provided to direct the flow. In a glass pilot vessel, full mixing is easily checked.

Especially to be checked is that there are no dead or stagnant spots where degradation and/or overheating could occur. If this issue is a real concern, then a CFD (computational fluid dynamics) could be made, showing flow patterns for both pilot plant and full-scale vessels. An experimental technique – process tomography – with electrodes inside the vessel provides a comprehensive picture. Tracer tests for any vessel are also revealing: An extraneous substance is added to the feed, and then samples are taken of the vessel output over time and analyzed for this tracer. The time-wise pattern is indicative of the internal flow pattern within the vessel. Of course, this only works for an existing vessel. If there are serious questions about the behavior of the full-scale vessel, then a full-sized mock-up can be built and tested.

4.3 Mathematical Modeling

Mathematical modeling is a very valuable supplement to the physical design of any plant, including a pilot plant. There is a widespread opinion that ordinary engineers no longer do math modeling but rather resort to one or another of the commercial simulators. Also, unfortunately, general-purpose languages are not always taught in undergraduate programs. However, commercial simulators have licensing fees and, because they are general purpose, may not be a good fit to the problem at hand. It is valuable to have a person on the design team who is able to construct models in a general programming language, e.g., FORTRAN or C. As ideas arise, the model can be adjusted to simulate them, possibly saving a lot of time and allowing more ideas to be tested. The existence of a well-documented math model is a great addition to the final report on the whole project. The building of a model takes time, so it should be started as soon as the broad outline of the process is known.

The power of math modeling may not be apparent to process designers. Many are aware of only the *steady-state* "flow sheet" type of model. In the example of Figure 2.1, this is the model that produced the flow rates in Figure 2.1.

The steady state equations, assuming chemical reactions are first-order, are presented for the process presented in Figure 2.1.

Author's note: A mass basis is used in this example rather than the more common molar basis. A different situation may require a modified approach:

$$A1 + A3 - \mathrm{k}\,M\,A2/(A2 + B2 + C2) + (\mathrm{k}/\mathrm{K})\,M\,B2/(A2 + B2 + C2) - A2 = 0 \tag{4.9}$$

$$A2 - A3 - A4 - A5 = 0 \tag{4.10}$$

$$\mathrm{k}\,M\,A2/(A2 + B2 + C2) - (\mathrm{k}/\mathrm{K})\,M\,B2/(A2 + \mathrm{B}2 + \mathrm{C}2) - \mathrm{kc}\,M\,B2/(A2 + B2 + C2) - B2 = 0 \tag{4.11}$$

$$B2 - B4 = 0 \tag{4.12}$$

$$\mathrm{kc}\,M\,B2/(A2 + B2 + C2) - C2 = 0 \tag{4.13}$$

$$C2 - C3 - C5 = 0 \tag{4.14}$$

where

M is the total mass in the reactor,
K is the equilibrium constant for $A \rightleftarrows B$ (dimensionless),
k is the forward rate constant for $A \rightarrow B$ (h^{-1}), and
kc is the forward rate constant for $B \rightarrow C$ (h^{-1}).

Solving these six equations requires knowledge of the values of M and K, k, and kc at the temperature of operation. It also requires the setting of one or more boundary conditions. In this case, it is sufficient to set:

the production rate of product B, i.e., $B4\,\mathrm{kg\,h^{-1}}$,

the purge rate of C, $C5\,\mathrm{kg\,h^{-1}}$, and

the ratio of A to C in the purge stream, $A5/C5$.

The process designer does not initially know M but rather seeks to determine what it must be. Another boundary condition, the ratio of B to A in the stream leaving the reactor, allows the reaction terms to be inverted and yield the value of M (this assumes the concentration of C to be small). The ratio of B to A is likely the specification that the designer wants to make since it defines the split between the reactor and separation column.

Steady-state modeling shows the flows and compositions that the designer hopes to see. However, process materials, being inanimate, do not read the flow sheet and therefore must be constrained, by process control, to adhere to the flow sheet. Control is therefore part of process design, and this aspect can be very much helped by creating a *dynamic* math model. Creating such a model should be very much within the capabilities of a skilled model builder.

Referring once again to the example of Figures 2.1 and 4.6, let M be the total mass of material in the reactor. Figure 2.3 shows this amount to be controlled by manipulating the flow of stream 1. A simple dynamic model is

$$\mathrm{d}[M]/\mathrm{d}[\phi] = F_1 + F_3 - F_2 \tag{4.15}$$

where ϕ is the time. Because of the nonzero derivative, we can numerically integrate the equation to follow how the process develops. However, unless the flows F_1, F_2, and F_3 rigorously follow the steady-state values, the reactor will either overfill or empty itself with time.

A controller function must be added to the model, and a proportional–integral control is selected,

$$f = \text{minus GAIN} \times \{(M - M^\circ) + \text{RESET}^{-1} \times \Sigma(M - M^\circ)\Delta\phi\} \tag{4.16}$$

where f is the correctional signal to the manipulated variable, F_1,

GAIN and RESET (or RESET time) are the controller constants, and

M° is the set point for level in the reactor.

For now, derivative action is being neglected,

$$\text{Instantaneous } F_1 \text{ (for use in Eq. (4.5)} = f + F_1 \text{ (from the steady} - \text{state equation))} \tag{4.17}$$

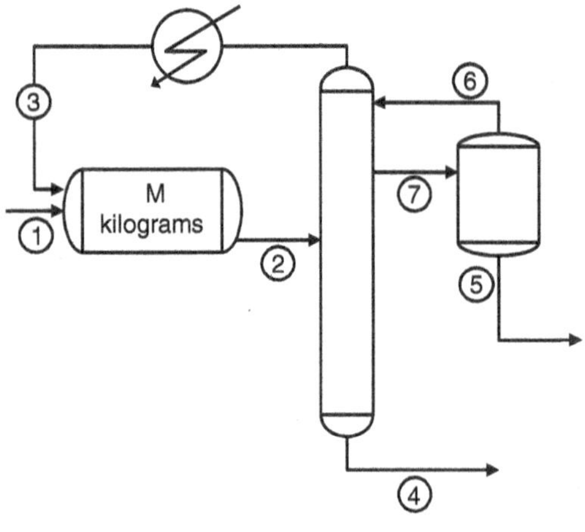

Figure 4.6 Reactor with recycle.

With this very simple model – Eqs. (4.15)–(4.17) – the fluctuations in F_1 and M can be tracked with time.

This model can be used to determine the best values for GAIN and RESET. In Eq. (4.7), a periodic (sinusoidal or square wave) fluctuation can be imposed on F_2 or F_3 and the effect can be seen in the resulting predictions of fluctuations of mass M. The values for GAIN and RESET can be varied to minimize the fluctuations in mass, thus indicating the best choices.

The simplicity of this example is owing to the simplicity of the task, namely (i) controlling the overall contents of a simple well-mixed vessel and (ii) using an active final control element – a pump. Creating a similar model for the control of temperature in the reboiler would require knowledge of the relationship between the liquid composition and temperature and also of the characteristics of the control valve.

References

Berg, D.A. (2011 Jul). Reduce piloting time and cost. *Chem. Eng. Prog.* 107 (7): 34–38.

Lockhart R.W. and Martinelli R.C. (1949 Jan). Proposed correlation of data for isothermal two-phase, two-component flow in pipes, *Chem. Eng. Prog.* 45 (1), 39–48.

Baker O. (1954). Simultaneous flow of oil and gas, Oil and Gas Journal, 53 (12), 185; 1957, 55 (45), 154.

Gmehling and Onken (2000), Gmehling and Onken series. https://www.ddbst.com/ddb-vle.html.

5

Operating Problems: Solution by Design

The objective in this chapter is to anticipate problems that may occur during full-scale operation and to suggest how initial design can help avoid them before they ever occur.

There will be enough problems that are not foreseen: Those residual problems will call for a very important skill in plant operation, namely troubleshooting. However, the goal of this chapter is to minimize the necessity for that skill.

Ten common potential trouble spots are examined here:

1. Buildup of extraneous substances
2. Corrosion
3. Erosion and cavitation
4. Flashing and phase separation
5. Excessive foaming and entrainment
6. Interaction between units
7. Liquid hammer and vibrations
8. Restrictions in piping systems
9. Scaling and fouling
10. Static buildup

5.1 Buildup of Extraneous Substances

Substances other than the desired product are often tolerable at low concentrations either in the product itself or at intermediate points. A substance may simply be considered as a product impurity. It may have entered with the feed stream or it may be internally generated chemically. At steady state, the goal is to have it expelled at the same rate as it enters or is generated and to have its *in situ* concentration low. Figure 5.1 shows a situation where this is not the case. The gas mixture which enters the condenser has a component which does not condense, at least under the conditions of the condenser. The mass of gas would gradually extend down into the vessel and interfere with its operation. The solution to the problem is also shown in the figure.

Figure 5.2 shows a distillation column (on the right) in which a minor component Z is trapped and reaches major concentrations. This happens when it vaporizes at the high temperature near the bottom of the column and condenses at the lower temperature near the top. The solution may be to withdraw a sidestream purge at the point where it has maximum concentration, or possibly

Practical Process Design for Chemical Engineers, First Edition. Keith Marchildon and David Mody.

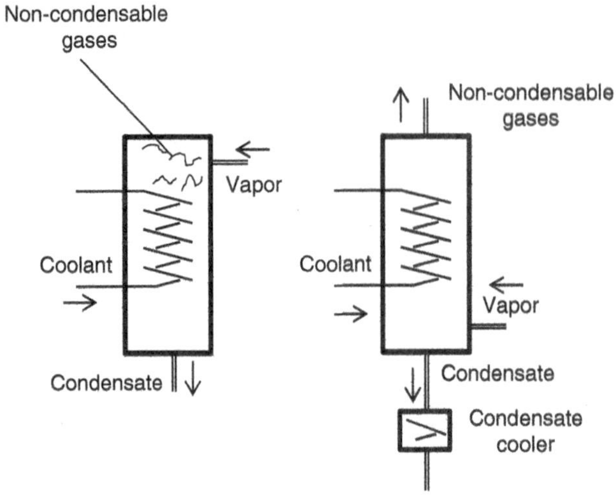

Figure 5.1 Non-condensables in a vapor condenser.

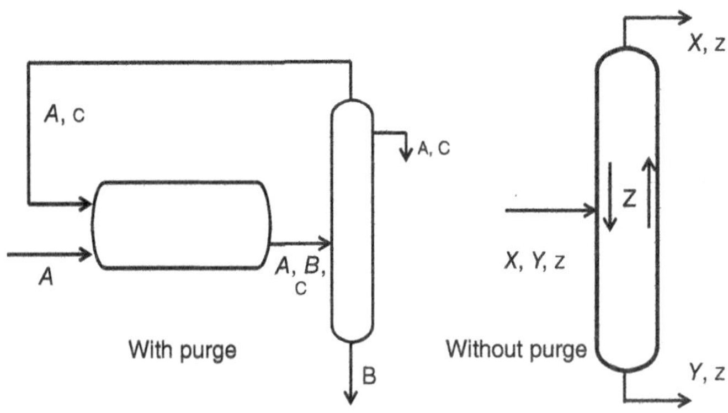

Figure 5.2 Entrapment and purge of components.

the column pressure could be changed, which alters relative volatility and affects boiling points, so that the component exits the tower. Figure 5.2 shows a purge stream of component *C* from a recycle system which helps keep its concentration in control. Such a stream unfortunately includes some wastage of useful components.

Dealing with component trapping requires knowledge of the chemistry of the process. It is also an effect that an integrated and comprehensive pilot plant needs to investigate.

Kister (2004) and Ramchandran (2006) provided more information on this topic.

5.2 Corrosion

The chemical breakdown of piping and containment vessels can take place both internally and externally. The latter occurs between insulation and the outer wall and is due to moisture invading that space. The prevention is to cover up outdoor insulation. To detect issues, potential trouble spots should be subjected to periodic visual, ultrasonic inspection, or X-ray inspection.

In choosing material for vessels and piping, the designer consults tables of material compatibility. If the process materials are complex, then testing should be done at conditions of eventual operation. Coupons of possible construction materials can be installed in a pilot plant, and their weight loss is measured over time.

The temperature of a heat transfer surface should be kept low by making it large and by providing a favorable coefficient of heat transfer. This minimizes the undesirable chemistry in both the surface metal and the process fluid next to it.

The quality of the construction material also plays a role. It should be checked (and rejected) for defects, sharp edges, crevasses, and dissimilar metals. The last condition can occur when an instrument of a different metal is installed.

Kane (2007) discussed corrosion monitoring, and Whitaker (2005) discussed corrosion under insulation.

5.3 Erosion and Cavitation

Over a long time, the interior of vessels and pipe will shed a few molecular layers due to friction with process fluids. Assuming no chemical attack, this erosion can largely be ignored if the flow velocity is kept to recommended rates [Ludwig (1999) provided additional guidelines]:

- 1.2–2.4 m s^{-1} for low-viscosity liquids,
- 25–40 for steam, and
- around 30 for air or gas.

The walls of pneumatic conveying pipes for solid particles are an obvious candidate for erosion, but the vendor/designer sizes these pipes with this in mind. A high-density solid material with sharp edges might warrant special attention.

Cavitation is the sudden appearance of vapor bubbles in an otherwise flow of liquid. When this happens in a centrifugal pump – which would be at the entrance hub where the local pressure is at a minimum – it can be damaging to the pump as well as reducing efficiency. Pump manufacturers specify a minimum net positive suction head (NPSH), for a given pump, namely how far above the fluid vapor pressure the upstream entry to the pump must be. This allows for the drop between the pump inlet pressure and hub pressure. That pressure difference is actually expressed in the "head" of liquid to allow for differences in liquid density. If a pump is run with an NPSH less than the manufacture-specified minimum, then cavitation may occur. It is recommended that the piping diameter and vertical sections be designed to keep the NPSH available at least 3 ft, or 20%, above the minimum required by the manufacturer.

The same effect can also occur for valves, as seen in Figure 5.3.

Ludwig (1999) provided guidance.

Figure 5.3 Cavitation in a valve. Source: Adapted from Ludwig (1999).

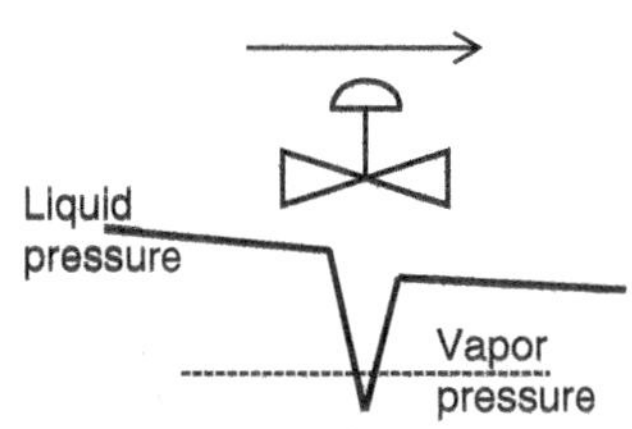

5.4 Flashing and Phase Separation

Some varieties of this phenomenon are as follows:

- A liquid stream splits into a vapor–liquid mixture.
- A liquid stream splits into two immiscible liquid phases.
- Solid particles appear in a liquid stream, or a coating starts to precipitate on solid surfaces.
- A vapor starts to condense into a liquid (or a solid). This is a particular problem when it happens in or before a compressor, which is designed to handle gases but is damaged by the presence of liquid.
- A solid being conveyed starts to liquefy or sublime.

To guard against these occurrences, it is fundamental to understand the phase equilibria and chemistry of the process materials. This lets the design maintain conditions of pressure and temperature to prevent the undesirable splits. In construction, cold spots and hot spots should be eliminated. The pipe or vessel may have to be traced with tempered water or hot oil or electrical heating. If high pressure is involved, seek out and consult an academic or other expert with a high-pressure viewing facility that allows the observation of phase behavior.

5.5 Excessive Foaming and Entrainment

As part of the process design, a determination should be made of the propensity of the process materials to foam. Organic compounds possessing polar groups are especially prone to foaming. The usual method of reducing this action is by the addition of an antifoam compound.

Whether a liquid is a foamer or not, it will expand if gas or vapor is passed up through it. See Figure 5.4. The key parameter is the superficial velocity of the gas U_S and of its related parameter the f-factor. A rule of thumb is to keep the superficial velocity less than 0.5 ft s^{-1} or 0.15 m s^{-1}. The f-factor is defined as

$$f = U_S/\sqrt{\text{liquid density}}$$

and it should be kept less than 0.1 in imperial units and 0.12 in SI units. This is done by making the cross-sectional area of the column adequate.

For distillation columns, the Souders–Brown criteria should be consulted.

The expansion of the mixture is related to the bulk density of the mixture,

$$H_{\text{expanded}}/H_{\text{liquid}} = 1/\text{bulk density}$$

where, for low viscosity liquids, the bulk density can fall to 0.6 at high U_S, and for high viscosity liquids, down to 0.35.

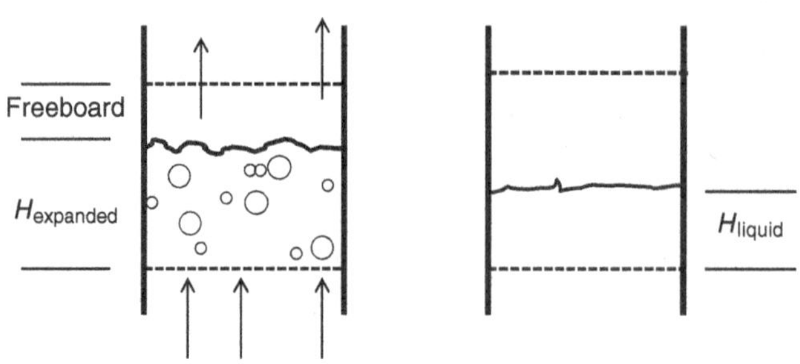

Figure 5.4 Expansion of a two-phase gas–liquid mixture.

Adequate freeboard must also be provided in order to allow liquid entrainment to settle back into the pool.

Koller (2004) discussed foam control in fermentation processes.

5.6 Interaction Between Units

Figures 5.5–5.7 illustrate some of the ways in which one part of a system inadvertently affects other parts. The system in Figure 5.5 shows two flows, *A* and *B*, fed by a common source. If the common line is too small, then the pressure drop in the common line imposed by the flow to *B* restricts the flow to *A*. The simplest solution is to make the common pipe bigger. A more expensive solution is to provide each inlet with its own separate pump.

Figure 5.6 shows flooding of a vessel because of low pressure. A solution would be to have a separate drainage system from the low pressure vessel. A cheaper solution would be to install a flow restriction (e.g. a valve) in the line between the two vessels so that the liquid from the second vessel sees a lesser downstream pressure (P_2 or less) and is able to flow.

Figure 5.7 shows a venting system where the driving force is inadequate: P_3 is too high.

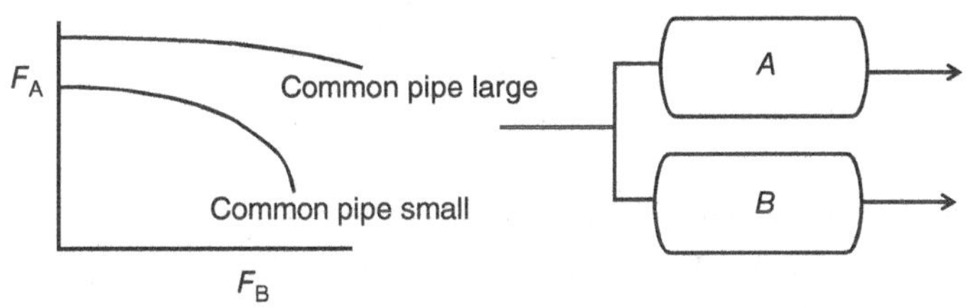

Figure 5.5 Interaction of parallel fluid flows.

Figure 5.6 Flooding due to pressure imbalance.

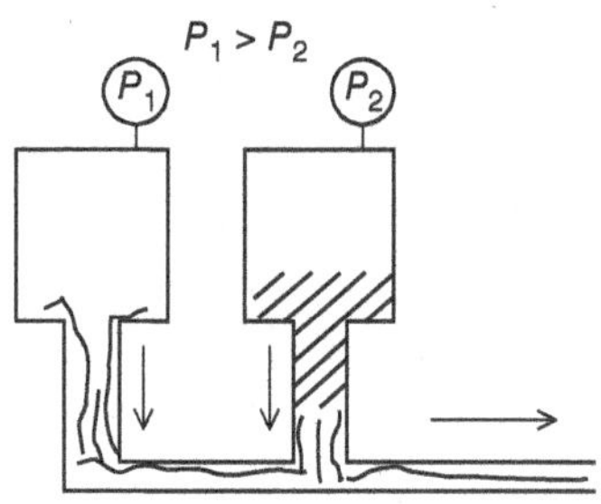

Figure 5.7 Venting inversion.

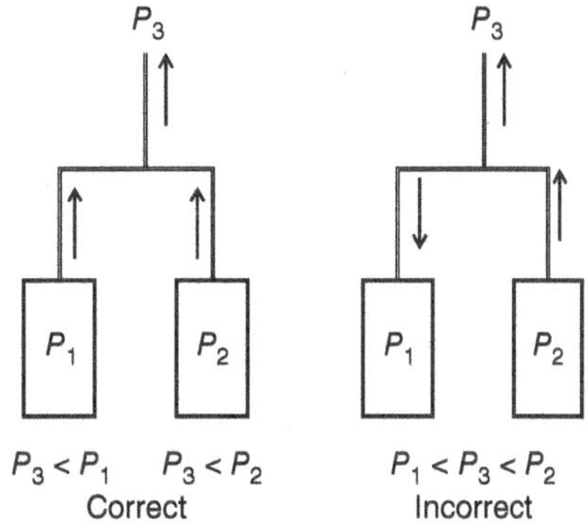

Pressure balances are often overlooked when chemistry, mass transfer, and heat transfer are the focus. Time should be scheduled for a separate review of this potentially troublesome aspect – which is generally fairly easy to correct. A "pressure profile" of the process can be generated to assist in understanding and eliminating such problems.

Other interactions can be prevented by providing buffer vessels between process units.

Structuring of control schemes requires consideration of unwanted interactions: avoid the tail wagging the dog.

5.7 Liquid Hammer and Vibrations

Two-phase gas–liquid flow can assume a number of forms, depending mainly on the ratio of gas flow to liquid flow and also on the velocity of the overall flow. Sketches of the varieties of flows and predictive methods are given in Chapter 12. For the present, the warning is to avoid the slug-flow region. This region is shown on the Baker plot for horizontal flow and on the Charles plot for vertical flow (Figures 5.8 and 5.9). In this type of two-phase flow, the gas and liquid travel in separate axial pockets. The liquid pockets, impelled by the high gas velocity, reach a speed such that at any change in direction, there is a large force on the piping and its support. This action is well recognized (especially by anyone in the vicinity of the steam hammer) and is considered potentially damaging and dangerous. Any situation where the two-phase flow is going to occur should be checked to make sure the flow regime is in a safe area of the Baker or Charles plot. If the desired flow is a gas and the liquid is simply a condensate, then installing a low-point drain in the pipe can help shift the flow regime into a safer territory.

"Hammer", also called water hammer or hydraulic shock, can also occur with only one phase (liquid). The mechanism is the sudden stopping of flow (e.g. closing of a valve) and the sudden change in fluid momentum. Significant damage to the piping can occur. Slowly closeling valves, providing surge suppressors, or piping changes can provide protection against liquid hammer.

The tubes in heat exchangers sometimes enter into vibrations. One such vibration mode occurs in sympathy with the frequency of periodic shedding of fluid from the wake behind a tube (Figure 5.10). A vendor would know about the conditions where this could occur, as well as conditions for other modes of vibration. It could affect the mechanical integrity of the heat exchanger.

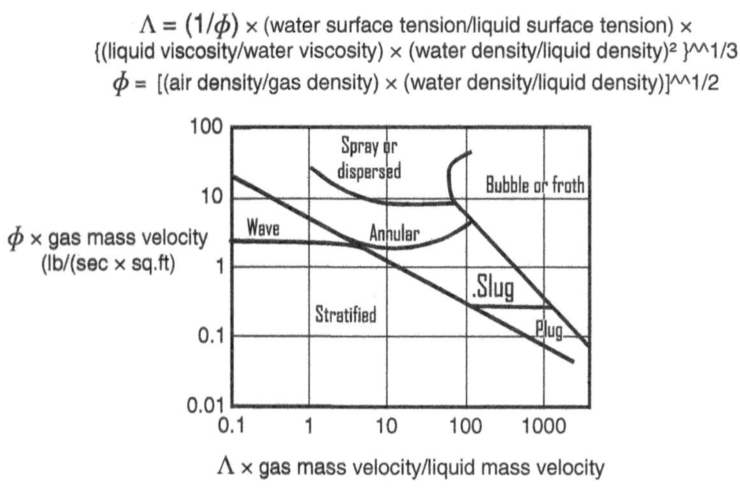

Figure 5.8 Generalized Baker plot.

Rv = volumetric flow rate of gas / volumetric flow rate of liquid

(Fr)tp = Froude No. = (two-phase velocity)^2 /(acceleration due to gravity × pipe diameter)

$\Lambda = (\mu_{liquid} / \mu_{water}) \times (\rho_{water} / \rho_{liquid})^A \times (\sigma_{water} / \sigma_{liquid})^B$ A = 0.5, B = 0.75

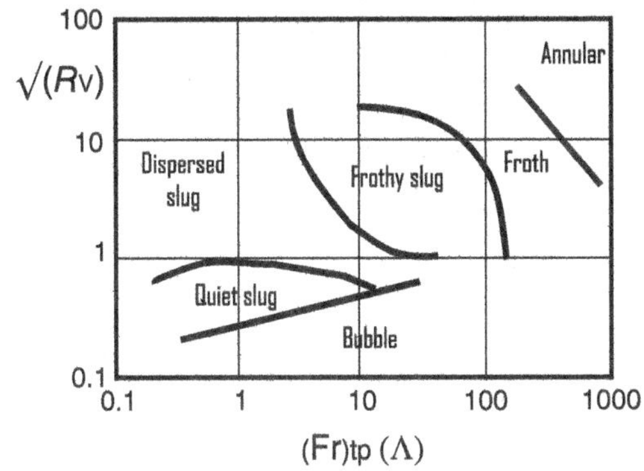

Figure 5.9 Charles plot for vertical two-phase flow.

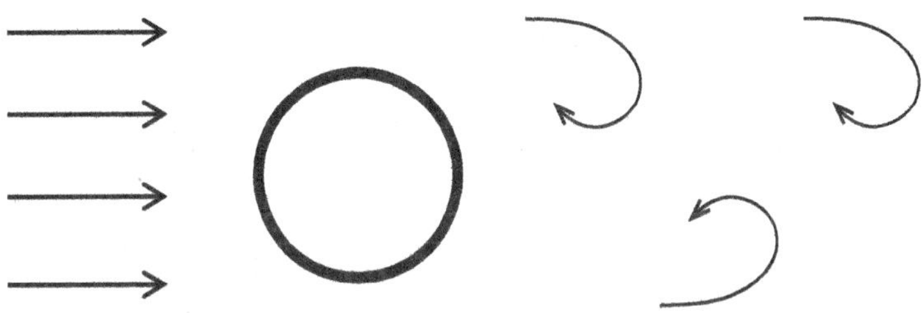

Figure 5.10 Wake shedding around cylindrical pipe.

5.8 Restrictions in Piping Systems

Among much more complicated design questions, inadequacies in piping should seldom happen. However, there are several situations that could arise.

1. It is desired to expand the plant, but the piping – sized as just big enough initially – imposes flow restrictions.
2. The pressure drop of fittings and valves was underestimated.
3. The piping got rougher because of erosion, corrosion, or precipitation, causing the friction factor to rise. Cavitation in the piping may not increase the pressure drop, but it may cause the piping or valves to fail prematurely.
4. A gravity-driven flow is impeded by vapor entrainment.
5. A gas flow, assumed to be incompressible and so calculated, turns out to have a Mach number greater than the 0.3 limit for incompressible-flow assumptions for calculations.

To minimize the occurrence of these difficulties, there are a few measures:

- Err on the generous side in pipe sizing, unless settling or stratification could occur. The larger size piping will impose less pressure drop initially and will facilitate plant expansions in the future.
- Use the most conservative predictions of pressure loss in valves and fittings.
- Use predictive methods for gravity flow, where entrained air or gas can accompany the liquid.

- If there is any doubt about whether a gas flow should be considered compressible (Mach No greater than 0.3), use the isothermal compressible-flow formula for pressure drop,

$$P_1^2 - P_2^2 = (4f\,L/D) \times [G^2\,\mathrm{R}\,T/(g_C\,M)] \times \{1 + [2/(4f\,L/D)]\,\ln(P_1/P_2)\} \tag{5.1}$$

  which reduces to the usual incompressible-flow formula at low Mach,

$$P_1 - P_2 = (4f\,L/D)\,[G^2/(2\,g_C\,\rho)] \tag{5.2}$$

where P is the pressure, f is the friction factor, L is the pipe length, D is the pipe internal diameter, G is the gas mass velocity, R is the gas constant (mass basis), T is the temperature, g_C is the flow-to-pressure conversion factor, M is the molecular weight, and ρ is the gas density.

5.9 Scaling and Fouling

Scaling and fouling show up gradually, typically in a loss of heat transfer across a surface or in an increase of pressure drop in the fluid flow. For heat transfer, these effects may be slowed down by providing adequate transfer area so that the surface temperature does not have to be excessive and by ensuring high fluid sheer rates (Nesta J. – Hyd. Proc. 2004). For fluid flow, provide access points where brushes or other cleaning tools can reach the scaled or fouled surface.

Passivating chemical agents may be able to keep "crud" from adhering to metal walls of pipes and vessels. See references for further applicable guidelines and articles on this.

5.10 Static Buildup

The sudden discharge of electricity between two unequally charged surfaces can be an unpleasant sensation when a home is at low humidity during the winter.

In an industrial plant, it can be the trigger of a fire or an explosion.

Electrostatic charge can develop whenever work is done on a liquid or solid. For a liquid, the work may consist of forcing it through a pipe or filter, agitating it, spraying it, and letting it fall or settle.

For solid pellets or powders, the work may consist of pneumatically transferring them through a pipe, blending them, grinding them, and classifying them.

Most of the hazards are eliminated by connecting vessels and piping to ground and by conductively bonding various parts of the system to one another.

There are several measures that can be taken to avoid this very serious situation of static buildup:

1. Ensure that filling systems operate at low velocity (less than 3 ft (1 m) per second if there is free fall of the fluid),
2. Avoid mist-generating situations,
3. Provide adequate and secure grounding and bonding,
4. For procedures where bonding is provided only periodically, allow time for the charge to "relax,"
5. For particles or powders that have been in motion, provide a charge-dissipation step before loading into nonmetallic containers, and
6. Refer to API Recommended Practice, "Protection against ignitions arising out of static, lightning, and stray currents".

References

Beain, A.; Heidari, J.; Gamble, C., Properly clean out your organic heat-transfer fluid system, *Chemical Engineering Progress*, 2001–2005, Vol.97 (5), p.74-77.

Bott, T. Reg, To foul or not foul: that is the question, *Chemical Engineering Progress*, 2001–2011, Vol.97 (11), p.30-36

Nesta, J. and Bennett, C.A. (2004). Reduce fouling in shell-and-tube heat exchangers. *Hydrocarbon Processing* 83 (7): 77–82. 5p.

Joshi, H.M., Mitigate fouling to improve heat exchanger reliability., *Hydrocarbon Processing*. Jan1999, Vol. 78 Issue 1, p93. 3p.

Kane, R.D. (2007 Apr). A new approach to corrosion monitoring. *Chem. Eng.* 114: 34–41.

Kister, H.Z. (2004 Aug). Component trapping in distillation towers: causes, symptoms and cures. *Chem. Eng. Prog.* 100: 22–23.

Koller, K. (2004 Aug). Foam control in fermentation processes. *Chem. Eng.* 111: 24–27.

Ludwig, E.E. (1999). *Applied Process Design for Chemical and Petrochemical Plants*, vol. 1. Gulf Professional Publishing.

Polley, G.T. (2002 Dec). Put fouling in its place. *Chem. Eng.* 46–49.

G. T. Polley; Wilson, D. I.; Petitjean, E.; Derouin, C., The fouling limit in crude oil preheat train retrofits, *Hydrocarbon Processing* Jul2005, Vol. 84 Issue 7, p71-80.

Turner, J.; Asquith, R. J., Stop foaming on hydrotreater 'hot' separator, *Hydrocarbon Processing* June 1999, Vol. 78 Issue 6, p113. 4p.

Ramchandran, S. (2006 Mar). Minimize trapped components in distillation columns. *Chem. Eng.* 113: 65–70.

Whitaker, T. (2005 Apr). Avoid pipe corrosion under insulation. *Hyd. Proc.* 84: 75.

6

Process Monitoring and Control

An important part of any chemical process is the instrumentation that is provided to monitor and control it. The common element of the two functions, monitoring and control, is measurement. Not every quantity that is measured gets controlled, but every quantity that is controlled must be measured. Many noncontrolled quantities are measured in order to understand how the process is functioning and to detect if it is drifting away from normalcy. From the point of view of the process operator, the more things that are measured the better, including sometimes hard-to-measure but important quantities. From the point of view of management and the project person who is installing the process, the fewer things measured the better because instrumentation costs money. It is often a tug-of-war between the two parties.

The control that is discussed in this chapter is *automatic process control*, which is the regulation of the process via quick-acting mechanisms that do not require human intervention except for the setting of set points (SPs). There is another type of control: *statistical process control*. Although not discussed in this chapter, it merits a description because it too contributes to the success of the process once in operation. It is "powered" by humans and takes place over time. The core activity is the time-wise gathering and statistical analysis of process data, including especially the characteristics of the final product. Each of these process quantities has a mean and an inherent variability about the mean. The data gathering aims to detect any departure from the mean or the usual variability. The aim is to detect incipient problems, determine the source, and correct them early.

The following topics of automatic process control are examined in this section:

- the options for measurement of control variables (CVs),
- combinations of controllers for specific purposes,
- causes of non-optimum control, and
- programmable controllers and distributed control systems.

6.1 Options for Measurement of Control Variables (CVs)

A single control loop, as shown in Figure 6.1, comprises at least three, generally five, and sometimes more elements. Three of these elements – the controller itself (LIC, level indicator and controller) plus signal converters for analog-to-digital and digital-to-analog – are generic nonmechanical devices. A person knowledgeable in controls specifies the details of these elements with input from the process designer who communicates desired ranges and relevant process details such as freezing points, the presence of solids, or trace corrosive components.

Practical Process Design for Chemical Engineers, First Edition. Keith Marchildon and David Mody.

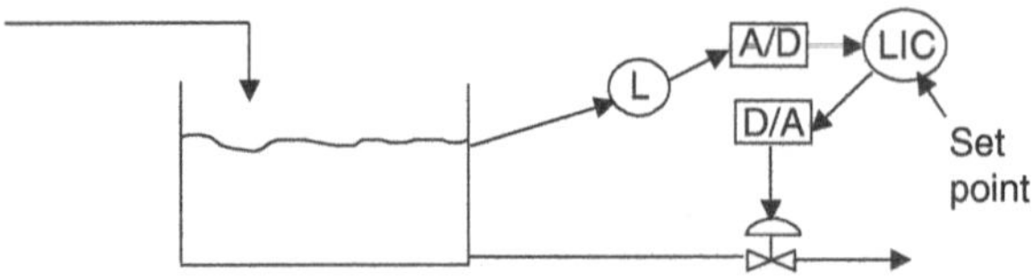

Figure 6.1 Single input, single output level control.

The fourth element is the final control device (a valve in Figure 6.1), and its choice may be a joint decision between a specialist and the process designer.

However, it is the fifth element that requires the greatest attention by the process designer: the device which continually measures the variable (the *CV*) which it desires to control. In the case of the control loop of Figure 6.1, the *CV* is the liquid level.

The other four components of the control loop are often left to the process control specialists, but choosing the method of measurement should involve the process designer and is therefore discussed here.

The following table shows most of the combinations of *controlled variables* and *manipulated variables (MVs)* that are found in chemical processes.

Other quantities that are controlled directly or by inference are compositions, including pH.

The *MV* is the "handle," whereby the controlled variable is being continually brought to its *SP*. It is seen that the same *CV* may be controlled by a choice of *MVs*.

In this section, measurement methods are reviewed for the four control quantities listed in the left-hand column of Table 6.1. The principle is described, and precautions are given that help avoid the "failure" of the instrument to deliver a reliable measurement. A fifth discussion concerns the inference of *composition* by the measurement of a physical quantity such as pH, refractive index, and viscosity.

Table 6.1 Common control loops in chemical processes.

Controlled variable	Manipulated variable
Temperature	Heating fluid pressure
	Heating fluid flow rate or valve opening
	Heating fluid temperature
	Heating fluid submergence
	Electrical current
	Electrical voltage
Fluid flow rate	Valve opening
	Pump speed
	Pressure or pressure differential
Liquid level	Rate of inflow or outflow
	Valve opening
	Pump speed
Pressure	Valve opening
	Pump speed or compressor fan speed
	Relief flow

6.1.1 Temperature

The simplest and most common device is the *thermocouple* (Figure 6.2). It makes use of the fact that when wires of two dissimilar metals are joined at two points of unequal temperature, a small voltage difference appears between the two ends. One end is the point where the temperature is to be measured. Theoretically, the other end should be kept at a fixed temperature, but, in practice, variation is compensated for electronically. The temperature at the measurement point is inferred from the electromotive force (EMF) at the other end.

The advantages of thermocouples are as follows:

- low cost and simplicity,
- high-temperature capability (750, 1250 °C depending on the wire materials), and
- potentially quick response.

The disadvantages are as follows:

- only moderate accuracy, e.g., +/− 2 °C,
- long-term instability,
- dependence on the measurement of small EMF, and
- susceptibility to extraneous electrical paths.

A thermocouple can be used bare-ended, but generally they are provided with a closed sheath to protect them from process fluids. The figure shows the thermocouple both insulated from the sheath and in contact with the sheath. The latter configuration gives faster response but requires that the sheath be insulated from the measurement circuitry. If the device is positioned in a thermowell, then it must make good contact with the end of the well.

Another electrically based instrument is the resistive temperature device or RTD (Figure 6.3). The principle is that the resistance of a metal changes with temperature. By making a length of wire (wound into a coil) subject to the temperature to be measured and building it into an automated Wheatstone bridge so that its resistance can be measured, the temperature is determined.

The wire coil occupies a finite space as opposed to the point junction of the thermocouple, so the measurement is not as localized.

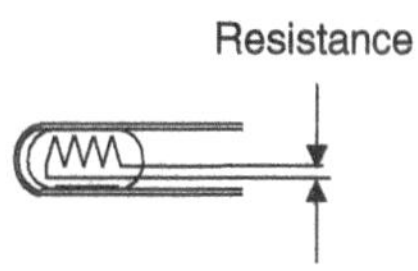

Figure 6.3 Resistive temperature device (RTD).

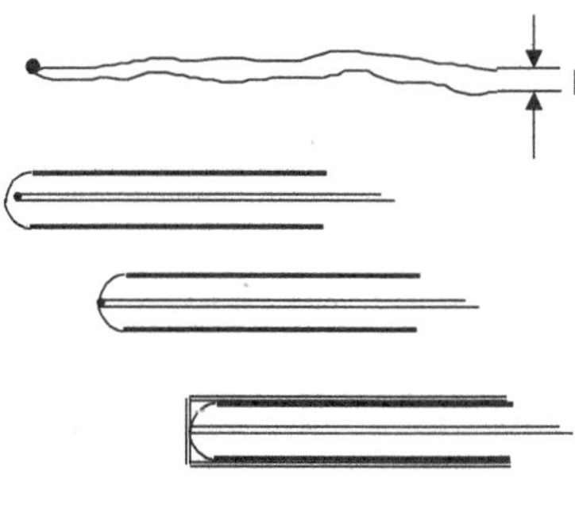

Figure 6.2 Thermocouples, sheathing and wells.

The advantages are as follows:

- accuracy, e.g., +/− 0.15 °C;
- long-term stability.

The disadvantages are as follows:

- cost,
- fragility, and
- upper temperature limit of about 500 °C.

Whereas the thermocouple must make good contact with the end of a thermowell, the RTD must have a tight fit along its sides.

6.1.2 Flow

One class of flow meter depends on the interconvertibility of pressure and momentum. In the *orifice* (Figure 6.4) and *venturi* meters, the flow is accelerated by passing through a constriction and the volumetric flow is inferred from the decrease in pressure.

A similar but more sophisticated device is the venturi tube (Figure 6.5), which is less prone to plugging or stagnation and which imposes less net drop in the pressure between upstream and downstream.

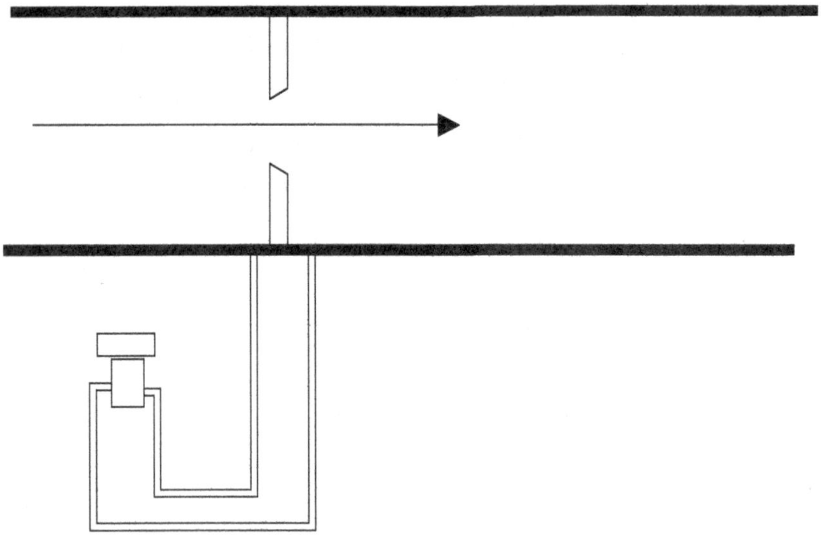

Figure 6.4 Orifice volumetric flow meter.

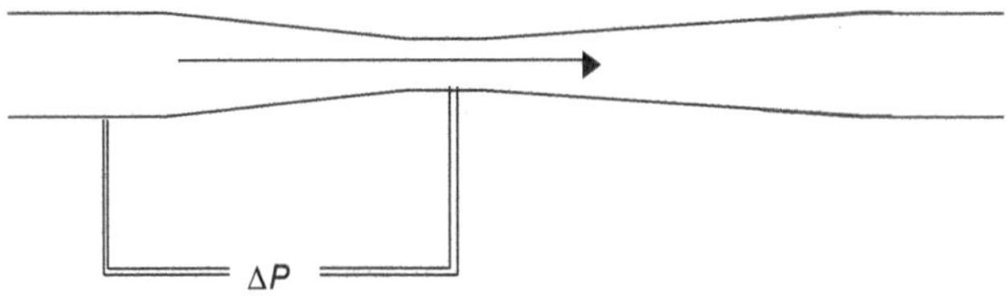

Figure 6.5 Venturi volumetric flow meter.

Figure 6.6 Pitot tube for volumetric flow.

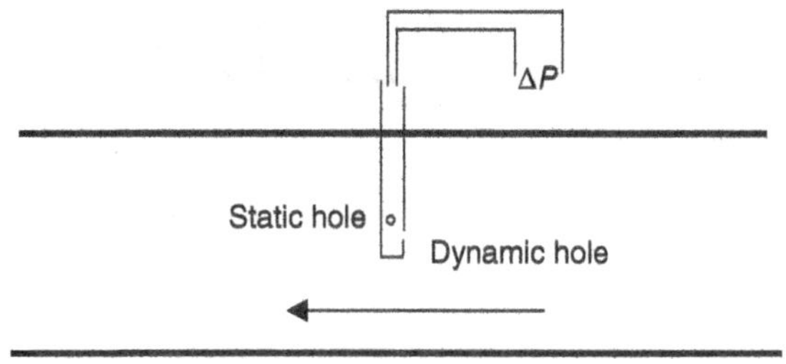

In the *pitot tube* (Figure 6.6), the flow is brought to zero at a point (the "dynamic hole") and the excess pressure above that at the "static hole" is a measure of velocity.

The pitot tube is even less obstructive to the flow but gives a local measurement of velocity which may not be representative of the total flow. The holes may also be prone to plugging.

An advantage of all these devices is their relative simplicity and low cost. A disadvantage is the fact that that the signal,

Pressure difference, ΔP is proportional to velocity^2

This relationship makes the devices inaccurate at flow rates below about 25% of full scale.

With all three devices, care must be taken to ensure zero or equal liquid head in the lines leading to the differential pressure cell.

Two instruments that provide a signal that is proportional to the first power of velocity are the *turbine meter* and the *vortex-shedding meter* (Figures 6.7 and 6.8). Both of them require an external electronic "pickup" to count either turbine rotation or shedding frequency.

The advantage of the turbine is good "turndown", about 10 : 1, and good accuracy, about 0.25%.

The disadvantage is susceptibility to fouling, especially of the bearings at each end of the turning shaft.

The advantages of the vortex-shedding meter are a turndown of 25 : 1, an accuracy of 1%, no moving parts, and resistance to fouling.

The disadvantages are the susceptibility of the electronics to miscellaneous noise and also the requirement of a Reynolds number range to ensure that the shedding frequency really is proportional to the velocity. This range is between 300 and 100 000 based on the diameter of the obstruction that generates the wake shedding.

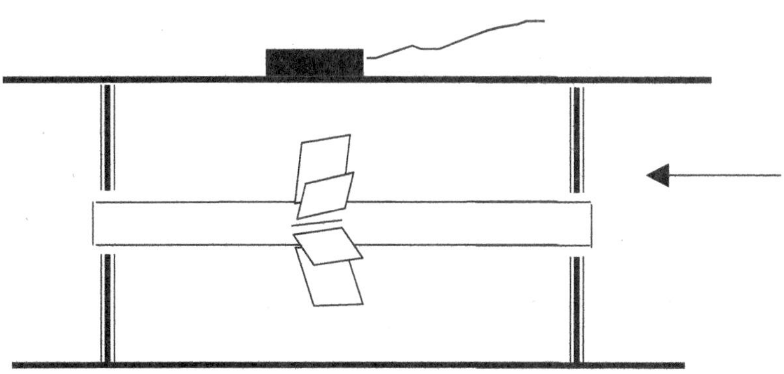

Figure 6.7 Turbine meter for volumetric flow.

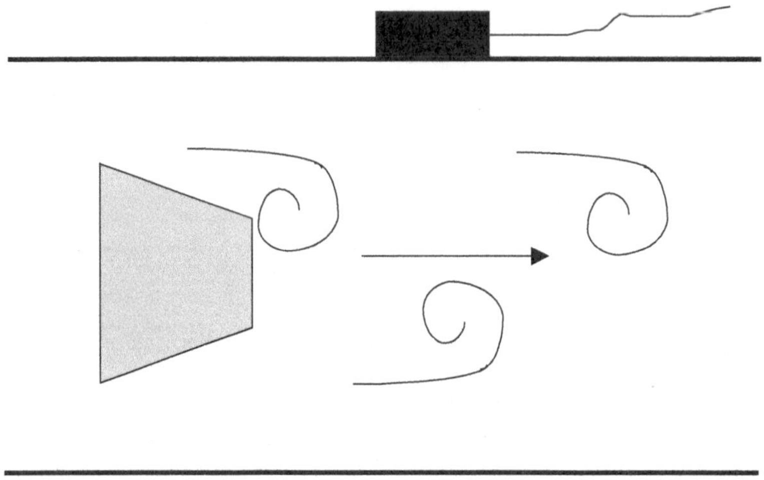

Figure 6.8 Vortex-shedding volumetric flow meter.

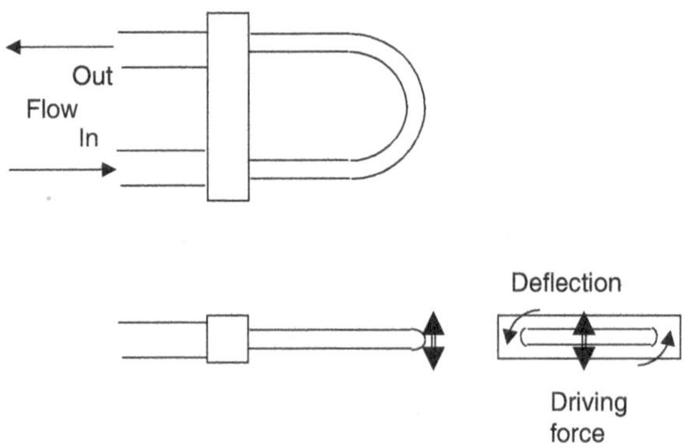

Figure 6.10 Coriolis mass flowmeter.

The *magnetic flow meter* or *magmeter* (Figure 6.9) works on the principle that a voltage (EMF) gets induced in a conductor moving through a magnetic field. In this case, the conductor is the flowing fluid, which must possess some conductivity and is therefore restricted to liquids. It is ineffective for demineralized water and nonconducting oils.

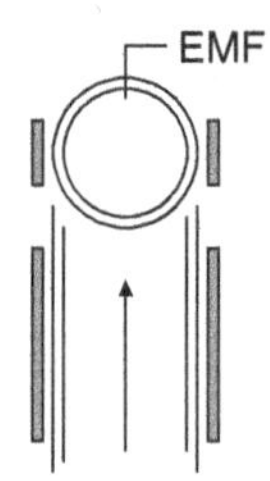

Figure 6.9 Magnetic volumetric flowmeter.

The advantages are a turndown of 30 : 1 and an accuracy of about 0.75%.

The disadvantages are as follows:

- restriction to conductive liquids;
- temperature limited to 200, 300 °C.

All of the above devices measure velocity or volumetric flow. The *Coriolis meter* or *mass flowmeter* (Figure 6.10) measures mass flow. The principle is that of Coriolis which is the rotational form of Newton's second law, Force = Mass × Acceleration. The fluid flows through a U-bend, the center of which is vibrated up and down. The result is a twisting of the U-tube by an amount proportional to the mass flow rate.

The advantages are as follows:

- high accuracy, 0.1%,
- high turndown, 100 : 1,
- applicable to both gases and liquids,

- sensitivity to mass rather than volumetric flow,
- pressure range up to 1500 psi, and
- temperature range −240 to +400 °C.

The disadvantage is cost.

All of the above measurements produce a signal related to instantaneous flow. An alternative is to accumulate and weigh material (M) over a time period (ϕ) and calculate the flow as M/ϕ. This could be a quick and simple method for bench-scale experiments. It is also a way of checking or calibrating the results of the above instantaneous methods. It has particular application for measuring the flow of solids, where the solid stream is periodically directed onto a load cell for fixed time intervals and the weight is measured.

6.1.3 Level

Two of the most common methods of measuring the level make use of the static-head pressure that a liquid exerts (Figure 6.11). In one case, the differential pressure between the vapor space above the liquid and a point below the surface is interpreted to determine the level. Seals can be provided to isolate the process fluid from the sensor to reduce the chance of freezing for plugging. In the second case (Figure 6.12), the pressure required to maintain a stream of gas into the lower levels of the liquid is measured.

The advantages of these two methods are simplicity and relatively low cost.

The disadvantages are as follows:

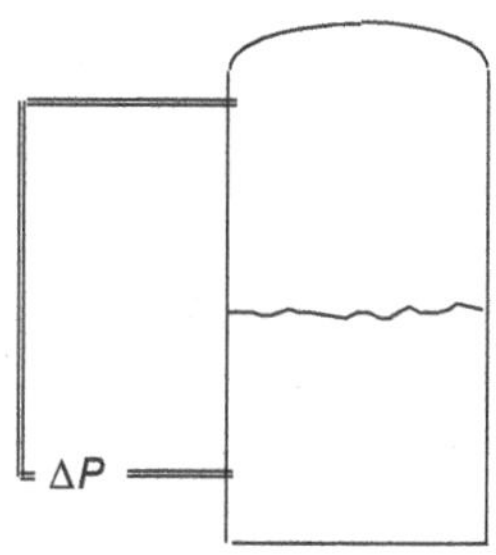

Figure 6.11 Liquid level by static head pressure differential.

- direct contact with the process liquid,
- dependence on the knowledge of density,
- restriction of the bubbler system to vented systems,
- necessity to keep lines free of condensate, e.g., condensate, and
- sweep gas reduces the possibility of plugging, but depending on the service, that possibility remains.

Figure 6.12 Liquid level by bubble-tube pressure.

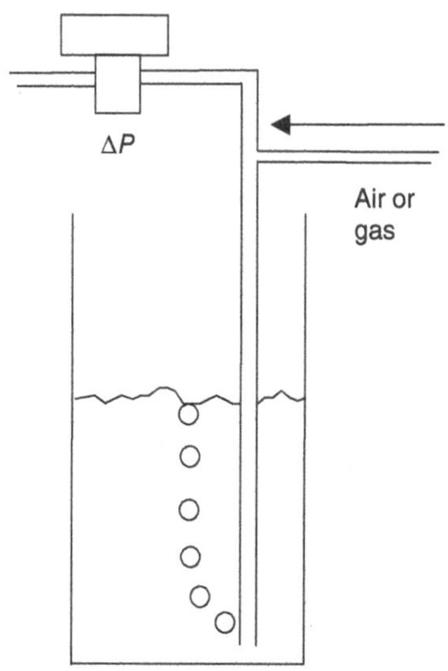

A better indication of the situation in the vessel is obtained by mounting a second measuring point at a different level in the liquid. The difference in static head between the two liquid-immersed points gives a measure of density, which can then be used in interpreting the height signal.

When it is desired to avoid intrusion into a vessel, for instance, if it is under high pressure or high vacuum or if the contents are hazardous, it is possible to use nuclear radiation to "see" what is inside (Figure 6.13). An expanding beam from a source on one side of the vessel is partially intercepted and attenuated by liquid in the vessel. The receiver on the other side picks up the residual radiation.

The nuclear source is usually the isotope Cesium 137 or Cobalt 60.

The advantages of this method are as follows:

- non-contact with process materials,
- effectiveness even through heavy-walled vessels,
- for existing vessels, no modification required, and
- low maintenance.

The disadvantages are as follows:

- stringent regulation,
- potential hazard to health if mismanaged,
- relatively narrow range of height, and
- confounding effect of foam and density.

The property of liquid capacitance can be used to detect the level (Figure 6.14). A probe is inserted into the liquid, and the capacitance is measured between an outer sheath and an inner core. The signal depends on how much of the surfaces are in contact with liquid.

The advantage is that such probes are in common use and therefore economical and well developed.

The disadvantages are as follows:

- full contact with the process liquid,
- capacitance sensitivity to the nature of the liquid, and
- possibility of surface fouling and therefore the change of response.

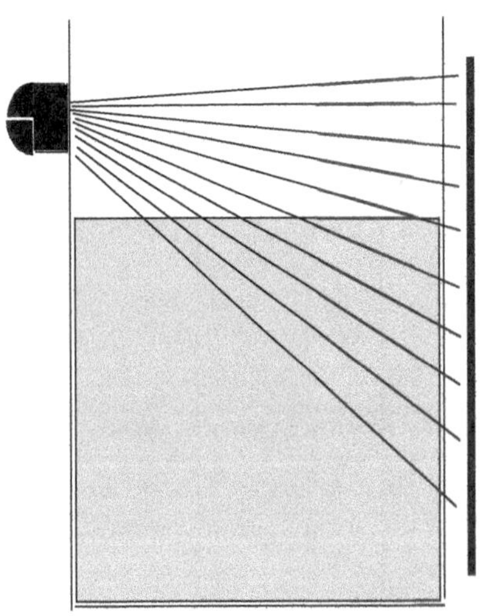

Figure 6.13 Nuclear level gage.

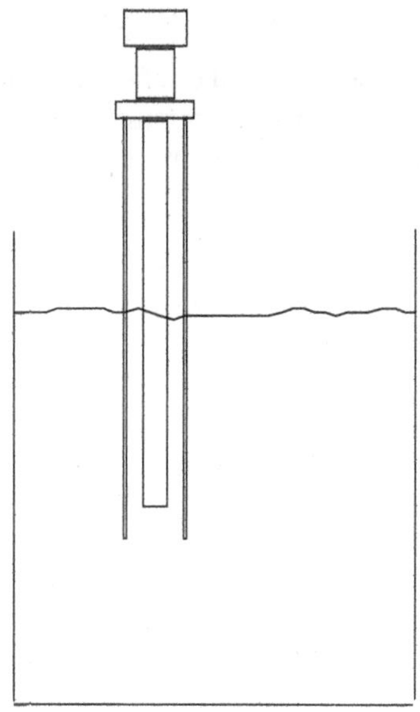

Figure 6.14 Capacitance liquid level probe.

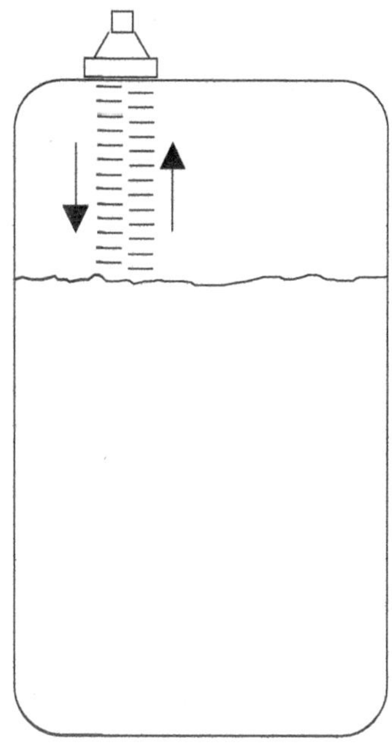

Figure 6.15 Ultrasonic level detector.

An ultrasonic source and detector mounted on the top of a vessel provides a measurement of level by measuring the time it takes for sound waves to travel down to the surface and reflect back (Figure 6.15).

The advantages are as follows:

- The intrusion into the vessel is at the top, where leakage and contamination are least likely.
- There are no moving parts.

The disadvantages are as follows:

- The signal can be affected by dust, foam, waves, and noise.
- The upper temperature limit is low.
- The instrument must be calibrated against an empty tank.

The level measurement of solids is complicated because of the fact that solids, unlike liquids, do not seek their own uniform level. A nuclear level gauge or an ultrasonic level detector could detect the peak of the solid pile. If only the amount of solids is required, then it can be weighed.

6.1.4 Pressure

Pressure is measured for its own sake and also as part of the measurement of flow and of liquid level. In many cases, the measurement is the difference of pressure between two points. The instrumentation is based on the movement of a flexible element, for instance, a bellows or a diaphragm, which is translated into a gauge reading and/or into a pneumatic or electrical signal for transmission to a controller (Figure 6.16).

If a differential pressure, say between "A" and "B", is being measured, the two pressures are connected to opposite sides of the instrument. If a single-point gauge pressure is desired, the instrument is put in "vented" mode, i.e., one side is left open to atmosphere. If single-point absolute pressure is desired, then one side of the instrument is plugged and evacuated (Figure 6.17).

In the case of differential measurement, the actual difference to be measured may be much smaller than the overall pressure of the system. In this case, the two sides of the device must be

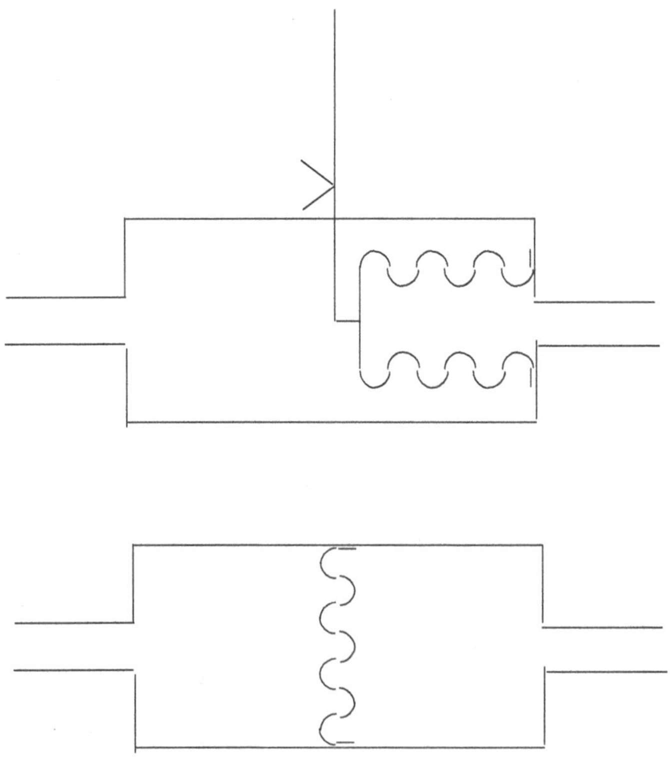

Figure 6.16 Bellows and diaphragms in pressure measurement.

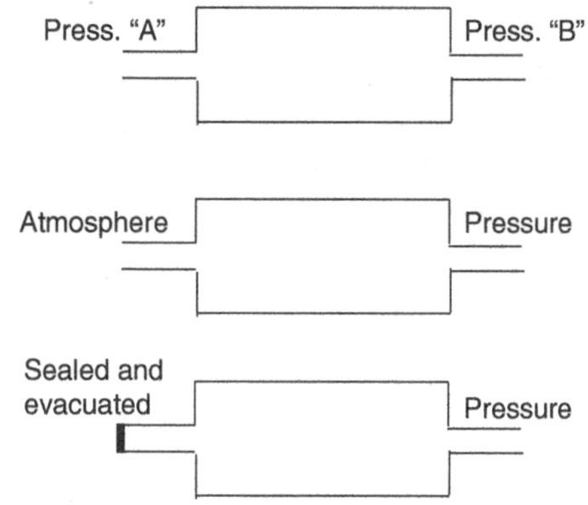

Figure 6.17 Pressure device configured for differential, gage and absolute measurement.

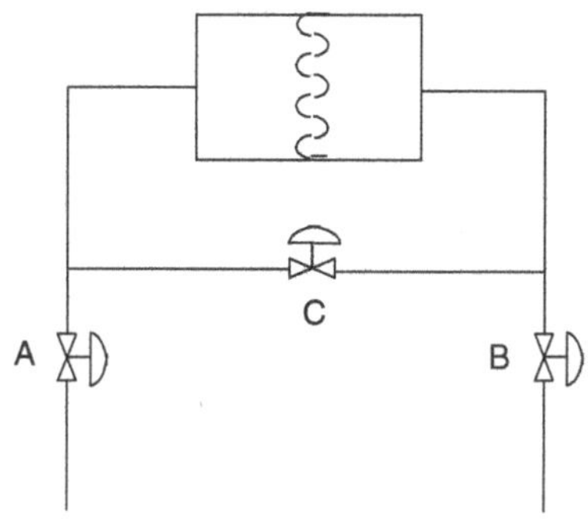

Figure 6.18 Differential pressure device with balance line.

equipped with a balance line which is open when the device is put in service to avoid having high unbalanced pressure exerted on one side of a delicate diaphragm or bellows (Figure 6.18). Valve "C" is then closed after both valves "A" and "B" are open.

6.1.5 Concentration

The most important information about a process stream is its content, and measuring that content is not as straightforward as measuring physical parameters like temperature and flow. The fallback is to take regular samples for analysis in a laboratory. The technique of *gel permeation chromatography* has been developed as an online instrument which can provide data for inspection and feedback within minutes. *Near infrared* can also detect key aspects of a stream and can be an integral part of a control loop. Both of these techniques require much time and expensive equipment, so they are not wide spread. However, they can make a process possible that would not otherwise be so.

The more common practice is to develop a relation between (a) some easily measured physical property and (b) the concentration of interest. Those physical properties include pH, refractive index, density, and viscosity. The relation must be robust in the sense that it is one-to-one and not influenced by extraneous factors such as temperature and the concentrations of other components.

6.2 Combinations of Controllers for Specific Purposes

The simplest configuration in process control is the single input/single output loop (SISO), as shown in Figure 6.1. The following illustrations expand on more complex possibilities.

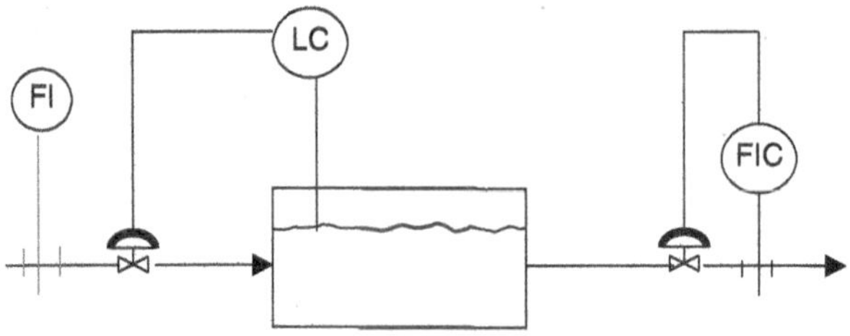

Figure 6.19 Level control and flow control.

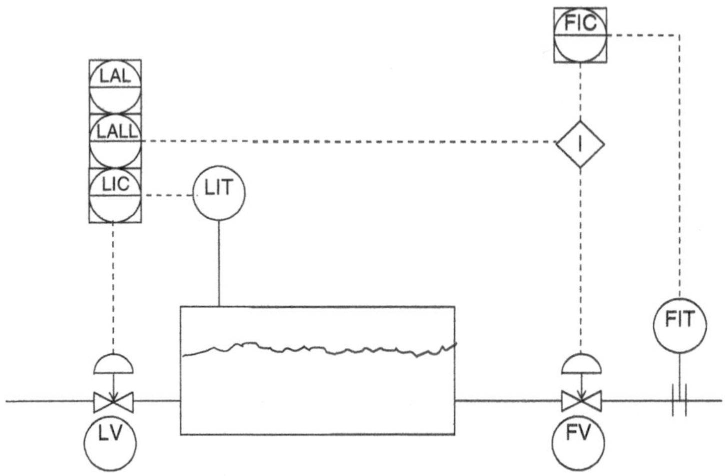

Figure 6.20 Alarm and interlock.

The two basic PMC units are *indicators* and *controllers* (and combined *indicator controllers).* Figure 6.19 shows two flows, both indicated (their values displayed), and one of them controlled, plus a liquid level which is controlled but not indicated.

These devices interact with each other only at the process level. The flow controller (on the right) causes the tank level to fluctuate, but that CV (the level) is regulated by its own loop (on the left).

The following schemes show some ways in which control loops are configured to work in tandem with one another.

Figure 6.20 shows an *alarm* is given if the contents of the tank drop below a prescribed level. A second alarm is given, and the outflow is shut off via an interlock if the tank level drops further to a second prescribed (low low) value.

Notes on symbology in Fig 6.20: The symbology in this diagram uses a dashed line per Instrument Society of America standard for P&ID's (standard ISA 5.1A). A dashed line represents a hardwired electrical signal connection between devices. The circle symbol (e.g. LIT (Level Indicating Transmitter)) is used to show instrument devices that are mounted in the field. The circle in a square symbol (e.g. LIC) shows a device that is located in a computer control system, usually a distributed control computer. The arrow pointing down on a valve stem indicates the valve is a fail closed valve.

Symbols and line types on PFDs commonly do not conform to ISA standards for P&IDs. The reader of a PFD is often expected to know whether a solid line type is intended to be a process line or a control signal (electrical, pneumatic, and software).

It is desired to control the tank temperature by using a steam heater (Figure 6.21). Instead of manipulating the steam valve directly to control the vessel temperature, we use *Cascade control.*

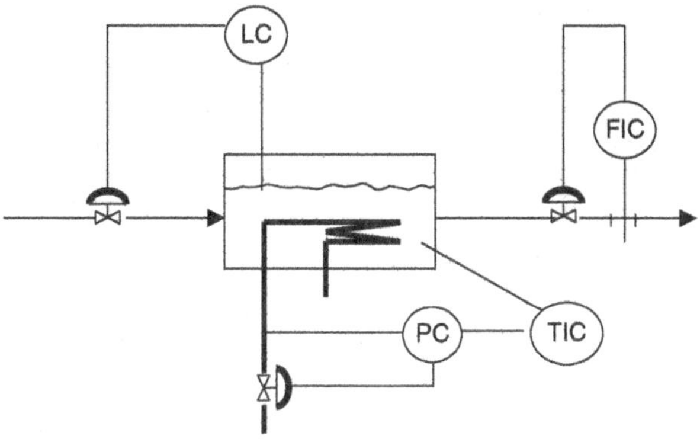

Figure 6.21 Cascade control.

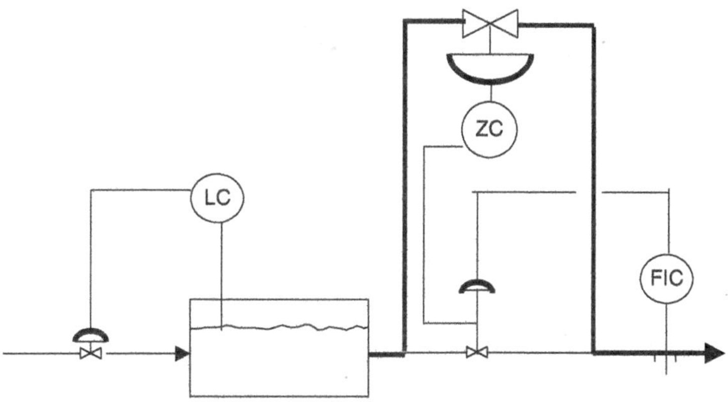

Figure 6.22 Balancing control or valve-position control.

The temperature is controlled by manipulating the SP of the steam pressure controller. The benefit is that the control is less sensitive to any fluctuations in steam upstream supply pressure. Note that the steam pressure is the *MV* of the outer control loop and is the *controlled* variable of the inner loop.

To obtain more precise control of the outflow, we provide parallel paths and do the immediate manipulation on the smaller flow (Figure 6.22). This valve is precise, but its range is small, so we add a slower loop to adjust the large flow to keep the small valve near the center of its range. This is called *balancing control* (in this case *valve-position control)*, and we are using the second loop to control the small-valve position, denoted by "Z".

It is desired to keep a constant composition of flow feeding a subsequent unit, so the sidestream flow is kept in a constant ratio to the main flow. This technique of *ratio control* (Figure 6.23) is a form of *feed-forward control*: instead of waiting for the final combined flow to show deviance from its desired composition and then correcting it, the correction is already made. In many cases, there is no online method of measuring composition, so *ratio control* is a necessity. Taking samples regularly and analyzing in the lab will ensure that the correct proportions are indeed being met.

In order to maintain closer control over the tank level, provision is made to respond early to changes in outflow SP setting. This is an example of *feed-forward control* even though, in this example, the direction of information flow is backward. The use of a *calculational block* is illustrated in Figure 6.24, combining the flow signal with the level signal and using a programmed algorithm to determine the correct position of the inflow valve.

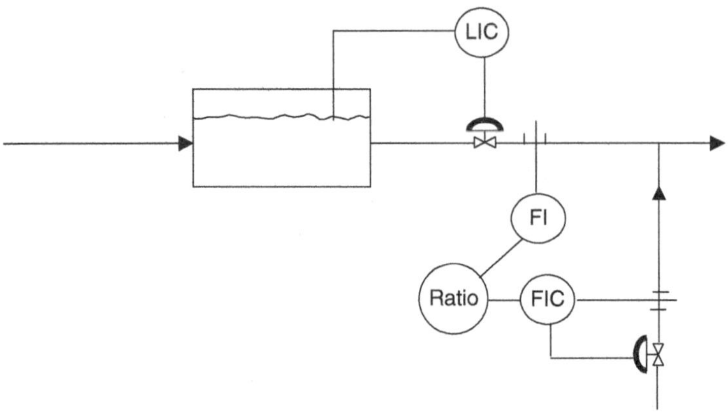

Figure 6.23 Ratio control.

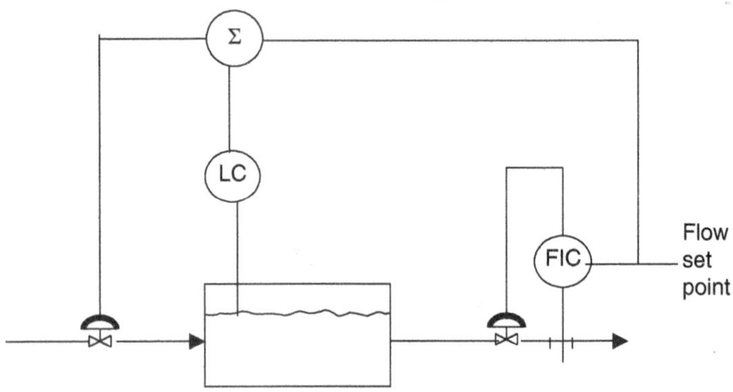

Figure 6.24 Control using a calculation block.

Sometimes there is an interaction between control loops: In this case, the heater may create bubbles which affect level and, at the same time, changes in level may uncover part of the heater and affect the ability to transfer heat. In Figure 6.25, "LIT" and "Temperature Transmitter (TT)" denote level-measuring and temperature-measuring devices. To deal with the interaction, *multiple input,*

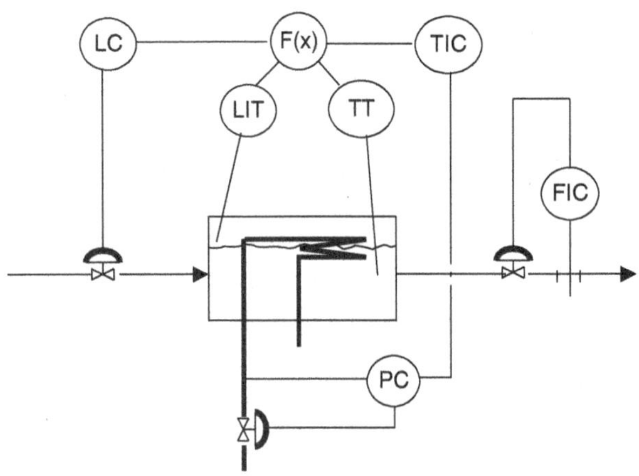

Figure 6.25 Multiple-input/multiple-output (MIMO) control.

multiple output (MIMO) control may be appropriate. This is not a common practice: 95% of control loops are said to be *SISO*.

As can be seen, there are specific avenues for reaching the accuracy and tightness of control that is required. For the operator of the process, it is important to know what is controlling what.

6.3 Causes of Non-Optimum Control

Some of the following difficulties are apparent only after the process is operational. However, many can be circumvented by proper attention during the process design stage.

Good and bad control is a very simple concept: How close does the *controlled variable* stay to the *SP*. It is understood that *CV* will always deviate somewhat because this is the nature of feedback control: A deviation is required before control action is taken. Only if disturbances are absent and if the SP is constant, will the *CV* stay exactly at the SP.

The key questions are as follows:

- How close is it really necessary for *CV* to be to its *SP*?
- How reliable is the measurement of *CV*?
- How fast and with how much cycling does *CV* return to its SP after a disturbance or a SP change?

The answer to the first question depends on the nature of the process. Sometimes the limits are wide: For instance, the level in a holding tank or surge tank might satisfactorily be allowed to range between almost empty and almost full although half-full may be the preferred level, i.e., the SP.

If the operator sees or suspects that the answer to the second or third question is not satisfactory, then there are four areas to examine.

The process itself may be difficult to control. It may be inherently unstable, in which case the control system has to deal with this fact. On the contrary, it may be suffering from disturbances that are correctable. For instance, it may be hard to control temperature in a vessel if a nearby door is opened from time to time to a winter wind. Sometimes control loops "fight" each other: In the example of Figure 6.25 showing MIMO control, if the level control and temperature control had been done separately (in two SISO loops), the interaction would have made one or both controls less than optimum. In Figure 6.26, if level control in the downstream vessel is aggressive, i.e., calling for large fluctuations in inflow, then it may be difficult to control the level in the upstream vessel.

Figure 6.26 also shows a design error that sometimes goes unnoticed across different PFD or P&ID sheets. When the same variable is being manipulated (i.e. liquid flow) by two different controller types, a dueling between controllers can occur.

Another situation where control may be difficult is for a loop that receives its SP as the output from another controller. For instance, it may be the "inner" loop in a cascade or balancing configuration. If the outer loop is tightly tuned, with high gain and rapid reset, it will cause the inner loop to work excessively.

The sensor, i.e., the measurement of the controlled variable, may be the source of trouble. There are several possibilities:

- It may be broken. This condition can manifest itself as a constant signal or as a rapidly fluctuating signal and is usually quite obvious.
- Its accuracy may be compromised: A thermocouple may be improperly installed in its thermowell, or a differential-pressure cell may have legs unevenly filled with liquid, or a flow meter may be operating at the low end of its range.
- It may be improperly calibrated, or it may have been replaced by a unit with different calibration.

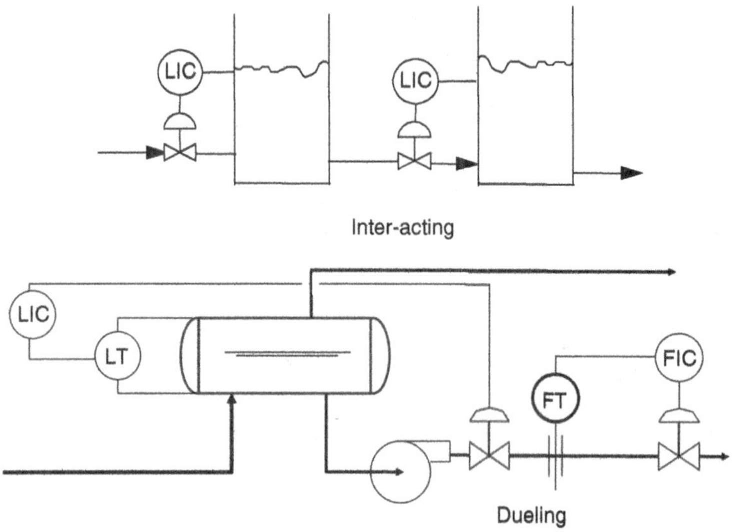

Figure 6.26 Inter-acting level controls versus dueling controls.

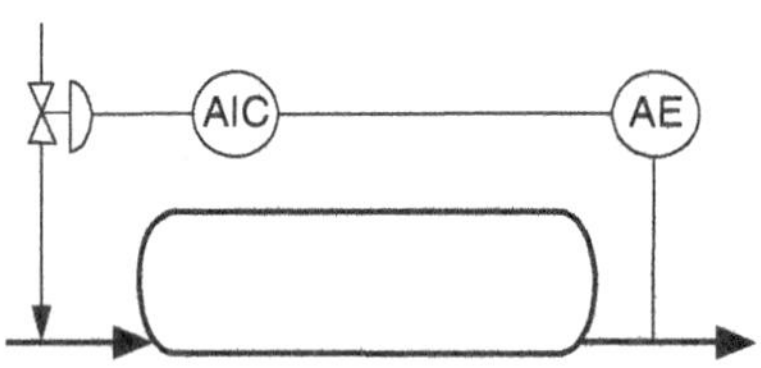

Figure 6.27 Difficult composition control.

It may be poorly located relative to the point at which the MV acts. In Figure 6.27, the presence of the vessel between the point of additive addition and the point of measurement makes it difficult to achieve good control ("A" denotes an analyzer measurement).

The final control element may suffer from malfunction. The most common final element is a valve, which may suffer from

- stickiness, causing it to move in jumps rather than smoothly with its input signal,
- hysteresis, causing it to have, for the same signal, a different opening depending on whether the opening is increasing or decreasing, and
- saturation.

The first two conditions can be checked by taking the valve offline and observing its position as the signal to it is changed manually. Proper maintenance can usually correct the problem.

Saturation is the condition where the element reaches the limit of its ability to affect the process. For instance, if the controller is calling for a flow that drives the valve wide open or to a state where further opening does not produce significantly more flow, then control is not achieved. This situation arises typically where either the valve was undersized in the first place or an increase in production rate causes a valve to become undersized. This is especially true of the controller that has an integral component to it, and a "integral anti-windup" feature should be considered.

The control system itself may be the cause of poor control. Control action is one of the three actions that affect the controlled variable, the other two being external disturbances and resetting

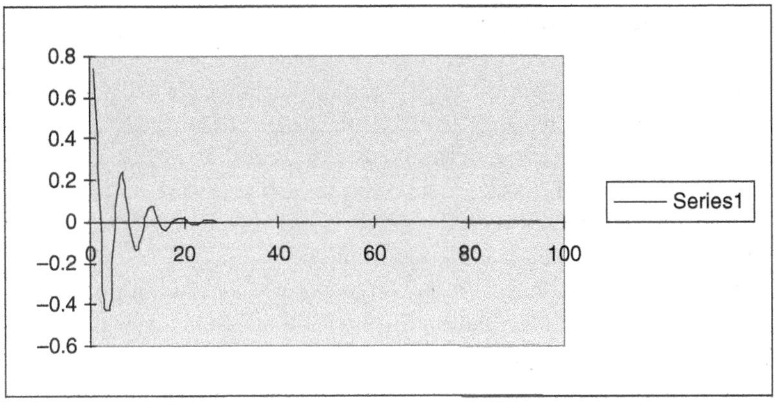

Figure 6.28 Rapid-response control.

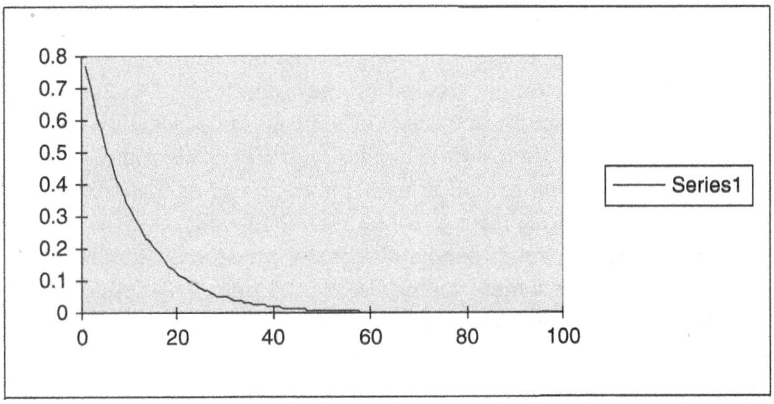

Figure 6.29 Slow response control.

the SP. Poor controller settings can make the controlled variable fluctuate and even make the system unstable. This happens when the process gain is set too high or when the reset (integral) action comes on too quickly. On the contrary, low gain and low integral action let the controlled variable wander loosely around the SP. Controller "tuning" is the art of setting gain, reset, and derivative action to the correct intensity for adequate control.

Figures 6.28 and 6.29 show the time-wise response of the *controlled variable* to a disturbance with fast-acting and slower-acting control. The former gets the *CV* back to its SP more quickly but in an oscillatory fashion with some significant undershoot. As long as the undershoot can be tolerated, this is a very satisfactory control action. If the oscillations persist and especially if they grow, then the control action is less ideal and possibly even destructive. Recall, the standard control equation for a control loop is

$$\mathrm{MV} = \mathrm{GAIN} \times \{\mathrm{CV} - \mathrm{SP} + (1/\mathrm{RESET}) \times \Sigma(\mathrm{CV} - \mathrm{SP}) + \mathrm{DERIV} \times \mathrm{d}(\mathrm{CV} - \mathrm{SP})/\mathrm{d}(\mathrm{time})\} \tag{6.1}$$

showing the proportional, integral, and derivative terms. Tuning the loop means choosing the optimum combination of GAIN, RESET TIME, and DERIV.

6.4 Programmable Controllers and Distributed Control Systems

A short history of the development of process control helps to put the current state in perspective.

The earliest process controllers were human beings, who read the gauges and turned the valves. For instance, in the manufacture of nitroglycerine, in the 1950s, the control was still carried out manually and at the process equipment itself. To save manpower, some plants were configured to bring the indicators and control elements into a central location: doing this required routing the actual process piping through that location. This was all very primitive.

As automatic controllers became available, they were still often grouped together in panels and in control rooms, but the measuring elements and final control elements stayed in the field. Rather than routing the process piping through the central location, it was necessary only to bring the information lines (pneumatic or electric) to the panel.

The advent of minicomputers led to the invention of the *Programmable Logic Controller* or *PLC* in the 1960s. It comprises one or more continuous control loops plus discrete control of events, relays, and interlocks. The device contains programmable software for carrying out these functions. It is built to be rugged so that it can be located at the process itself. *PLC*s are everywhere in industry today.

Historically, a totally different concept in computer-based process control was *DDC* or *Direct Digital Control*. Because the individual pneumatic or electronic controller is designed to operate according to a mathematical expression (i.e. the control equation, Eq. 6.1), it seemed natural to turn this function over to a large central computer since mathematics is a thing that computers do very well. The computer would communicate through cables with the field-located measuring devices and the final control elements but would handle all control functions within itself. It could be programmed with control algorithms more sophisticated than those of an electrical or pneumatic device. However, there was reluctance to put all the control "eggs in one basket" since a computer sometimes fails and a whole plant might fail along with it.

A less daunting approach was *supervisory SP control*, in which traditional controllers were used but their SPs were set by a computer. This approach is commonly used today in labs, pilot plants, and other small operations, with a Personal Computer (*PC*) supplying the calculational power.

In the mid-1970s, *distributed control systems* (DCS) were introduced (i.e. Honeywell TDC2000), which left the individual controllers and *PLCs* in or near the field but which provides communication with them and communication with the frontline human operators. The *DCS* provides valuable enhancement to the whole control scheme, including "hot" backup to individual controllers that occasionally fail.

Computerized central systems like *DCS*, *PLC*, and *PC*-based configurations provide greatly enhanced capability to help the operator run a stable and well-observed process:

- simple and sophisticated control algorithms,
- reliability through redundancy,
- easy ability to reset SPs as well as controller settings,
- programmability of varying difficulty,
- graphic display of current information,
- tabular and graphical records of past operation, i.e., a data historian,
- alarms,

- logging of events and alarms, and
- recipe handling.

The details of these capabilities are decided by the process designer and by the future operators of the process, in consultation with the vendor of the *DCS*. The document containing this information is called the *functional specification* and, if a distributed control system is part of the project, is a significant part of the process design.

7

Design for Safety and Health

A process design, no matter how efficient in carrying out its commercial purpose, would be completely unacceptable if it was prone to cause injury, health impairment, equipment damage, or detrimental change to the environment. It is part of the responsibility of the process designer to make sure these events do not happen.

The following three steps are needed:

1. Assess hazards of the process materials, the process steps, and the process area.
2. Design the process to deal with all hazards while operating in its intended manner, but add reactive measures for possible component failures.
3. Assemble a team of mixed competency colleagues to review both the process and the adequacy of Safety and Health measures. Document the assessment as well as the changes made as a result.

The US Occupational Safety and Health Administration (OSHA) provides some guidance to the process designer in carrying out the above tasks. They have formulated a *Process Safety Management* (PSM) program, to which US industries that handle hazardous materials must adhere. It is a fourteen-part program that extends over the life of the facility, with five of those parts of particular relevance to design and start-up of *new* facilities:

- Process safety information (#1)
- Process hazard analysis (#2)
- Operating procedures (#3)
- Training (#4)
- Pre-startup safety review (#13)

The following account largely follows the PSM requirements for these five parts.

It would also be wise, as part of the design, to refer to the recommendations of Chapter 5.

7.1 Identification of Safety and Health Hazards

At the top of the list of hazards are substances that, in themselves, are hazardous. They may cause injury to the skin (whether cold or hot), they may be acutely or chronically poisonous even in small doses, and they may promote cancer. The OSHA website lists 254 substances that are considered highly hazardous (Note: there are 137 substances as of the 2019 edition of OSHA 1910.119 App A). Your jurisdiction may have guidelines you should consult. The table contains, for each substance, the name, the CAS (Chem Abstracts) number, and the threshold quantity (pounds), above which there is potential for severe damage and/or catastrophe. It is also advisable to check the

Practical Process Design for Chemical Engineers, First Edition. Keith Marchildon and David Mody.

AEGLs (Acute Exposure Guideline Levels), ERPGs (Emergency Response Planning Guidelines), and TEELs (Temporary Emergency Exposure Limits) for any chemicals that may be present, and guidelines for exposure concentrations.

Next on the list of hazardous substances are substances that, although not inherently deleterious to personnel and equipment, are ones that can react with the environment or with other substances to create hazards. For example, substances that decompose at high temperature, or a hydrocarbon mixed with air, or a polymer monomer exposed to an excess of catalyst are all candidates.

Apart from substances, the existence of high pressure can be a hazard. Chemical processes very often operate at elevated pressure, and the hazard is a pressure that lies outside the *Maximum Allowable Working Pressure (MAWP)* of the equipment. Containment within the MAWP range is assured by mechanical and pipe specialists, who calculate the wall thickness by formula. The MAWP is chosen as some safety factor above the *design pressure*, which is stated by the process designer as the maximum pressure the process device will ever reach in normal operation. The vessel is stamped with the MAWP. Relief valves are set at the design pressure since they may be protecting other elements in a process.

The hazard of pressure can also apply to subatmospheric vessels. A vessel containing ammonia vapor suddenly receives an influx of water, which absorbs all the ammonia and leaves a large vacuum. The vessel will collapse if it is not designed for higher pressure on its outside than on its inside. Commonly process vessels and storage tanks are specified to have a vacuum rating. Process vessels are often designed for full vacuum, but storage tanks usually are not.

One cause of overpressure is a blockage somewhere downstream of a pump, which should show up in pressure and/or flow measurements.

Overpressure can be the result of over-temperature. It is much more serious if it takes place where a liquid is the sole phase confined in a vessel rather than in a two-phase, vapor–liquid, mixture. In the latter case, pressure is constrained to follow the vapor–liquid-equilibrium relation. In the liquid-only case, there is no vapor to compress, and the liquid has very little compressibility, so the pressure rises precipitately. For water, the rule of thumb is 100 psi per °F or 12 bar per °C. Thus, this situation requires very careful monitoring, possibly interlock protection, and protection via pressure safety relief devices (relief valves, rupture disks, etc.).

High temperature is ubiquitous in chemical processes, producing products which are then stable at temperatures of human use. Some hazards of high temperature are as follows:

- Insufficient insulation to protect the human person. OSHA restricts human contact to 60 °C and, at this temperature, for only up to five seconds.
- Insufficient insulation to maintain process flows from freezing in cold weather.
- Interaction with pressure as noted above.
- Acceleration of chemical reaction and, if exothermic, start of a runaway situation if temperature is not controlled.
- If excessive, weakening of the strength of containment vessels and piping.

The area in which the facility is being located may already have hazards created by prior facilities. In North America, areas are classified by the National Electrical code (US) or the National Fire Protection Association as

- Division 1, where an atmospheric substance is present or likely to be present in normal operation and could be ignited;
- Division 2, where such a substance is not present in normal operation but could be present in abnormal operation.

Each of the two divisions are further classified according to the nature of the substance

- Class 1, gas or vapor;
- Class 2, dust;
- Class 3, fiber.

The knowledge of the class and division is needed when purchasing equipment and controls. Division 1 requires items that will not spark.

None of these hazards can be allowed to jeopardize people and plant, and so must be controlled or eliminated by proper process design.

7.2 Process Design for Hazard Control: Equipment

A process designer may be unlucky enough to have to deal with an inherently hazardous substance, but there are a number of practices and devices to reduce the risk. One approach, known as inherent safety, is to substitute the material for something more safe or minimize the inventory. If large amounts are needed over time, then most of it can be stored in facilities that are specially designed for secure storage. Piping systems will be chosen to minimize leakage: noncorroding seamless-wall tubing with minimal openable connections. Where connections are needed, they should be with compressible fittings; fittings with vacuum capability are considered most secure against leakage. To keep the fluid from escaping up the stem in a leaking valve, *bellows-sealed valves* should be used, where the stem resides inside the bellows and is attached to the bottom of the bellows. The bellows is attached and sealed at its top to the body of the valve. See Figure 7.1.

Rather than using a centrifugal pump to move an inherently hazardous liquid, the better choice is a seal-less centrifugal pump, which keeps process liquids away from leaking mechanical seals. Other choices include *diaphragm* and *peristaltic pumps*, which separate fluids from the surroundings and can be easier to clean: See Figures 7.2 and 7.3, but consider the effects of the diaphragm or peristaltic pump failing.

Figure 7.1 Bellows-sealed valve.

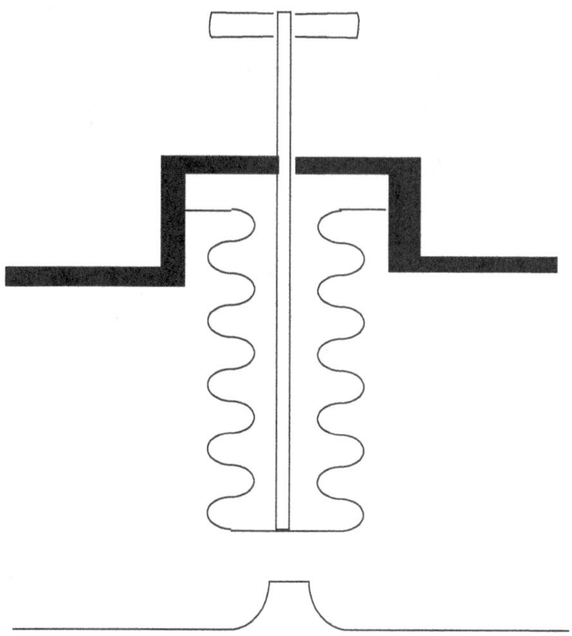

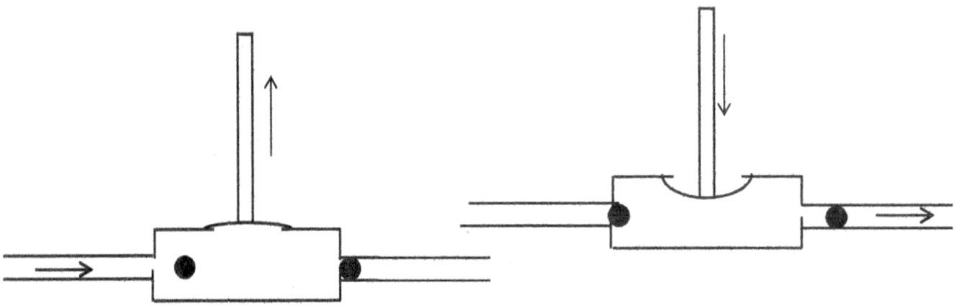

Figure 7.2 Diaphragm pump.

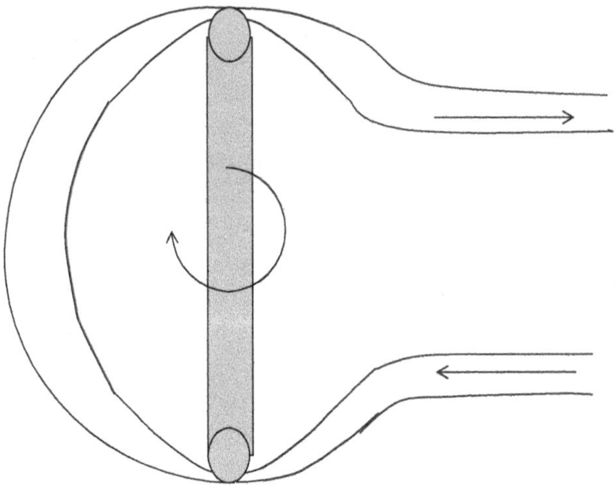

Figure 7.3 Peristaltic pump.

For the measurement of temperature, a thermowell containing either a thermocouple or a resistance temperature device is suitable. At other points, a thermocouple can be welded or soldered onto the exterior wall of a pipe or vessel and covered up with insulation. The objective is to prevent leakage and to prevent contact between process substances and hardware.

Various commercial devices measure pressure without mechanical contact with the fluid. The pressure is generally conveyed across a membrane, but you must consider the effects of that membrane failing either into the process or to the surroundings.

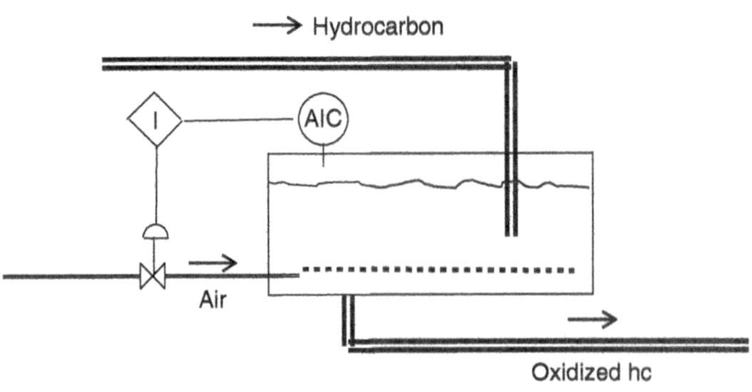

Figure 7.4 Controlled oxidation of a hydrocarbon. Source: Adapted from Santo (2023).

For substances that are less of a hazard, many of the above measures are still advisable. These are compounds that are not objectionable in themselves but must be controlled in their interactions with other substances. Typically, a flammable material must be kept from air. However, sometimes a controlled oxidation is part of the process itself. See Figure 7.4 where air is used to produce a mild oxidation of a hydrocarbon. A close check is kept on the oxygen content (AIC, Analyzer Indicating Controller) of the vessel and triggers an interlock on the air supply if the level rises above a predetermined safe amount.

Santo (2023) discussed inherently safer design concepts to existing facilities.

7.3 Process Design for Hazard Control: Instrumentation

When pressure, temperature, flow, or composition depart from the values or limits intended in the design of the process, one or more of the consequences need to result according to the "layers of protection" as follows:

- early indication to the personnel operating the process,
- alarms,
- interlocks,
- relief systems, and
- plant and community emergency response systems.

This layered protection approach may require significant study since nonconformity may come from different causes. Safety is such a paramount imperative that there should be no issue with time or expense for making everything safe.

Alert operators regularly and systematically scan the indicators of the operating variables and can generally tell if there is something unusual. One of the outputs from a controller is the value of the *manipulated variable*, which is the signal sent by the controller to the final control element. However, this value does not necessarily tell what the final element is doing. A sticky or broken valve does not show up. Attaching a physical indicator to a valve stem lets the operator check. These valves may be in a process-flow line or in a heating-fluid line. Some advanced flow controllers already have this capability. See Figure 7.5. It is also possible to use limit switches which are inputs to the computer control system to ensure valves have properly opened and closed.

Likewise, it is a good practice to install a confirmation of flow using a pressure switch, a speed indicator (tachometer) on a pump, or a flow measurement after the pump. With two pumps in parallel, one in use at a time, any unplanned change in flow can be checked out by switching pumps. Pumps need servicing anyways. See Figure 7.6.

While conscientious operators will notice the gradual trending of controlled and manipulated variables, sometimes the deviation from normal occurs quickly. In this case, an alarm needs to show up, either on the operator control room screen or audibly. Controllers should have indicators

Figure 7.5 Position indicator on valve.

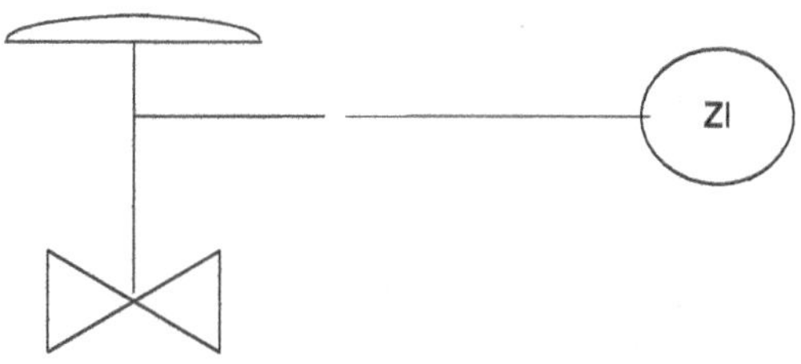

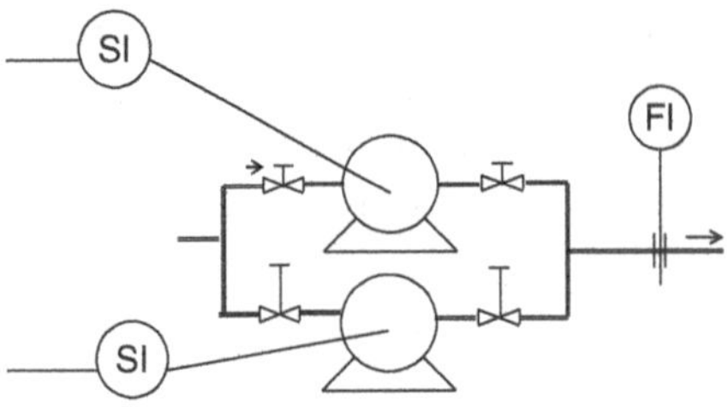

Figure 7.6 Spare pump – manual switch.

for small and large deviations of controlled variables. For example, a level controller would have a level alarm low (LAL) for small deviations below the set point and level alarm low low (LALL) for larger deviations. They can show up on the screen, and the LALL would be accompanied by an audible alarm. In critical situations, two or three measurement devices may be provided, and possibly with different underlying measurement techniques to protect from failures that may have common modes. For example, a level sensor may become plugged and isolated from the process in a failure scenario, so a different technology (i.e. capacitance switch or ultrasonic level) that would resist the failure scenario might be used in addition to the level sensor.

Alarm management should be thought through. Too many alarms cause operators to lose focus. Alarms for small deviations can interfere with alarms for more serious situations. The science of alarm management in general has been studied and reported on.

An example of a tank level control is shown in Figures 7.7 and 7.8. On the right, a valve is manipulated to control a flow rate out of the tank. On the left, a valve is manipulated to maintain the level at its set point in the tank. The controllers for both loops are digital, with analogue/digital converters interposed with the analogue signals from the measuring devices and to the final control element (P&IDs often do not show the a/d converters because they are built into modern transmitters or valves). The level control loop has high and high–high warnings, plus a digital interlock (or a "trip") to shut off the feed valve (left-hand valve) and keep the tank from overflowing. The right-hand valve also has an interlock to keep the tank from running dry.

The warnings and interlock on the incoming flow are activated digitally within the level controller (LIC). The interlock on the outgoing flow is hardwired using either the computers software (shown) or analogue hardwired signals (not shown) which are either pneumatic or 4–20 mA electrical or, in critical situations, both computer and hardwired. This avoids any computer delay or computer malfunction.

In cases where there is no tolerance for uncertainty about the controlled variable, which could happen if a single measuring element malfunctions, a second parallel measuring point can be installed (Fig 7.8). However, if the indicators start to disagree there is no way to know which is right. Hence, a third measuring point is added and the correct value of *CV* is taken according to the two indications that agree. An even surer way to gain the "truth" is for the three measurements to be made by different principles, so as to avoid common-cause errors. The severity of the situation may require a separate isolation valve, and the signal from the level transmitters and the logic solver for the voting system be built into a separate computer to ensure a reliable interlock system. The reader is referred to the topic of Functional Safety, Safety Instrumented Systems, Safety Integrity Level design in standards such as IEC 61508 and 61511, and Coleman (2020).

In spite of alarms and interlocks, unsafe conditions can still arise, sometimes suddenly, and they may require relief flow from a vessel. Relief valves must be provided per the legal jurisdiction the plant is located, and usually that means following ASME code requirements. The difficulties around relief valves include:

- Sizing them. The fluid which they must handle can be of varied composition – gas or vapor, vapor and liquid, and liquid and solid. There are sizing methods for each of these combinations.
- Maintaining them so that they function at the rare time when needed.
- Handling the effluent.
- Predicting the effect on the rest of the process caused by their sudden action.
- Possibly designing a system of relief valves if relief is needed at different locations and different design pressures in the process.

The process designer generally leaves the choice of the details of relief to a specialist or someone with more experience.

Safety systems are sometimes summed up in a scheme called *layers of protection*, as shown in Figure 7.9. The objective for the process designer is to build a process that never has to climb very far up the ladder.

VanCamp (2016) discussed alarm management; Coleman (2020) discussed avoiding SIL (Safety Integrity Level) misconceptions, and Schmidt (2007) discussed tolerable risk – the missing link to complete risk assessments.

7.4 Process Reviews for Safety and Health

The process designer (or the process design team) is responsible for formulating a safe process, but they then bring in a review team to assess and augment their work. This team should include

- the designers themselves,
- a facilitator,
- a control specialist,
- a mechanical or pipe specialist,

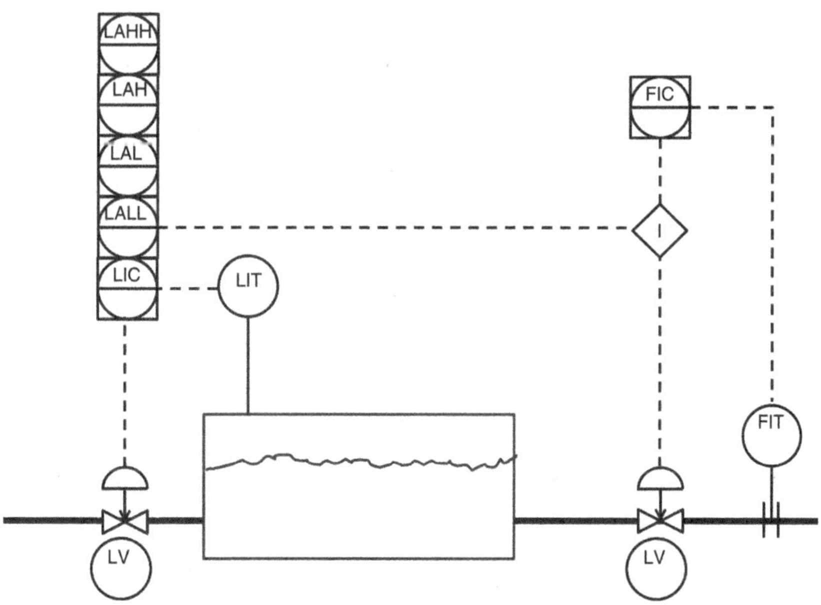

Figure 7.7 Alarm and hardwired interlock.

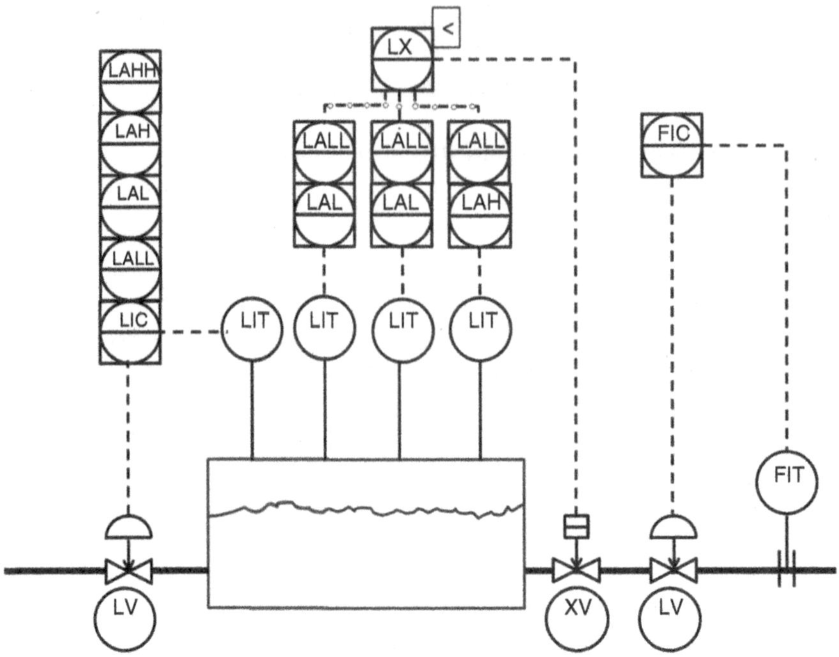

Figure 7.8 Alarm voting Hardwired interlock.

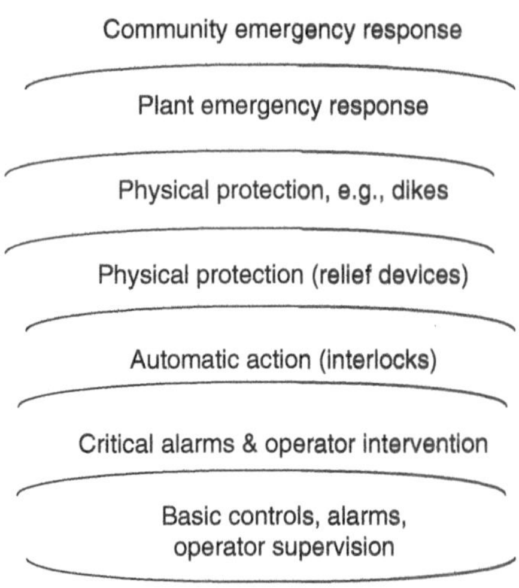

Figure 7.9 Layers of protection.

- the project manager, and
- one or more of the people who will operate the process.

The review may take several sessions and will require full wide-awake participation by all participants. These people have tasks of their own, so there needs to be efficiency both at the sessions and in providing documents and drawings (PFDs, P&IDs, and equipment drawings) for preliminary study.

Ideally, a project would divide the review into phases. The first phase, occurring early in the design, would examine the inherent hazards, such as the chemicals used, conditions present, and the PFD. At this stage, significant design changes can be made with minimal (possibly positive) impact on the schedule and project cost. Subsequent phases, featuring progressively more engineering detail (e.g. P&ID and varying degrees of plant layout completion – even a cubic 3D model in the form of blocks is useful), can delve into the process in greater detail, but changes will be at greater cost in budget and schedule impact the later the review is done. It is worth noting that the plant's location can influence the risk levels of some events, so having that information available early on is useful.

Depending on the available information, different review formats might be suitable. The review will adhere to a specific format, with potential methods including:

- what if questions,
- checklist,
- hazard and operability study (HAZOP),
- failure mode and analysis (FMEA),
- fault tree analysis (FTA), or
  Bow Tie analysis, and
- Layers of Protection Analysis (LOPA).

The "What if" scenario and the FMEA approach are similar, in that a failure of a component (large or small) is postulated and the consequences are then thought through. It depends very much on the experiences of the participants: What they have experienced themselves and what they have observed over the course of their career. The FTA approach is in the opposite direction: A truly bad occurrence is studied and the conditions that would lead up to it are identified.

The most common – and time-consuming – approach is the HAZOP. If the process is complex, it is broken down into nodes and every aspect is examined:

> flow, pressure, temperature, level, time (batch, start-up, shutdown, draining, venting, inerting), agitation, reaction, utility failure, DCS failure, and vibrations.

The questions asked regarding flow, for instance, are the causes and effects of its being too high, too low, not flowing at all, flowing in reverse, and being contaminated. Since most of the conditions are unlikely but must be included, a HAZOP study can be quite extensive. However, it is the "gold standard" as a review. An experienced facilitator will mix in a smattering of "What if" and FMEA at crucial points in the process.

A scribe writes down the conclusions after the discussions on each issue. It will include action items that arise. At the next session, the process designer(s) reports on what has been done for each item or not done if they have, on reflection, come to the conclusion that the situation needs no changes.

The review of processes is mandatory in the United States according to process safety management. A lax review or an undocumented review leaves the designer and the company open to legal proceedings if there is an accident. From the moral standpoint, a good process review is a must.

7.5 Training and Operating Procedures (PSM #3, #2)

These two activities generate a document, but those documents are greatly different in content and style.

The *Training Document* has at its core the process description, set at a technical level between process operators and engineers. It does not hurt operators to learn a little of chemical engineering, and will enhance their familiarity and judgment in supervising the process.

The document should also have a long-term mission by having added to it the design calculations and any mathematical modeling. It will serve as the basis for future engineers to troubleshoot the process and to change capacity or product type.

Operating procedures also require a document but of a different sort. It comprises step-by-step procedures for start-up, normal shutdown, nonscheduled shutdown, emergency response to a variety of occurrences, and normal supervision of the operation. This document needs to have all physical elements of the process numbered and labeled, initially just on the P&IDs, and then tagged on the equipment when built. It is a laborious task to prepare these instructions but is essential to avoid having operators missing some crucial step.

7.6 Pre-Startup Safety and Health Review

Safety is always paramount. In the earlier part of this chapter, the objective was to build safety and health into the *design* of the process. When the equipment is installed and ready to be started up, the review is aimed at safe *operation*. There are obvious perils like tripping hazards or an electrical panel with no cover. At a deeper level, the control personnel should check that the wiring from device to device is properly connected, that measurement devices are sending plausible signals, and that the final control elements are responding to signals.

References

Coleman, A. (2020 Jan). Avoiding SIL misconceptions. *Chem. Eng.* 127 (1): 45–47.

Santo, R. (2023 Aug). Apply inherently safer design concepts to existing facilities. *Chem. Eng. Prog.* 119: 43–49.

Schmidt M., Tolerable risk: while determined risk is generally well understood, tolerable risk can be the missing link to complete risk assessments, Access Intelligence, LLC *Chemical Engineering* (New York) 2007 Sept, 114(9): 69.

VanCamp, K. (2016 Mar). Alarm management by the numbers. *Chem. Eng.* 123: 50–55.

8

Protecting the Environment

Protecting the environment is the logical off-site extension of protecting Safety and Health within the plant.

Like safety and health, it has four aspects:

Morality	Legality	Technology	Economics

Most ideally, the plant or process should have the least possible "footprint":

- It should draw only the minimal required materials and energy to make the product.
- It should yield only the desired product and pure streams of side components like water and air.

The design and construction of a plant or process usually requires approval from regulatory authorities, particularly if there are deviations from the ideal footprint. It is part of the process designer's responsibility to provide conditions and equipment to minimize deviations. A caution is that environmental rules change with time, always to tighter standards, so the process designer needs to consider jumping ahead of current rules. An example of this is the concept of a circular economy which will require corporations to look at their products and processes in a new way (Glavic 2021).

Many articles have been written on this topic. Savage (2009) asked "What does it mean to be green?"; Anastas & Zimmerman (2003), Allen (2007), and Anonymous (2002) listed general principles and approaches. Contreras & Bravo (2011) and Mendez (2007) have valuable suggestions. Curran (2015) proposed the idea of a *life cycle assessment* in making decisions. Jenkins (2010) cautioned that likely increases in *carbon pricing* should be factored into economics. Williams & Dunwoody (2012) recommended the use of mathematical models to quantitatively design for environmental protection. Ternes (2012) reviewed the intricacies of the regulatory process, and Jennings (2008) provided a good background on the topic of climate change. Glavic (2021) provided a review of process design and sustainable development in Europe. Other references are Allen (2007), Anonymous (2002), Chin & Gillespie (2010), and Contreras & Bravo (2011).

The present chapter deals with the two aspects of the "footprint":

1. Consumption: The use, by the plant or process, of material substances and of energy;
2. Emission: The emissions of substances other than the intended product.

The objective is to minimize all of these various flows.

Practical Process Design for Chemical Engineers, First Edition. Keith Marchildon and David Mody.

8.1 Consumption

To the outside world, the main entities which a plant consumes are raw process materials, water, air, and energy. The energy may appear in the form of delivered fuel or it may be direct as electricity or water-powered. Minimization of these consumables often presents opportunities for savings in operating costs, which of course have to be measured against capital cost to achieve.

The three components of consumption are now considered in turn.

8.1.1 Raw Materials

The very minimum amount of the principal raw materials is obtained from the reaction equation,

$$A \rightarrow B \tag{8.1}$$

where A is the raw material and B is the desired product. However, two types of eventuality interfere with this simple relationship.

One is the presence of another process, chemical or physical, so that a second consumption of A occurs in parallel

$$\begin{array}{lll} A \rightarrow B & \text{or} & A \rightarrow B \rightarrow\rightarrow\rightarrow \\ \quad\searrow C & & \quad\searrow C \end{array} \tag{8.2}$$

Lab studies or pilot plant runs can identify the chemistry and hopefully identify conditions where the parallel process is minimized. Assuming that the final product needs to be substantially free of the co-product C, there may be loss of A or B in separating out component C. Capital spending will likely be needed especially if the optimum temperature for best yield is low and requires a larger process unit. Capital may also need to be spent in obtaining a clean separation of impurities.

The other is wastage of components A and B, as explained here. Potentially, one of the largest, although intermittent, waste is at start-up, shutdown, and changes in the product type. Here is where an unsteady-state mathematical model can help by identifying a time-wise transition process path that keeps the product close to specification for as long as possible during transitions. Figure 8.1 illustrates such a transition. In this simple but illustrative case, it is desired to double the concentration of component C in the outlet of the vessel. The quick action of a forced transition is seen.

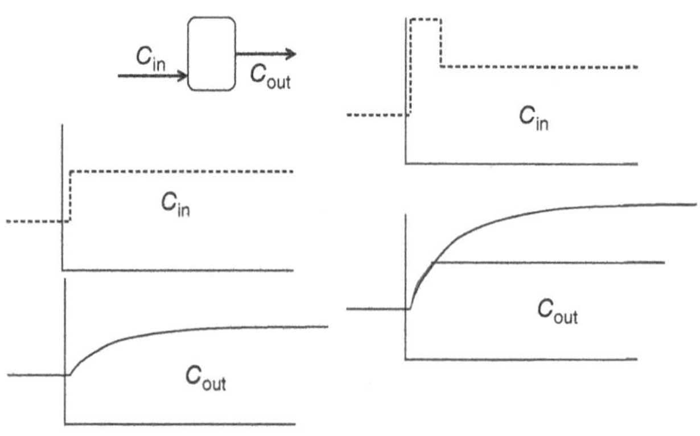

Figure 8.1 Forced transition.

In this case of a well-mixed holding tank, the outflow composition of component *C* is desired to be doubled. This is done by doubling the composition of the inflow. If the inflow is just doubled, then the outflow takes an exponential asymptotic increase to its new value, leaving a long period of transition. However, if the inflow is tripled for a short period, the outflow achieves its desired new value very quickly.

Transition waste may also be reduced or eliminated by adding holding vessels to the facility, to which nonconforming product can be diverted during transition. It may be possible to gradually blend this material back into the product stream when the transition is over. Holding tanks cost money, but they are much prized by operating personnel to deal with upset conditions as well as transitions.

8.1.2 Consumption of Water

Water is a ubiquitous material in most processes where substances are being consumed, transformed, or produced. Muller et al. (2013) suggested optimization of energy use in cooling systems. Willa (2005) suggested proper distribution in cooling towers. Schultz (2008) urged water reuse and conservation. Sutikno (2007) and Tanthapanichakoon (2012) proposed energy savings around steam generation.

Facilities are often situated near lakes and rivers or over aquifers. The supply may be plentiful, but nevertheless application must be made to "take water". Sometimes water is not so plentiful, and it is required to recycle and reuse it within the process. These measures may also be necessary to avoid hydraulic overload of the treatment system for water being returned to the environment. Within-process reuse of water requires capital investment, so the cost/necessity will be examined in process design.

Figure 8.2 summarizes the common paths of water within a plant.

The uses to which water is put dictate the sort of treatment which is applied.

1. Because its naturally occurring temperature is generally lower than process temperatures, it is an excellent coolant. It is also an excellent heating medium in the form of steam. These two applications are often the only ones in many processes. The separation between the water and the process fluids keeps the water relatively contamination free. Treatment is needed only to keep the water courses suspension-free and to minimize surface scaling. Treatment #1, on the supply side, would likely be softening and filtration (at least to keep the fish out). Water purity is especially important for uses in food processing and other human contact applications such as cosmetics and bio/pharma or health-related industry, in electronics, and in feed to high-pressure

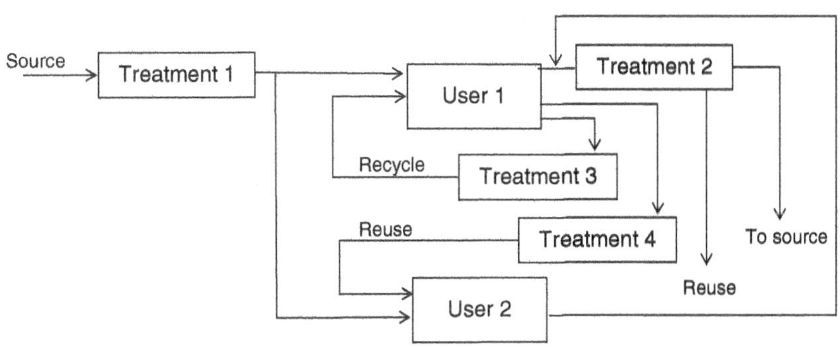

Figure 8.2 Adventures of process water.

steam production. Process equipment metallurgy and process materials such as catalysts may also be sensitive to water quality. Fortunately, there are companies that specialize in providing water treatment and water quality services.

2. The RECYCLE loop shown with User #1 and Treatment #3 could represent a steam boiler and a process unit, where the condensate from the process is used as the boiler feed water. Treatment #3 is the addition of anti-scaling, anticorrosion, and anti-bacteriological compounds to maintain levels obtained in Treatment #1.
3. The same RECYCLE loop could also represent cooling of the process followed by cooling of the water so that it can be used again. The latter task commonly takes place in a cooling tower, where air is used to partially vaporize and to thereby lower the water temperature. Hairston (2003) discussed cooling towers, and Huchler (2016) suggested the measures to eliminate *legionella* bacteria from cooling towers. Puckorius (2013) considered recycled water for cooling tower makeup.
4. Treatment #4 represents the use of water outflow from one process unit as the feed for a second, unlike, process unit. This is a very broad categorization depending on the properties of the effluent and the acceptable requirements of the influent. The discrepancy may simply be temperature, but could be impurities in the effluent. Some ways in which water is used and may get contaminated (i.e. in Unit #1) are as a diluent or as a solvent in extraction. Techniques for cleaning up this stream are suggested in the literature: chemical softening, reverse osmosis, coagulation followed by ultrafiltration or RO, membrane filtration, oxidation, ozonation, peroxidation, ultraviolet radiation, electrodialysis, ion exchange, heat, and biotreatment – the choice depending on the nature of the objectionable substances to be reduced or eliminated. Cost is also a factor.

Authors who have listed and described these techniques are Aerts & Tong (2009), Cartwright (2006), Hairston (2003), Harfst (2010), Jenkins (2013, 2015), and Uribe et al. (2015). Harfst particularly focused on nonchemical treatments.

5. Treatment #2 is the interface between the process and the world around. It must be effective in order for the plant to legally operate. Some of the above techniques may form part, or all, of "Treatment 2", but the most common sequence is the mechanical separation of insoluble impurities, followed by a biological digestion of soluble impurities. This operation is examined in Section 8.2.2, where we also look at the other aspects of the process footprint. Zero discharge plant practices are becoming the goal of operations.

The above considerations of use, recycle, reuse, and treatment could constitute a major sub-design project in its own right. The complexity lends itself to mathematical modeling.

Cartwright (2006) identified three treatments of in-house water: to get it up to required quality for use and to keep it there. Amminudin et al. (2008) and Durham & Patria (2006) recommended a survey of existing water systems in a plant, in search of opportunities for improvement. Huchler (1999) focused on getting a record of the use of utility water. Foo et al. (2006) suggested the application of the techniques of *pinch analysis*. Nolen (2016) also suggested the application of energy conservation principles to water conservation. Nystuen (2011) and Putra (2015) recommended the integration of the treatment of wastewater with the conservation of water – that is the joint consideration of the above four "treatments". Pakzadeh & Zbacnik (2015) and Schultz (2008) suggested incidental sources of water that may be acceptable, at least after some treatment or "polishing". Anonymous (2015B) looked for sustainability. Parkinson (2005) suggested about getting the most out of water. Da Silva & Goodman (2014) were helpful. Durham & Patria (2006) showed that wastewater can sometimes be a reliable source. Frenkel (2002) discussed the pretreatment of water. Harfst (2010) and Jenkins (2013) discussed the nonchemical treatment of water. Jenkins (2015) covered water treatment technologies in general, and Jenkins (2020) suggested about the use of ion-exchange resins. Dalan (2000) provided advice on zero liquid discharge design. Uribe et al. (2015) listed advanced technologies for water treatment and reuse.

The manufacturer applies to the regulatory body for permission to establish these flows. They receive either acceptance or a set of requirements to reduce or alter the streams. Generally, there are negotiations between the parties to arrive at a path forward which is environmentally acceptable and minimally costly and with an eye to future tightening of regulations.

8.1.3 Consumption of Energy

The two energy streams that feed most plants are fuel (natural gas as opposed to oil or coal) and electricity. The former is used for the generation of heat, the latter for powering mechanical devices. Electricity is too expensive to be degraded into heat, except in some specific cases where the load is small or high temperature is needed, or where it is awkward to use a heating fluid.

Natural gas is often compared unfavorably with electricity because of its carbon dioxide combustion product, although it is the least CO_2 producer of all the fossil fuels. However, electricity also has some social flaws in its production. It may come from innocuous sources like isolated hydroelectric sites or from nuclear reactors. However, it may also come from a fossil-fuel burning generating station or a hydroelectric site that despoils the land around where people live.

The starting point is best to distinguish the above distinction between sources, i.e., natural gas and electricity and to look for ways to transfer energy from one to the other where there is otherwise going to be wastage.

Alternative sources of energy could be explored. Wood chips and the remains of dead trees have been studied. A small nuclear reactor seems daunting but may become practical and manageable in the near future: SMRs (Small Modular Reactors) in the range of around 300 MW (see Figure 8.6) are being developed. Geothermal heating is also coming to the fore, although that possibility depends on the siting of the plant and what lies beneath it. Unfortunately, the surroundings around the installation become exhausted of heat because of slow diffusivity of heat in the soil.

Many of the referred articles assume that a plant already exists and can be explored for energy savings. Some general references are Rossiter (2006), Hardin (2007), Harrison (2008), Chin & Gillespie (2010), Elshout & Marchant (2010), Rossiter & Venkatesan (2012), Reddy et al. (2013), Scheihing (2014), Rossiter & Davis Jr (2014), Blume (2014), Scheihing (2014), Rossiter & Jones (2014), Rossiter (2015A–C), Rossiter (2015A), and Anonymous (2015A). There are a multitude of suggestions – some major but mostly minor, which do add up. These references show the process designer what the final plant might look like. For an existing plant, Griesbach et al. (2013) recommended conducting an energy design review. Any of these concepts could form part of the predesign planning, and a mock audit could be done during a HAZOP review.

Among the suggestions are those that indicate measurement devices that will help in reducing the energy load. Some of the devices are as follows:

- Sampling of exhaust gases from fossil-fuel combustion (e.g. combustibles, carbon monoxide and nitrogen oxides) to ensure that the combustion process is as complete as possible.
- Characterization of streams, by way of measuring flow, of sampling for contents and concentration – preferably with quick online measurements such as gas chromatographs.
- Provision for passing information to online optimizing computers.

Fully implementing all of these devices may be controversial to the funders of the project, so at a minimum the sampling and measure points should be installed and plugged off until further equipment is eventually bought. This all requires planning for the final state. Some references for in-plant information and automation are White (2010), Blume (2014), and Parisi et al. (2015).

The general objective is to fully use the heating value of the fuel [see Axon (2009)] and to recover waste heat. Recovery may be simply transferring heat from a hot waste source into a lower-temperature useful stream - perhaps for domestic and comfort use within the plant. This application requires efficient transfer of heat. An article, Anonymous (2015A), described a commercial device the *Compabloc heat exchanger* by Alfa Laval. It is claimed that it can provide 3–5 times the rate of standard devices and is also easy to keep clean.

Reddy et al. (2013) reviewed the opportunities and methods of waste heat recovery. Labrecque & Kiari (2006) also provided suggestions. Flue gases are a likely source, see Bourji et al. (2010).

One use for waste heat is to run turbines for the production of electricity. This may mean the purchase of turbines solely for this purpose, see Ganapathy (2009) and Kaupert et al. (2015). Obviously, the saving of a particular amount of energy must be sufficient to justify the capital expense. The operation is governed, in the abstract, by the Rankin Cycle. Bourji & Winstead (2013) described a Rankin Cycle based on an organic liquid of higher volatility than water, to which the waste heat has been transferred and which causes it to vaporize.

A combined cycle gas turbine uses hot combustion gas to generate electricity and continuously follows it up with the transfer of heat from the exit gas into steam, which powers another generator.

Many plants have installed, or are installing, *cogeneration* facilities. As described in the literature, a common heat source drives the production of steam, some of which is immediately applied to electricity-producing equipment or other rotating machinery. Claims for the conversion of incoming fuel energy into plant-useful energy can be as high as 85%. This can be much higher than if the two operations were done separately.

The objective is to wring as much useful energy as possible out of whatever fuel is expended.

Some other pointers may be useful. Add heat only when it needs to be added. In general, proper instrumentation will tell that. If a vessel is to supply a hot fluid, there is no use of heating the whole vessel; rather heat the outflow as it emerges. Both heating and cooling require energy: use "pinch" or other methods to transfer heat between streams; however, start-up and shutdown, as well as process operation dependencies, require careful thought in a system with heat integration. Remember that energy can constitute a major operating expense, especially in jurisdictions where electricity is expensive.

8.2 Emission of Waste

This is the other facet of plant footprint, the uncalled-for, unwanted emission of extraneous substances into the public domain. It is also the area where the plant may get into trouble, so requires attention early in the design phase. Regulators may lack urgency of action, so can hold up approval or plant start-up. Four areas are looked at here (noise is not considered here, but can be an issue, and it is regulated in jurisdictions):

- liquid waste,
- gaseous waste,
- solid waste, and
- carbon dioxide and other climate-affecting substances.

Raaphorst (1998), Lad (2002), Jones & Rankin (2015), Shah (2016), and Shah (2016) alerted managers and process designers to the strict mandates around this topic. Powell (2008) presented "best practices", a guide to managers and technical personnel. Emissions monitoring is insisted by regulators, see Malosh (2008). Siegell (2010) suggested auditing emissions inventory. Moftah (2000), Reinemann (2006), and Khaqan (2011) recommended statistical assistance in getting the data. Neimeier (2004) recognized that emissions testing costs money, so get the most from tests. Scheier (2001) focused on completeness in assessing the potential to emit. Schuster (2015) reminded that what we can measure, we can improve. Ulrich (2007) stressed considering pollution control during the design phase. White (2010) recommended generous use of automation.

Provision for dealing with waste is highly specialized and not merely considered as a typical part of process design. There is too much at stake and too many options that require specialized help.

8.2.1 Dealing with Liquid Waste

A chemical plant or a plant using chemicals to produce materials is bound to have significant liquid waste. This waste could be a concentrated nonaqueous substance, which would require a specialized company to haul it away and dispose of it. Losing such a stream could be costly, so a process would undoubtedly be examined to avoid or minimize the loss. See Mueller & Cipullo (2008).

Very typically, this waste is in dilute form as part of an aqueous stream. The object is to remove the contaminant from the stream, so that water can be sent back into the environment often first through a public wastewater system. This problem is discussed here.

The general topic of wastewater was discussed by Gucciardi (2003) and Doble & Geetha (2011). Huchler (2010) explained the essential step of obtaining a permit to emit water. Putra (2015) suggested improvements for wastewater plants.

8.2.1.1 Biological Treatment of Wastewater

With regard to technology, the *activated sludge process* is the most-used process, being based on the remarkable ability of certain microorganisms to grow by extracting organic materials and biodegradable compounds from water. Information on this process is contained in Carson (2002), Kleerebezem & Macarie (2003), Schultz (2005), Cartwright (2006), Doble (2006), Marshall (2007), and Sustarsic (2009). This process was devised in the early 1900s in England, and various forms have been implemented worldwide.

Most activated sludge processes operate with oxygen (usually air-oxygen) assisting in bringing about the chemistry of assimilation. This need depends on the particular microorganisms. The physical form is for the organisms to be suspended and swirled in the wastewater. Two references give a good picture of this *aerobic* process: Sustaric (see above) and Ataei (2010). The other detail which makes this process possible is the clumping together of the organisms, so the Arntsen at they separate out after the absorption.

Figure 8.3 shows schematically how a suspended-solid aerobic system operates. Some of the peripherals are included, including a holding tank for containing water resulting from unusual conditions that might upset the treatment system.

The *aeration tank* is the heart, with agitation of some sort keeping the solid flocs of organisms in suspension and also entraining and distributing the oxygen. The physical mixture continually flows out of the aeration tank into the *clarifier*, where the microorganism flocs settle out as sludge. Part is sent to sludge disposal; the rest is recycled back to the aeration tank to maintain concentration. A key consideration is the mutual dependence of the organisms and the water, i.e., there must be sufficient contamination in the water to feed the hungry "bugs". If the water feed is cut off or drops to a low content, then supplemental feed must be administered.

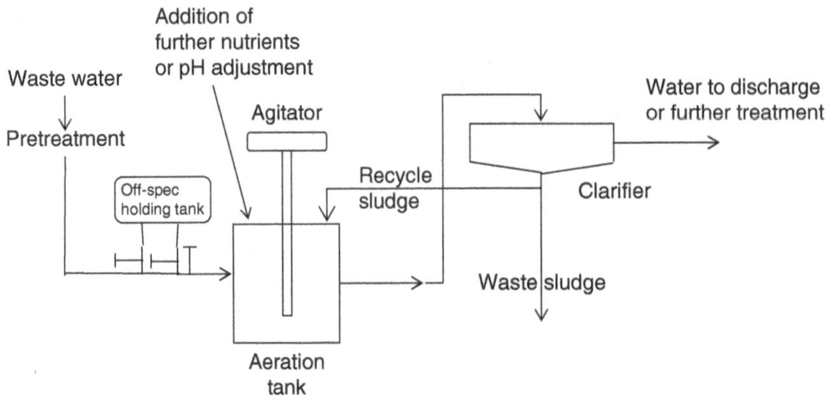

Figure 8.3 Suspended-solid aerobic water-waste treatment.

A variant of the process is to conduct it without oxygen. Again, this possibility depends on the particular combinations of microorganisms and water contamination. Kleerebezem and Macarie (see above) explained this process, giving its advantages and weaknesses. Marshall (2007) presented a list of contaminants that lend themselves to *anaerobic* treatment. A third variation, *anoxic* low-oxygen condition is a possibility. In general, the omission of oxygen makes the process less expensive to operate, energy-wise.

The above distinctions apply to oxygen use. A second distinction applies to the form of the microorganisms. Instead of being on the loose in the wastewater, they are attached to, and grow on, solid surfaces, which may be fixed or may float about in the waste. These reactors are referred to as *biofilm*. A fixed bed of such surfaces is known as a *trickling filter*. Flocs do break loose from the solid surfaces but not to the extent that they need recycling back to the aeration tank. The gases from anaerobic processes may be combustible (unlike the CO_2 from aerobic processes) and could be tapped for energy. Gas from solid landfill may also be combustible and only needs trapping.

A very instructive account of the use of the two configurations, biofilm and suspended solid, in tandem with each other is given by An (2013). The biofilm in this case was contained on wafers, which floated in suspension, so that the process was called MBBR – *Moving Bed Biofilm Reactor*.

8.2.1.2 Other Treatments of Wastewater

Wastewater may contain components unsuitable for bio treatment, which need to be removed independently. Such are volatile components (VOCs) in significant amounts. Zygula (2008) presented a trayed steam-stripping column, where contaminated water enters at the top, flows downward, and is met by steam flowing upward. With proper design the steam carries VOCs up and out of the device. Stripped water flows out the bottom for release or for a further type of treatment. The top stream contains the VOCs along with some steam, all of which is condensed and disposed of. It may be partially funneled back to the plant process. The optimal pressure for such a system is close to atmospheric.

Dissolved air flotation (DAF) is a process for floating solid material out of water, typically flocs of microorganisms out of bio treatment sludge. Water, flowing upstream, deposits solids near the top where they are withdrawn. The water flows back down a parallel channel where it is withdrawn. This process is described by Vlyssides et al. (2006). Carson (2002) provided tips on the handling of sludge.

Carr & Vaughn (2003) alerted to the efficacy of activated carbon in many applications. It has very high absorption powers and, for instance, is good at removing such heavy metals as arsenic, cadmium, chromium, lead, and mercury. Other absorbents are zeolites and other molecular sieves. Vardhan et al. (2019) suggested heavy-metal pollution and remedial measures.

An objective of *zero* liquid discharge has been adopted by some enterprises. Dalan (2000) and Shaw & Brosdal (2008) advised on how to approach this happy state. Evaporation and possibly crystallization are useful.

Ataei (2010) showed that wastewater treatment can present energy-conservation opportunities.

Sometimes the contamination is in the form of ground water. Beal & Faircloth (2002) and Adams et al. (2006) discussed ISB, *in situ bio remediation* for cleaning up this form of pollution.

Arntsen (2020) described a recent installation where membrane filtration is used to clean up a stream prior to aerobic treatment.

8.2.2 Dealing with Gaseous Waste

Again, this is a specialized area of design and not to be approached without engaging specialists. The objective is to reduce VOCs (volatile organic compounds), HAPs (hazardous air pollutants such as sulfur dioxide, benzene, perchloroethylene, methylene chloride, and a whole host of others), NOx (nitrogen oxides), and carbon monoxide to permissible levels for emission from the plant. This often means reducing them by 99%. An overall survey of methods is provided by Moretti (2002) and Karell (2013). These methods are discussed in general terms here. Moretti provided a map of methods suitable for different conditions.

Treatment of waste gas streams is often referred to as "end of pipe". A better alternative is to modify the process or procedures so that these contaminants do not show up at the end of the pipe or are at least minimized. That sort of development may have too long a time frame for the project at hand. Buchanan (2007) helped with the efficiency of air pollution control devices. Gendron et al. (2004) discussed perimeter air-monitoring programs.

8.2.2.1 Thermal Oxidation and Thermal Catalytic Oxidation

This group of methods is very aggressive and effective for most waste gas impurities.

Air is used to combust with the unwanted components and convert them into the innocuous products water and CO_2. References are Klobucar (2002), Venkatesh & Woodhull (2003), Mack (2003), Blocki (2005), Goldshmid (2005), Sims (2007), Baranski & Underwood (2014), and Armstrong & Predatsch (2019). Baranski & Underwood (2014) specifically discussed the catalytic approach. Goldshmid is particularly useful in providing a map of which oxidizer is suitable for combinations of waste flow and waste concentration.

The simplest version is the *direct-flame thermal oxidizer*. It consists of three parts: a burner, which mixes air plus gas plus a fuel such as methane, a mixing zone, and a retention zone for complete combustion. It is flexible relative to flow rates and composition, but expensive to operate because of the need for fuel.

The *recuperative thermal oxidizer* conveys the heat generated by the combustion of impurities to heat and boil other streams. The heat exchangers are an essential part of the device, but there may be flexibility in their design.

The *regenerative thermal oxidizer* contains within itself two or more beds of ceramic media to alternatively store and then release the heat. The flow is cycled back and forth. Three beds are often used for smoother operation and can give heat recovery of typically 93–97%. Essentially, the heat is transferred to the incoming waste gas which passes through the bed on its way to combine with air.

Generally, no additional heat (i.e. natural gas) is needed. Recuperative and regenerative systems are obviously less expensive to operate, but they cost more. Some typical temperatures are 850–1050 °C. Destruction efficiency (DE) of contaminants can be 90%, even as high as 99.5, hopefully enough to satisfy regulators.

Catalysts for oxidation are also used (as in motor vehicles) and are also shown in Figure 8.4. There is a reduction in temperature requirement of 100–200 °C, although if part of the contamination is halogenated, that requires more temperature. There are the usual difficulties with beds, e.g., "blinding" with particles, poisoning with sulfur, silicon, arsenic, phosphorus, and the periodic need for refurbishing the bed. The catalyst may be a precious metal, e.g., palladium or platinum, or a metal oxide (e.g. vanadium or titanium), and it may be in the form of a monolith or a packed bed. Sims (2007) discussed minimization of particle interference.

Figure 8.5 shows the concept of the two cycles of the regenerative oxidizer. It is controlled by the action of the valves. In the diagram, the closed valves are shown in black. The arrows show where flow takes place. The heat recovery in this device is probably the best that can be had.

Figure 8.5 shows the cycles of the *regenerative* oxidizer. Combustion gases are cooled; waste stream is heated. This is all internal and an integral part of the system. The switch is made automatically through valves that are alternatively open and closed. The valves that are closed are shown in black in the diagram. Needless to say, such high-frequency high-temperature valves must be of good quality and leak-proof.

Some quantitative characteristics are that total residence time in the combusters is generally 0.75–2 seconds. A flow of gaseous waste is typically 1000 scfm, with 5000 being considered high. A volumetric concentration of 10 ppm is considered low. A *rotary concentrator*, as described by Herraiz et al. (2020), may be used to increase concentration before thermal or other treatment. That application is for CO_2 but can apply to other gases. Moretti (2002) claimed a 1000 : 1 concentrating is possible.

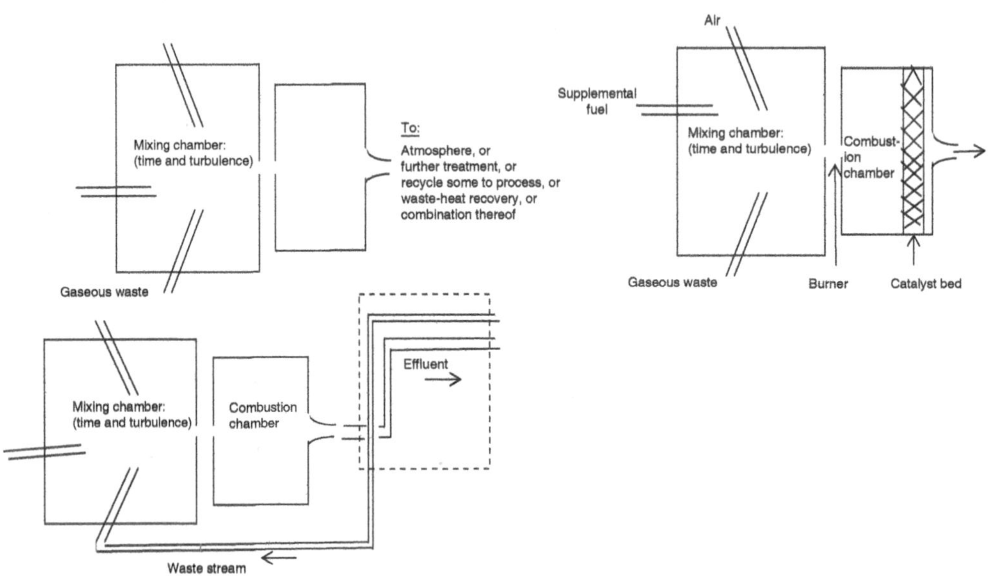

Figure 8.4 Direct-flame oxidizer, recuperative thermal oxidizer, catalytic thermal oxidizer.

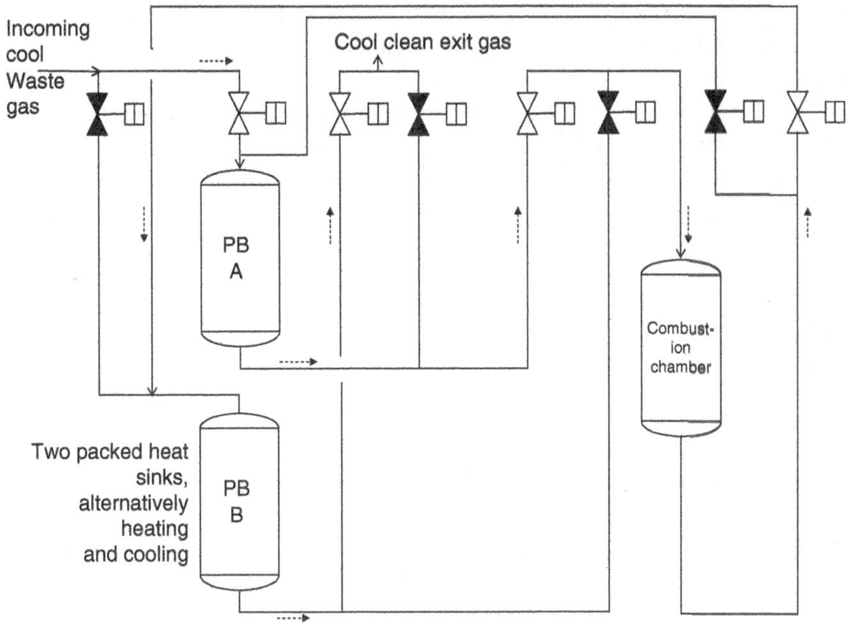

Figure 8.5 Regenerative thermal oxidizer – cycles.

Finally, the option of thermal oxidation is effective but extreme. The high temperatures are hazardous to personnel and a source of fire. A flame arrester is sometimes installed on the hot waste gas before combustion since the gas may be a collection from various sources some of which would be compromised by blowback. A safety rule is that the waste stream be maintained at 25% or less of its LEL, lower explosive limit.

8.2.2.2 Other Treatments for Waste Gases

Moretti (2002) is a good source of alternatives (along with thermal oxidation) for dealing more mildly with gaseous emissions. These are described here.

1. Adsorption: The adsorbing material is one of the known substances with affinity for organics, i.e., activated carbon, zeolites, or polymeric adsorbents. Two or more beds are needed so that gas can be directed through a clean one, while the other(s) are cleaned. Efficiency is claimed up to 95%.
2. Absorption: The waste gas stream is scrubbed with a suitable solvent – often just water, or mineral oils, or other nonvolatile petroleum oils. Of course, the contaminant must be soluble. The solvent plus contaminant must be continuously separated so that the solvent can be reused and the contaminant can be sent for further treatment or back into the plant process.
3. Condensation: Low temperature is generally used. Along with its beneficial action, the scrubbant must undergo cleanup, whether continuous or periodic, which still leaves the CO_2 on hand, to be disposed of.
4. Flares: Liquids are rigorously kept out of the stack in order to keep hot burning particulates from falling on people or on the ground. A flame arrester prevents blowback, especially if feed to the flare falls too low. Bader et al. (2011) advised on selecting the proper flare systems.
5. Biofiltration: The filter bed may consist of natural materials (e.g. compost, soil, or bark), supporting and nourishing microorganisms that ingest contaminants. The system must be

continuously kept wetted. Contaminants that contain oxygen, such as aldehydes, ketones, alcohols, ethers, esters, and organic acids, are good candidates. Such filters have surprisingly long lives – in years.

6. Membranes: A solvent dissolves and carries in the waste gas, which then permeates the membrane, leaving the contaminant up to 20 times more concentrated than the original feed. This makes it more easy to dispose of, whether treated or returned partially to the plant process. Startin & Elliott (2009) suggested controlling emissions with ceramic filters.

8.2.3 Nitrogen Oxides (NOx)

The term *nitrogen oxides* incorporates a collection of compounds containing the nitrogen–oxygen link. Most of the emissions accompany combustion, i.e., from burners and smoke stacks. Another source can be the emission from thermal oxidizers. NOx is a source of smog, especially in geographical terrains that allow this unhealthy nuisance to develop. The necessity to control NOx is especially concentrated in vicinities of high population.

The production mechanism and rate of production of nitrogen oxides are still being studied. McGowan (2004) suggested that the predominant mechanism is described by the Zeldovich equation:

$$N_2 + O \rightarrow NO + N$$

$$N + O_2 \rightarrow NO + O$$

$$\text{Rate of NOx formation} = A\,[N_2] \int_0^t \exp(-b/T)[O_2]\,d\theta \tag{8.3}$$

Consequently, the objective is to keep temperature low and excess oxygen low as well. The latter requirement runs into the need to use excessive oxygen to gain complete consumption of fuel. A well-instrumented combuster, especially if it includes quick or online analysis, can optimize the balance between the two needs.

Anonymous (2020B) deals with NOx absorption. NOx emissions and their reduction are discussed by McGowan (2004), Dakshinamoorthy (2009), and Colannino (2020). Colannino listed 9 (!) measures to alleviate NOx emissions, of which one is to *stage* the adding of fuel and air in boilers, to keep both temperature and oxygen low.

8.2.4 Fugitive Emissions

The many valves and seals around chemical plants all have the potential to leak vaporized process fluids and unwanted contaminants into the surrounding. Parkinson (2007), Adkins & Ehlers (2008), Drago (2010), Wilwerding (2011), and Khiani (2017) wrote about this difficulty. Some of these references actually discuss the making and characteristics of a low-leak valve. Others detail the interaction of regulators (e.g. the EPA in the United States) with commercial plants – with everyone having a common goal. A 2008 estimate is that global fugitive emissions are greater than one million metric tons per year.

Much of the discussion has to do with measuring emissions. The original procedure had personnel taking "sniffer" samples around points of possible leakage, generally around valves, of which there are many in a chemical plant. Valves are estimated to cause 65% of fugitive emissions. This plan was very costly in people effort. Over the years, the techniques of *image analysis* have been worked upon, in which scanning can reveal the extent and composition of leakage. However, the goal of LDAR – leak detection and repair – is always underlying. A fixed-mounted scanner can monitor a range of potential leak points and can alert if an adverse event occurs.

A vacuum system also has the potential for fugitive emissions, as explained by Peress (2002).

None of these measures have anything to do with internal flow past a valve when it is closed, which is a plant problem.

8.2.5 Odors

Odors may seem like a minor problem among other technical challenges, but residents outside the plant will not long put up with the nuisance; and rightly so, even if the plant pre-dated their moving in. If ignored by the company, the problem gets referred to the political sphere and soon a restraining order and plant shutdown occur. In some jurisdictions, odor is actually considered a pollutant.

Neuman (1999), Tetley (2002), and Gans (2006) considered the problem. The consensus is that dilution along with mixing technology can be effective.

8.2.6 Greenhouse Gases

This is a situation when The Tragedy of the Commons could take place, where a commonly owned domain is exploited solely for local interests and eventually despoiled. That situation does not seem to be the case with greenhouse gases and climate change, and those considerations are slowly rising to the top of human concerns. Reducing the emission of greenhouse gases will become the future dominant mandate, both for nations and for industries.

The measures taken by governments are obviously of interest to citizens, but measures that could be taken by industry are the focus for the process designer. Here, we look at some remedial measures, as well as the details of reporting.

Other discussions of greenhouse gas control or mitigation are given by Parisi et al. (2015), Sullivan & Oliva (2007), Princiotta (2007), Jennings (2008), and Balicka (2018).

8.2.6.1 Carbon Dioxide

While CO_2 is not the only greenhouse gas – there are many, including water and methane, its steady release to the atmosphere is commonly seen as driving climate change. Reduction of CO_2 is the number one objective. Methane is a much more potent greenhouse gas but is not as ubiquitous and is being dealt with separately. Water, with its heat-trapping action, is essential to keeping the planet warm enough for human habitation.

The single forward step for a plant is to use, as a fuel, natural gas – this to replace coal or oil. It is only an intermediate step because burning still generates carbon dioxide. Comparison of fuels shows that emissions of CO_2, in pounds per million btu's of heat, are

anthracite coal	228.6
diesel and heating oil	161.3
natural gas	117.0

Natural gas has many different compositions depending on its source, but generally contains methane at 75% or more, with hydrocarbons up to C^5 and higher hydrocarbons making up the rest. Impurities such as carbon dioxide, nitrogen, helium, and water also subsist.

Until an adequate alternative source of energy is developed, industry is stuck with reducing CO_2 emissions. Axon (2009) offered suggestions on improving the combustion system efficiency.

One measure is to extract as much energy as possible from the burning of natural gas. A combined cycle was alluded to in the conservation of energy (Section 8.1.2). A combined cycle gas turbine uses hot burning combustion gas to generate electricity and continuously follows it up with the transfer of heat from the exit gas into steam, which powers another generator. The hot wastewater or steam from the steam turbine can be further used.

The topic of using condensing economizers for added heat recovery from flue gas frequently arises. Although there is often reluctance to adopt this method due to concerns about corrosion and equipment lifespan, there have been successful implementations of the technology. Thus, the energy value of the natural gas is almost fully used. However, this will still not bring about the *major* reduction in carbon dioxide emission that is sought.

8.2.6.2 Alternative Sources of Energy

Carbon dioxide is generated within a plant when natural gas is burned for heat or for powering gas turbines, and outside the plant as part of the mix in the generation of electricity. Obviously NG, for all its low-CO_2 propensity, must eventually be replaced or dealt with. The technologies are being developed, and a company or plant might want to work with a promising and pertinent developer to accelerate and test new methods.

Here are some current suggestions.

1. Symbolic: Install racks of solar panels on the sunny side of the plant; install small windmills on the windy side of the plant; plant CO_2-absorbing trees all around. Even taken together, these actions will likely come nowhere close to canceling out the detrimental emissions. What they do is to advertise to the neighbors and the world that this plant takes climate change seriously. This is a contribution to the world effort. Since it costs capital funds to do this, the power from the panels and windmills could be sold to the grid for income.
2. Geothermal: When one looks at the heat and energy in the earth's core, it appears that it would more than satisfy our consumption on the surface forever – without carbon dioxide. Two factors militate against this drawing up of the energy:
   (1) access must be made far under the earth (say, 10 km) to get at the hot spots;
   (2) the ground around the equipment (e.g. piping) gradually loses its temperature which takes a long time to recover due to the poor thermal conductivity of the soil.

   Typically, in continental regions, in the first 3–5 km down, the *geothermal gradient* is 25 °C km^{-1}. However, the plant builder should check on his/her own particular site, which may have a more favorable gradient. At the very least, there may be opportunities for heat pumping. Geothermal to Rankin Cycle electrical generators typically convert about 11% of the geothermal heat to electricity; thus, the symbiotic use of low temperature heat is often investigated along with the generation of electricity in these systems.

   Plants often buy up large amounts of land around them, partly for future expansion and partly to keep odor-sensitive and noise-sensitive neighbors at bay. Geothermal stations thrive on lateral separation, so they could be established around the property. The whole system could be phased in, with new wells being dug or deepened, until a truly new meaningful energy stream was built up.
3. Carbon capture and utilization: This approach, because of its expense, is more suited to major CO_2 emitters, like gas turbine power plants. It is not a popular concept since it stores but does not eliminate CO_2. Carbon dioxide is captured by a solvent, typically a methyl amine, although

more effective solvents are being sought. The solvent must give up its capture since it needs to be used again. And so there is the carbon dioxide, although now in a purer form that may enhance its usefulness. Anonymous (2020A) discussed carbon capture.

A more acceptable term is CCUS, where the "U" stands for utilization. Unfortunately, the production of carbon dioxide greatly exceeds its needs. If the plant location is anywhere near an oil field, then some CO_2 can be used for *Enhanced Oil Recovery, EOR*. There are many studies where carbon dioxide can form a useful, stable combination with other materials. Because it is at the bottom of the energy chain, it is hard to turn it into anything useful.

4. Burning of on-surface natural products: A fortuitous combination may be a plant near a dead or dying forest or other natural product that could fuel a boiler. Since the natural decomposition of the wood would release CO_2 over the near time, burning it just accelerates the eventuality.
5. Small-module nuclear reactor: The few, rather disastrous accidents with large nuclear stations have dampened the enthusiasm for this form of power. However, it is such an effective solution to the CO_2 problem that it is resurging. For plant applications, there is much development of small reactors. For instance, a 15-MW reactor is under development for remote communities, and there are many more developments, bigger or smaller. The use of nuclear power may be daunting, but it is a technology that can be learned. The new reactors have enhanced safety features. Many people are turning in this direction. The negatives of this technology are well recognized and are being worked on: disposal of waste fuel, need for recharging, and shutdown during overheating.

Figure 8.6 reminds designers of the scale of the problem and the remedial measure.

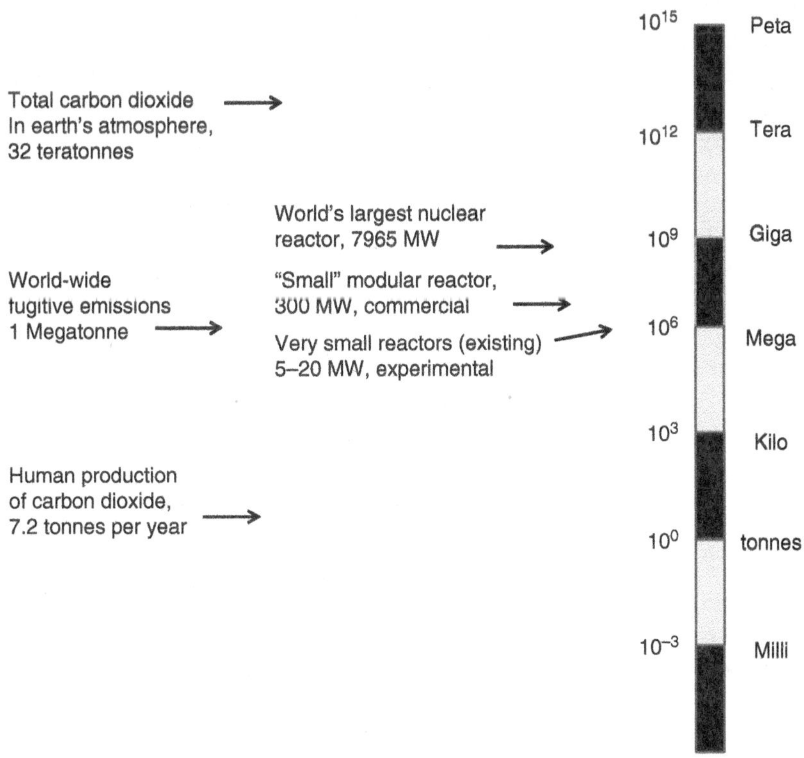

Figure 8.6 Scales.

8.2.7 Auditing and Regulation of Greenhouse Gases (GHG)

Regulators are requiring reports on GHG emissions, although not necessarily forcing reductions. Wintergreen & Sandler (2004), Ritter et al. (2005), and Baranski & Ternes (2010) advised on preparing such reports, and they described "the monitoring plan as a living document that, at a minimum, describes

- the calculation methods that will be used for reporting GHG emissions
- the data required for those calculations
- the data-gathering equipment's calibration, maintenance, and repair procedures
- the party who will be responsible for gathering each data component".

Consistency is key here, with details differing from plant to plant.

8.2.8 Handling of Solid Waste

Solid-waste disposal methods vary depending on local regulations, environmental considerations, and available infrastructure. Here are some common methods for disposing of solid waste:

1. Recycling: Recycling involves collecting and processing materials like paper, plastics, glass, and metals to manufacture new products. Recycling helps reduce the volume of solid waste and conserves resources.
2. Hazardous waste disposal through contracting: Handling and disposing of hazardous wastes, such as chemicals and toxic materials, require specialized methods to protect both people and the environment. In cases where specific expertise is needed, or due to scale and other considerations, it is often prudent to contract this task to qualified companies. These firms may employ methods like incineration, secure landfilling, or other specialized treatments.
3. Landfilling: This is one of the most common methods worldwide. Solid waste is buried in designated landfill areas. Landfills are engineered to minimize environmental impact and control leachate and gas emissions. However, landfills can still contribute to environmental problems if not properly managed.
4. Sanitary landfilling: This is an engineered landfill with measures to minimize environmental impacts, including leachate collection systems and gas extraction. While both landfilling and sanitary landfilling involve burying solid waste, the key difference is the level of engineering and environmental safeguards in place to protect the environment and public health. Sanitary landfills are considered a more advanced and environmentally responsible form of landfilling compared to traditional landfills.
5. Composting: Organic waste like food scraps and yard waste can be composted, turning it into nutrient-rich soil conditioner. Composting reduces waste and enriches soil.
6. Mechanical Biological Treatment (MBT): MBT combines mechanical sorting and biological treatment (composting or anaerobic digestion) to process mixed waste streams.
7. Bioremediation: This method uses microorganisms to break down or detoxify hazardous waste in soil and water.
8. Incineration: This method involves burning solid waste at high temperatures, with air or oxygen to reduce its volume and convert it into ash, flue gas, and heat energy. Modern incinerators are equipped with emission control systems to minimize air pollution. Energy can also be recovered from the process.
9. Pyrolysis and gasification: These are advanced thermal treatment methods that operate in the absence of oxygen (or with limited oxygen). They break down the waste into syngas (a mixture of hydrogen and carbon monoxide), char, and other byproducts.

10. Waste-to-Energy (WtE): Some facilities use waste as a fuel to generate electricity or heat. This method reduces the volume of waste and produces energy.
11. Dumping and open burning: These are informal and often illegal methods used in some areas, particularly in regions with limited waste management infrastructure. These methods can be harmful to the environment and public health.

The choice of disposal method depends on factors like waste composition, local regulations, environmental concerns, and available resources. Many places are shifting toward waste reduction, recycling, and sustainable waste management practices to minimize the environmental impact of waste disposal.

References

Adams, D.J., Faris, B., and Vlassopoulos, D. (2006 Feb). Evaluating in situ bioremediation for groundwater cleanup. *Chem. Eng. Prog.* 102 (2): 20–28.

Adkins, M. and Ehlers, P. (2008 Jan). How to choose a ball valve that curbs fugitive emissions. *Hyd. Proc.* 87 (1): 119–123.

Adler, S.F. (2001 Apr). Biofiltration: a primer. *Chem. Eng. Prog.* 97 (4): 33–42.

Aerts, P. and Tong, F. (2009 Sep). Strategies for water reuse. *Chem. Eng.* 116 (9): 34–39.

Allen, D.T. (2007 Dec). Green engineering and the design of chemical processes and products. *Chem. Eng.* 114 (13): 36–40.

Amminudin, K.A., Sofian, M.N., Hawari, H. et al. (2008 Dec). Consider water reuse options for your facility. *Hyd. Proc.* 87 (12): 99–102.

An, L. (2013 Oct). Biological wastewater treatment: selecting the process. *Chem. Eng.* 120 (10): 38–43.

Anastas, P.T. and Zimmerman, J.B. (2003 Mar). Design through the 12 principles of green engineering. *Environ. Sci. Technol.* 1: 94A–101A.

Anonymous (2002 Jun). Ten steps to consider. *Chem. Eng.* 109 (6): 79.

Anonymous (2015A Feb 20). Multi-faceted approach on energy efficiency methods. *Hyd. Proc.* 94 (2): 85–87.

Anonymous (2015B Oct). Creating sustainable water treatment systems. *Chem. Eng. Prog.* 111 (10): 17.

Anonymous (2020A May). CO_2 capture. *Chem. Eng.* 127 (5): 10.

Anonymous (2020B Jun). NOx adsorption. *Chem. Eng.* 127 (6): 10–11.

Armstrong, P. and Predatsch, E. (2019 Oct). Use new oxidation technologies to achieve near-zero emissions of NOx and VOC. *Hyd. Proc.* 98 (10): 33–37.

Arnt, en, B. (2020 Sep). Designing an integrated wastewater-treatment system. *Chem. Eng.* 127 (9): 39–44.

Ataei A., Wastewater treatment: energy-conservation opportunities, *Chem. Eng.* 2010 Jan, 117*(1)*, 34–41.

Axon, B. (2009 Aug). Improve combustion system efficiency. *Chem. Eng. Prog.* 105 (8): 41–46.

Bader, A., Baukal, C.E. Jr., and Bussman, W. (2011 Jul). Selecting the proper flare systems. *Chem. Eng. Prog.* 107 (7): 45–50.

Balicka, I. (2018 Sep). Carbon capture, utilization, and storage: a plan for greater energy efficiency. *Chem. Eng. Prog.* 114 (9): 23.

Baranski, K. and Ternes, M.E. (2010 Mar). Submitting your first GHG report. *Chem. Eng. Prog.* 106 (3): 65–70.

Baranski, K. and Underwood, B. (2014 Jun). Mitigate air pollution with catalytic technology. *Chem. Eng. Prog.* 110 (6): 30–34.

Beal, D.R. and Faircloth, H. (2002 Mar). Tackling tough groundwater contaminants. *Chem. Eng.* 109 (3): 91–94.

Belenchia, V. (2010 Dec). Chemical lifecycle management. *Chem. Eng.* 117 (13): 44–47.

Bhatia, S.C. (2014). *Advanced Renewable Energy Systems*. CRC Press.

Blocki, S. (2005 Jul). Choose the right emissions-control option. *Chem. Eng.* 112 (7): 44–47.

Blume, A. (2014 Mar). Real-time infosystem measures energy use. *Hyd. Proc.* 93 (3): 18.

Bourji, A., Barnhart, J., Winningham, J., and Winstead, A. (2010 Sep). Recover waste heat from flue gas. *Chem. Eng.* 117 (9): 37–40.

Bourji, A. and Winstead, A. (2013 Jan). Optimizing an organic rankine cycle. *Chem. Eng. Prog.* 109 (1): 35–39.

Buchanan, L.S. (2007 May). Determine the efficiency of air-pollution control devices. *Chem. Eng. Prog.* 103 (5): 37–42.

Carr, S. and Vaughn, R. (2003 Apr). Taking stock of activated carbon's many talents. *Chem. Eng.* 110 (4): 76–80.

Carson, J.W. (2002 Jan). Don't get stuck on sludge handling. *Chem. Eng.* 109 (1): 79–81.

Cartwright, P. (2006 Mar). Process water treatment–challenges and solutions. *Chem. Eng.* 113 (3): 50–56.

Chin, W. and Gillespie, J. (2010 Mar). Breaking down the silos between energy and asset management. *Hyd. Proc.* 89 (3): 35–38.

Colannino, J. (2020 May). Low-cost techniques for NOx reduction. *Chem. Eng.* 127 (5): 30–36.

Contreras, C.D. and Bravo, F. (2011 Aug). Practice green chemical engineering. *Chem. Eng.* 118 (8): 41–44.

Cooper, H.W. (2010 Feb). Producing electricity and chemicals simultaneously. *Chem. Eng. Prog.* 106 (2): 24–32.

Curran, M.A. (2015 Oct). Life cycle assessment: a systems approach to environmental management and sustainability. *Chem. Eng. Prog.* 111 (10): 26–35.

Dai, W. (2005 Nov). Modeling accidental releases. *Chem. Eng. Prog.* 101 (11): 33–39.

Dakshinamoorthy, D. (2009 Sep). Consider advanced modeling to control air emissions. *Hyd. Proc.* 88 (9): 81–86.

Dalan, J.A. (2000 Nov). 9 things to know about zero liquid discharge. *Chem. Eng. Prog.* 96 (11): 71–76.

Da Silva, A.K. and Goodman, A. (2014 Apr). Reduce water consumption through recycling. *Chem. Eng. Prog.* 110 (4): 29–37.

Doble, M. (2006 Jun). Biological treatment of VOCs. *Chem. Eng.* 113 (6): 35–41.

Doble, M. (2008 Aug). Green reactors. *Chem. Eng. Prog.* 104 (8): 33–42.

Doble, M. and Geetha, V. (2011 Oct). Understand the fundamentals of wastewater treatment. *Chem. Eng. Prog.* 107 (10): 36–42.

Drago, J. (2010 Jun). Containing fugitive emissions. *Chem. Eng.* 117 (6): 43–46.

Drago, J. (2015 Oct). Emissions regulation and control. *Chem. Eng.* 122 (10): 56–60.

Duckett, J.E. (2006 Jun). Right-sizing environmental controls. *Chem. Eng. Prog.* 102 (6): 43–47.

Durham, B. and Patria, L. (2006 Oct). Wastewater: a reliable water resource. *Chem. Eng.* 113 (10): 50–54.

El-Halwagy, M.M., Lovelady, E.M., Abdel-Wahab, A. et al. (2009 Feb). Apply process integration to environmental impact assessment. *Chem. Eng. Prog.* 105 (2): 36–42.

Elshout, R. and Marchant, L. (2010 Feb). Doing an energy audit. *Chem. Eng.* 117 (2): 34–36.

Foo, D.C.Y., Manan, Z.A., and Tan, Y.L. (2006 Jul). Use cascade analysis to optimize water networks. *Chem. Eng. Prog.* 102 (7): 45–52.

Frenkel, V. (2002 Jul). Use pretreatment to improve process-water treatment. *Chem. Eng.* 109 (7): 54–58.

Ganapathy, V. (2009 May). Rethink planning for heat-recovery systems. *Hyd. Proc.* 88 (5): 47–51.

Gans, C. (2006 Aug). Consider mixed-flow impeller technology for odor management. *Chem. Eng. Prog.* 102 (8): 43–47.

Gendron, L.J., Sacco, A.M., and Chaudhuri, I. (2004 Oct). Implementing perimeter air-monitoring programs. *Chem. Eng.* 111 (11): 42. 64–66.

Glavic, P. (2021). Process design and sustainable development—a European perspective. *Processes* 9: 148.

Goldshmid, Y. (2005 May). Destroying organic wastes: thermal oxidizer basics. *Chem. Eng. Prog.* 101 (5): 40–47.

Griesbach, R., Corthay, E.T., Storey, R., and Dover, M. (2013 May). Conducting an energy design review. *Chem. Eng. Prog.* 109 (5): 40–43.

Gucciardi, J.M. (2003 Jun). Wastewater treatment. *Chem. Eng. Prog.* 99 (6): 25.

Hairston, D. (2003 Apr). In cooling towers, it's the water that matters. *Chem. Eng.* 10 (4): 25–28.

Hannah C. C., Perry T. W., Understand the installation of steam tracing for long-term application success-Part 1 and Part 2, *Hyd. Proc.* 2018 Sep & Oct. 97 85-87 & pp. 67–70.

Hardin, D. (2007 Dec). Industrial power management: what plants need to know about the future electric system. *Hyd. Proc.* 86 (12): 95–99.

Harfst, W.F. (2010 Apr). Non-chemical water treatment. *Chem. Eng.* 117 (4): 66–69.

Harrison, J. (2008 Aug). Making energy efficiency pay. *Chem. Eng.* 115 (8): 34–40.

Herraiz, L., Palfi, E., Sánchez Fernández, E., and Lucquiaud, M. (2020 Dec). Rotary adsorption: selective recycling of CO_2 in combined cycle gas turbine power plants. *Front. Energy Res.* 8: 482708.

Huchler, L.A. (1999 Nov). Are you properly monitoring your utility water in process units? *Hyd. Proc.* 78 (11): 127.

Huchler, L.A. (2010 Jan). Wastewater discharge permits: what should you know? *Hyd. Proc.* 89 (1): 82.

Huchler, L.A. (2016 Jan). Minimize the risk of *Legionella* contamination in industrial cooling towers. *Chem. Eng. Prog.* 112 (1): 46–48.

Jenkins, S. (2010 Jul). Conservation economics: carbon pricing impacts. *Chem. Eng.* 117 (7): 25.

Jenkins, S. (2013 Jan). Non-chemical water treatment. *Chem. Eng.* 120 (1): 27.

Jenkins, S. (2015 Dec). Water treatment technologies. *Chem. Eng.* 122: 32.

Jenkins, S. (2020 Jun). Water treatment: ion exchange resins. *Chem. Eng.* 127 (6): 24.

Jennings, P.A. (2008 Feb). Dealing with climate change at the local level. *Chem. Eng. Prog.* 104 (2): 40–44.

Jones, F.V. and Rankin, S. (2015 Jul). The integral role of waste management in capital projects. *Chem. Eng.* 122 (7): 43–49. *109(5)*, 29–33.

Karell, M. (2013 May). Develop an effective air-pollution-control strategy. *Chem. Eng. Prog.* 109 (5): 29–33.

Kaupert, K., Rull, R., and Iles, R. (2015 May). Maximize energy recovery with small steam turbines. *Hyd. Proc.* 94 (5): 51–54.

Khaqan, A. (2011 Apr). PEMS: the low-cost alternative to emissions monitoring. *Chem. Eng.* 118 (4): 28–31.

Khiani, G. (2017 Oct). Valves: emission standards and current practices. *Chem. Eng.* 124 (10): 60–63.

Kleerebezem, R. and Macarie, H. (2003 Apr). Treating industrial wastewater: anaerobic digestion comes of age: anaerobic treatment systems offer important advantages over conventionally applied aerobic processes for removing organic pollutants from water-based streams.(Cover Story). *Chem. Eng.* 110 (4): 56–64.

Klobucar, J.M. (2002 Feb). Pollutant destructions: comparing thermal oxidizer designs. *Chem. Eng.* 109 (2): 62–67.

Labrecque, R. and Kiari, G.B. (2006 Oct). Get the most out of waste heat. *Chem. Eng.* 113 (10): 40–43.

Lad, V.N. (2002 Jul). Pollution prevention: don't neglect it. *Chem. Eng.* 109 (7): 59–63.

Mack, S. (2003 Sep). VOC control: thermal vs catalytic oxidation. *Chem. Eng.* 110 (9): 58–60.

Malosh, K.M. (2008 Jul). Develop and maintain your environmental management system. *Chem. Eng. Prog.* 104 (7): 53–56.

Marshall, R. (2007 Jan). Wastewater treatment. *Chem. Eng.* 114 (1): 31.

McGowan, T.F. (2004 Oct). Charting a path for cost-effective NOx control. *Chem. Eng.* 111 (11): 36–41.

Mendez, M. (2007 Dec). The role of chemical engineers in green engineering. *Chem. Eng.* 114 (13): 41–44.

Moftah, K. (2000 Aug). Apply statistical rules to determine emissions. *Chem. Eng. Prog.* 96 (8): 61–63.

Moretti, E.C. (2002 Jun). Reduce VOC and HAP emissions. *Chem. Eng. Prog.* 98 (6): 30–40.

Mueller, J. and Cipullo, M. (2008 Jan). Minimize wastes from specialty chemical processes. *Chem. Eng. Prog.* 104 (1): 30–34.

Muller, M.R., Muller, M.B., and Rao, P. (2013 Nov). Optimize energy use in industrial cooling systems. *Chem. Eng. Prog.* 109 (11): 18–23.

Neimeier, R.M. (2004 Nov). Get the most from your emissions testing buck. *Chem. Eng. Prog.* 100 (11): 31–35.

Neuman, V.A. (1999 Nov). Consider dilution for odor control. *Chem. Eng. Prog.* 95 (11): 81–84.

Nolen, S. (2016 Apr). Leveraging energy management for water conservation. *Chem. Eng. Prog.* 112 (4): 41–47.

Nystuen, D.L. (2011 Apr). Reduce your plant's water footprint. *Chem. Eng. Prog.* 107 (4): 28–33.

Ondrey, G. (2004 Jan). A way to cut the cost of CO_2 recovery. *Chem. Eng.* 111 (1): 13.

Pakzadeh, B. and Zbacnik, R. (2015 Sep). Treating wastewater for industrial reuse. *Chem. Eng.* 122 (9): 56–61.

Parisi, P., Staroselsky, S., and Hoop, E. (2015 Aug). Optimizing turbomachinery energy savings with control technologies. *Hyd. Proc.* 94 (8): 71–74.

Parkinson, G. (2005 Feb). Wringing more out of water. *Chem. Eng.* 112 (2): 21–25.

Parkinson, G. (2007 Dec). A smarter way to detect fugitive emissions. *Chem. Eng. Prog.* 103 (12): 7–10.

Peress, J. (2002 May). Estimate emissions from vacuum operations. *Chem. Eng. Prog.* 98 (5): 40–46.

Powell, I. (2008 Aug). Best practices in hazardous-waste management. *Chem. Eng. Prog.* 104 (8): 49–51.

Princiotta, F. (2007 Nov). Mitigating global climate change through power-generation technology. *Chem. Eng. Prog.* 103 (11): 24–32.

Puckorius, P.R. (2013 Feb). Consider recycled water for your cooling tower makeup. *Chem. Eng. Prog.* 109 (2): 24–29.

Putra, Z.A. (2015 Dec). Improve performance and reduce costs of wastewater treatment plants. *Hyd. Proc.* 94 (12): 75–78.

Raaphorst, H. (1998 Jul). Crack down on emissions. *Chem. Eng.* 105 (7): 151.

Reddy, C.C.S., Naidu, S.V., and Rangaiah, G.P. (2013 Jan). Waste heat recovery methods and technologies. *Chem. Eng.* 120 (1): 28–38.

Reinemann, P.S. III, (2006 Jul). The maturation of a technology: predictive emissions monitoring. *Chem. Eng.* 113 (7): 50–55.

Ritter, K., Nordrum, S., Shires, T., and Lev-on, M. (2005 Sep). Ensuring consistent greenhouse gas emissions estimates. *Chem. Eng. Prog.* 101 (9): 30–37.

Rossiter, A. (2006 Dec). Make your plant more energy efficient. *Chem. Eng. Prog.* 101 (12): 31–36.

Rossiter, A.P. and Venkatesan, V. (2012 Dec). Easy ways to improve energy efficiency. *Chem. Eng. Prog.* 108 (12): 16–20.

Rossiter, A.P. and Davis, J.L. Jr., (2014 Sep). Identify process improvements for energy efficiency. *Chem. Eng. Prog.* 110 (9): 53–58.

Rossiter, A.P. and Jones, B. (2014 Dec). Achieving excellence in energy efficiency. *Chem. Eng.* 121 (13): 44–49.

Rossiter, A.P. (2015A Mar). Energy efficiency: getting in early pays off exponentially. *Hyd. Proc.* 94 (3): 21.

Rossiter, A.P. (2015B April). *Energy Management and Efficiency for the Process Industry*. AIChE/Wiley Publishers.

Rossiter, A.P. (2015C May). The four pillars of industrial energy efficiency. *Chem. Eng. Prog.* 111 (5): 40–44.

Savage, P.E. (2009 Sep). What does it mean to be green? *Chem. Eng. Prog.* 105 (9): 30–34.

Scheier, J. (2001 Dec). Correctly calculate the potential to emit. *Chem. Eng. Prog.* 97 (12): 50–52.

Scheihing, P. (2014 Sep). Save energy through the superior energy performance program. *Chem. Eng. Prog.* 110 (9): 48–51.

Schultz, T.E. (2005 Oct). Biological wastewater treatment. *Chem. Eng.* 112 (10): 48–51.

Schultz, T.E. (2008 Sep). Water reuse and conservation in the CPI. *Chem. Eng.* 115 (9): 44–50.

Schuster, D. (2015 Jan). Sustainability: what we measure we can improve. *Chem. Eng. Prog.* 111 (1): 34.

Shah, G.C. (2016 Nov). What every manager should know about hazardous waste management. *Hyd. Proc.* 95 (11): 43–48.

Shaw, W.A. and Brosdal, J.D. (2008 Oct). Strategies to minimize wastewater discharge. *Chem. Eng.* 115 (10): 60–68.

Siegell, J.H. (2010 Jan). Audit your emissions inventory. *Chem. Eng. Prog.* 106 (1): 44–49.

Sims, P. (2007 Nov). Thermal oxidizers aid in pollution control. *Hyd. Proc.* 86 (11): 69–70.

Smith, R. and Varbanov, P. (2005 Jul). What's the price of steam. *Chem. Eng. Prog.* 101 (7): 29–33.

Startin, A. and Elliott, G. (2009 Jan). Controlling emissions with ceramic filters. *Chem. Eng.* 116 (1): 35–39.

Sullivan, J. and Oliva, M. (2007 Aug). Greenhouse gases – an effective strategy for managing GHG emissions. *Chem. Eng.* 114 (8): 34–40.

Sustarsic, M. (2009 Nov). Wastewater treatment: understanding the activated sludge process. *Chem. Eng. Prog.* 105 (11): 26–29.

Sutikno, T. (2007 Aug). Optimize energy savings on steam generation. *Hyd. Proc.* 86 (8): 105–108.

Tanthapanichakoon, W. (2012 Jan). Saving energy in multilevel steam systems. *Chem. Eng. Prog.* 108 (1): 27–32.

Ternes, M.E. (2012 Feb). Environmental law for chemical engineers. *Chem. Eng. Prog.* 108 (2): 36–43.

Tetley, P.A. (2002 Apr). For wastewater odors, dilution may be the best solution. *Chem. Eng.* 109 (4): 83–85.

Uhlman, B.W. and Saling, P. (2010 Dec). Measuring and communicating sustainability through eco-efficiency analysis. *Chem. Eng. Prog.* 106 (12): 17–26.

Ulrich, G.D. (2007 Jun). Predesign for pollution prevention and control. *Chem. Eng. Prog.* 103 (6): 53–60.

Uribe, I.A., Mosquera-Corral, A.N., Rodicio, J.L., and Esplugas, S. (2015 Oct). Advanced technologies for water treatment and reuse. *AICHE J.* 61 (10): 3146–3158.

Vardhan, K.H., Ponnusamy, S.K., and Panda, R.C. (2019). A review on heavy metal pollution, toxicity and remedial measures: current trends and future perspectives. *J. Mol. Liq.* 290: 111197.

Venkatesh, M. and Woodhull, J. (2003 Aug). Destroying gaseous emissions: pick the right thermal oxidizer. *Chem. Eng.* 110 (8): 67–70.

Vlyssides, A., Barampouti, E.M., and Mai, S. (2006 Jan). Simplify wastewater treatment process design. *Chem. Eng. Prog.* 102 (1): 42–46.

White, D.C. (2010 Jan). Save energy through automation. *Chem. Eng. Prog.* 106 (1): 26–33.

Willa, J.L. (2005 May). Ensure proper distribution in cooling towers. *Chem. Eng. Prog.* 101 (5): 29–33.

Williams, A. and Dunwoody, K. (2012 July). Evaluating green projects: modeling improves economic benefits. *Chem. Eng.* 119 (7): 34–43.

Wilwerding, J. (2011 Jun). Fugitive emissions from valves: update. *Hyd. Proc.* 90 (1): 81–83.

Wintergreen, J.T. and Sandler, L.M. (2004 Apr). Preparing for GHG inventory verification. *Chem. Eng. Prog.* 100 (4): 35–38.

Zygula, T.M. (2008 May). Designing steam stripping columns for wastewater. *Chem. Eng.* 115 (5): 52–56.

9

Capital Cost Estimating and Economic Analysis

Most chemical engineers are familiar with and have utilized some form of factored equipment estimate (FEE) method and economics in the undergraduate capstone design courses. This chapter is for those who want to expand on what was taught at school in a way that will help them understand and create better estimates. The references section contains the original sources of estimating procedures, some quite old, but well worth obtaining and reading.

This chapter supports engineers and others in developing capital cost (CAPEX) estimates and the project economic analysis for the front-end stage gate decisions.

9.1 What Is an Estimate

When we say an estimate, we usually mean a capital project estimate, the dollar figure representing the total cost for engineering, procurement, construction, and commissioning of a chemical plant. It is also known as the project CAPEX, or the depreciable asset value when it includes the owner-associated project costs. The estimate will have a stated accuracy range and a date the estimate was prepared. Along with the estimate values are several supporting documents explaining the basis for the estimate, the most important being the scope of work. The scope explains in words what the proposed project will provide and explains assumptions, inclusions, exemptions, and details that the numbers in the estimate cannot. Depending on the type of estimate, differing amounts of engineering will be required, resulting in a collection of engineering documents to show the design basis and details.

A CAPEX estimate contains a list of items that fit into the following categories:

1. Direct costs: All costs that are directly attributed to installing equipment. This includes the equipment, bulk materials (piping, valves, and instruments), and labor to install this. It also includes what is known as "Off-sites" which includes pipe racks, rail spurs, and utility facilities such as waste water, steam boilers, and so forth.
2. Indirect costs: This category includes costs that are spread across many pieces of equipment, such as engineering, freight, insurance, overhead, and construction supervision.
3. Contingency, allowance, fees, etc.: This includes those items, the first fill of the plant, taxes, duties, and escalation.

Practical Process Design for Chemical Engineers, First Edition. Keith Marchildon and David Mody.

9.2 Why Estimate

Chapter 10 describes the business steps of bringing a new chemical product from discovery to final product launch. Section 10.5 and Figure 10.5 illustrate how the estimates fit within the project execution framework for a typical chemical engineering project. Section 10.6 delves into the front-end engineering in further detail. In the context of a stage gate review process, each gate typically culminates with an estimate review as part of the decision-making process to continue.

There are numerous reasons why an estimate is needed in a project, and understanding why you are preparing an estimate is key to understanding how much effort must be put into the task. Estimates and the subsequent economics analysis support that business process by providing insight into the financial implications and risks of the project at regular intervals. However, estimates cost money, not just in preparing the cost values but in the engineering that leads to those values. Thus, there are numerous approaches tailored to minimizing the effort costs while providing a required level of accuracy (Lagace 2006).

A stage gate business process is said to reduce the overall cost of innovation of a company by stopping projects that are uneconomical before significant engineering costs are wasted on a dead-end design. This chapter is intended to help a person understand the deliverables and steps involved in developing the capital and operating costs (OPEX) for a design and making economic decisions through various stage gates. Later in the life of a project, the "detailed" estimate is used to monitor and control the construction so that overruns in cost and time can be identified and possibly avoided before they become serious obstructions.

9.3 The What and Why of Economic Analysis

Economic analysis is one aspect of deciding whether to approve and continue with a project. The analysis utilizes the capital estimate and adds the OPEX to determine the cost of manufacturing a product. If revenues are available, they can be added, and a year-over-year analysis of cash inputs and outputs can be built to provide a discounted cash flow (DCF) model. Subsequently, the following metrics can be developed: payback period, net present value (NPV), ROR (Rate of Return), IRR (Internal Rate of Return), and others which support decisions to:

- continue, recycle, or discontinue a project based on corporate benchmarks;
- compare other investments more easily for ranking and prioritization;
- include projections in forward-looking business documents, such as forecasts of future earnings;
- assess the factors contributing to economic risks in the project;
- employ Monte Carlo analysis methods to estimate probabilities for a range of values.

Above is the usual understanding and use of DCF analysis regarding project approval. Christensen (2008) presented an alternate approach to project approval. The premise he states is that the stage gate process is flawed, that conventional DCF analysis methods require assumptions that can be intentionally, or unintendedly, biased in a way that would mislead decision-makers. His solution is called "discovery-driven planning", and it is to work backward to find the assumptions that must be met to achieve the necessary financial hurdles. He refers to this as a reverse income statement. The method determines the most important assumptions in ensuring a project's success and works toward ensuring their success. A checklist can be used to help. The team "revises its strategy until assumptions are all plausible. If no set of plausible assumptions will support the case for success, the project is killed". Regardless of the approach, many of the tools require working with an economic analysis.

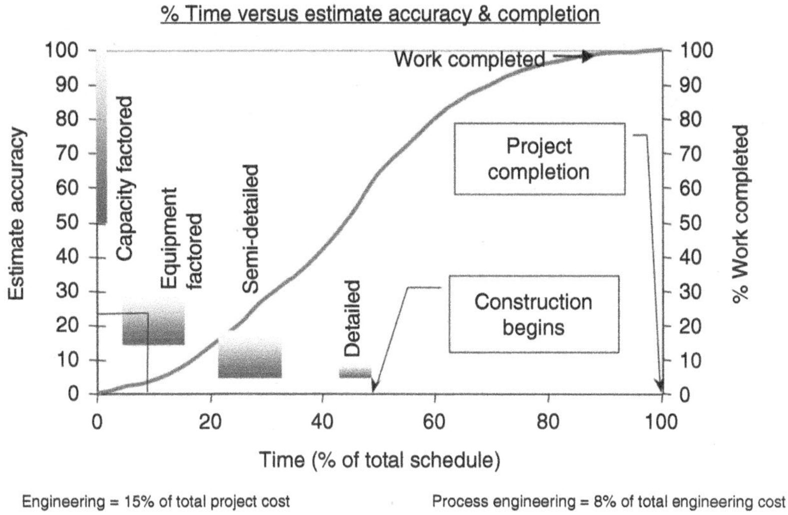

Figure 9.1 Estimate and project timeline.

9.4 A Process Engineer's Role in Estimating

In smaller projects, or ones that are very early in development, the process engineer may perform most, or all, of the tasks in creating the estimate and analysis, given some guidelines from the business roles. On larger projects, or later in the stage gate process, specialists in estimating, as well as other engineering disciplines, and business decision and risk analysis people will be more involved. In those cases, the process engineer provides an engineering lead role and helps to ensure all the right people have the information they need; people understand the assumptions that are present, and that important details are not missed.

Early-stage estimates aim to support the decision-making process of whether to continue a project. Figure 9.1 shows the progression of different estimates, their accuracy, and the project completion timeline. The approach is always to minimize the cost of creating the estimate while providing a desired level of accuracy. The estimate classes are defined by numerous people, such as the Association for the Advancement of Cost Estimates (AACE), but are described below with the accuracy and information required for chemical engineering projects by Lagace (2006) (Table 9.1).

9.5 Estimate Types and Methods

9.5.1 Order of Magnitude

One of the issues every project grapples with is having an idea of the CAPEX before a project begins. The following describes fast but low-accuracy methods for determining the approximate cost of a plant. Unfortunately, people often believe the first number they hear is the correct number, probably because it is the lowest value they will see, and that can hurt credibility later in a project, so caution is advised in using the methods below.

Experienced engineers and business people have the experience to guesstimate the CAPEX of a process remarkably well.

Table 9.1 Estimate types and required information.

Estimate description	Order of magnitude	Preliminary	Budget	Control	Definitive
Class	V	IV	III	II	I
Also called	Curve quality or capacity factored	Equipment factored or conceptual estimate	Software conceptual estimate	Project estimate	Detailed estimate or check estimate
Purpose	Early cost indication/ planning	Studies/planning/ evaluate alternate processes	Studies/planning	Cost-plus bids	Lump sum bids
Method of preparation	Cost-capacity curves, historical or published data (i.e. dollars per barrel)	Total installed cost factored from equipment cost using Lang/Guthrie factors	Estimator with perhaps minor amount of manual estimating. Some vendor quotes as necessary	Combination of computer and manual methods Fewer Quotes than definitive	Primarily manual using in-house programs
Accuracy	+50%/−30%	+35%/−20%	+25%/−15%	+15%/−10%	+10%/−5%
Minimum information needed					
Plant capacity	X				
Plant location and expected date	X	X	X	X	X
Narrative scope of work				X	X
PFDs		X			
Plot size/plan		X Size	X Plan	X Plan	X Plan
P&IDs			X	X	X
Sized equipment list		X	X	X	X
Equipment layouts				X	X
Equipment data sheets and specs			X	X	X
Equipment pricing		Generated in the estimate	X	X	X
Electrical equipment list		X	X	X	X

See Lagace for further details relevant to Class I through III estimates. Source: Adapted from Lagace (2006).

There may be some information available such as a plant cost per ton of product made, or per $ of sales (turnover ratio). See Garrett (2012).

You might consider back-calculating a range for an acceptable CAPEX using the observation that the CAPEX and OPEX combined must be paid by the difference between revenue and raw material costs (gross margin). In a sense, this is a simplified variation on a payback period.

Prior experience with similar processes can provide you with a CAPEX:OPEX ratio that can be used, or failing that by utilizing a wide range of values leads to a rough range of values for the CAPEX as follows:

- Gross Margin = Revenue – Raw Material Costs.
- Gross margin is divided up to pay for CAPEX + OPEX, ignore profit, taxes, fixed expense, etc., for the benefit of simplicity.
- CAPEX is recovered over multiple years.
- If a plant is dominated by OPEX, the CAPEX:OPEX ratio per year could be 10 : 90 (1/10th).
- If a plant is dominated by CAPEX costs, the CAPEX:OPEX ratio per year could be 90 : 10 (9/10th).

Example: If the Revenue – Raw Materials costs are $100MM per year, and the life of the plant is 10 years, the CAPEX might range from:

$$\$100\text{MM} \times 1/10 \times 10\,\text{years} = \$100\text{MM to}$$

$$\$100\text{MM} \times 9/10 \times 10\,\text{years} = \$900\text{MM}$$

While this is a large range, it does provide some useful information very quickly.

Unfortunately, published CAPEX:OPEX ratios for various technologies are not readily available, so they need to be developed by industry or company for a more accurate assessment.

Other approaches to back calculating an approximate range of allowable CAPEXs can be used:

- Return on Investment (ROI), where the % return is assumed, and the expected cash flow per year is used to provide an initial investment value.
- Return on Assets (ROA) is a value that can be determined from a corporation's publicly traded financial statements. Net income, also known as profit (income after all expenses, depreciation, interest, and taxes has been deducted from revenue), is divided by total assets. This % can be used to determine a corporation's typical asset value based on the income you expect it to provide. An alternative to this would be the ratio of gross profit to asset value. Gross profit includes the costs to produce the product such as raw materials, labor, and utilities, but not taxes.
- Asset turnover ratio is the revenue divided by the asset value. These values can also be derived from the corporate financial documents' values more specific to the company's business.

9.5.2 Licensor Estimate

If a process is to be built from licensor information, initial estimates of economics can be obtained from them. On the other hand, licensor estimates can be difficult to obtain without signing nondisclosure documents, which can increase the time and effort to obtain them.

9.5.3 Capacity Factored Estimate

The capacity cost curve type estimate, also known as the 6/10th's law, is a very useful one. Where a larger multiunit process is being estimated, it might resemble other technologies which have been previously been built and operating, and for which cost information is available. The cost of a new plant based on that prior plant can be determined using a ratio of the two plant capacities to an exponent, adjusted for inflation between years, and if necessary, adjusted for differences in locations.

Cost of new plant = Cost of old plant (capacity new plant/capacity old plant)^exponent

The exponent often used is 0.6 to 0.7 but varies. Dysert (2003B) discussed capacity factored estimates and presented a table for different technologies. Perry had a list of exponents by technology.

Since the old plant and the new plant are being built at different times, inflation can be accommodated by utilization of a "plant cost index" value available for different years, and an educated guess for the value when the new plant will be built. Plant cost index values are available from sources such as the magazine Chemical Engineering (CEPCI index value) and others. See Dysert (2003B), Ulrich (2009), and Williams (2021). See also the Cost Index Section 9.5.4.2.

If the regulatory environment has changed since the older plant was built, a hybrid estimate composed of a capacity factored estimate and a more detailed method described below may achieve the accuracy needed with minimal engineering effort.

There are numerous sources of information for these types of estimates. It is likely that a large company has a database of these projects that can be relied on, but an hour in the library or searching for similar projects is an excellent starting place. One source of process and technology feasibility studies, and capacity factored cost information is the National Renewable Energy Laboratory (NREL) website.

While the estimates below contain more detail, and thus more perceived accuracy, a capacity factored estimate has some advantages in that it may include equipment and issues that other estimate methods could miss until significant engineering effort has been invested. Merrow (1979) warned, "A carefully implemented time-consuming estimate based on uncertain plant characteristics is little better than an initial estimate performed quickly".

To compensate for inaccuracies in this method, contingency is added (as it is to all estimates). Contingency is discussed well by Lagace (2006) and in the chapter below. Methods for calculating contingency that are said to achieve estimate accuracy to within +/−10% regardless of the engineering development performed are presented by Merrow (1981). This method is further discussed below (section 9.8.2.4).

9.5.4 Factored Equipment Estimate (FEE)

There are three types of estimates that are used that fit in this category. All use the equipment costs and multipliers to determine the overall plant cost. They require that you have the equipment determined, sized, and to some extent costed. For this reason, there is substantially more engineering costs and time involved in developing an estimate this way.

The first of these estimates, the Lang factor method, sums up the equipment costs and multiplies this value by 5 (or some value similar to this based on experience) to reach a total installed cost (TIC). Despite the simplicity of this method, a very large proportion of companies use this method ubiquitously. The second, pioneered by Guthrie and others, uses a more complex method, which can simultaneously determine costs for the equipment based on physical design parameters like size. This is the preferred method used by engineering companies who have a budget to prepare the engineering required for it.

If you miss equipment in a FEE, it can impact the estimate, possibly significantly. Ensuring enough engineering design is complete to use this method along with judicious use of decision and risk analysis techniques is advised.

It was stated by Lagace (2006) that in almost all cases, estimates are more accurate than the scope used to develop them, and that most inaccurate estimates are caused by things we forgot to include, things we decided to leave out, wishful thinking, and things in the realm of "known unknowns". Since the FEE methods rely on pricing all the equipment (or at least all the expensive equipment), before diving into the method, we will look at the issues a process engineer should understand that contributes to inaccurate FEE estimates.

Many FEE estimates have a process simulation to support them. Understandably, for the sake of completing the engineering in a reasonable amount of time, simulations (and hand material balances) do not necessarily model the entire process and usually have assumptions built in such as:

- Not all the chemical components that could exist in a process are in the model.
- There is zero pressure drop between equipment, and gravity or hydraulic head effects are not considered.
- The process is at steady state, and start-up or shutdown is not considered in the model.
- A fictitious unit-op (i.e. "component splitter") can somehow be turned into a piece of equipment later. Models can use black box unit-ops where an actual process cannot.
- Process storage and storage tanks for feeds and products may not be represented in a model, and these can have significant cost implications.
- Feed streams magically appear, and waste streams magically disappear. These assumptions can require extensive engineering and subsequent capital and/or OPEX.
- Control of the process, and product quality, is achievable.

The design process usually requires an engineering step after the simulation to generate a more detailed PFD that is adequate to base an estimate upon. The engineer and estimator must ensure all the necessary pumps, compressors, and fans, start-up heaters/coolers, and so forth are included. They must satisfy themselves that they have accounted for the frictional, process pressure differential, and gravity requirements to move material around the plant. They should also ensure equipment for start-up or shutdown has been included. An approach for waste streams must be defined, so the estimate can have associated capital equipment or OPEX included.

An approach to ensure a higher quality FEE is to prepare preliminary P&IDs for the plant. P&IDs will inevitably uncover some equipment that was missing from the design. The P&IDs can also provide information about the complexity of the control system which can be included in the estimate.

9.5.4.1 Obtaining Equipment Prices

FEEs rely on equipment prices and factors to determine the TIC. Obtaining equipment prices is therefore key but getting vendor prices can be a lengthy and expensive task. Thus, there are fast-track methods using correlations, or experience to determine the equipment prices, as described in Figure 9.2.

Vendor quotes require the most engineering effort and time to obtain but provide the benefit of accuracy and a price "guarantee" for a period of time. That guarantee can be useful in the risk analysis for the project. If only a "budget quote" is requested, a vendor can use their own experience to provide an estimate in a shorter period without resorting to their own engineering, but without the price guarantee. There is a balance between accuracy for minimum risk, time, and cost.

Engineering experience from recent jobs can provide a reasonable level of accuracy in lieu of equipment quotes. An experienced person may just know that a fan or pump will cost roughly $10 000, and if, relative to the overall estimate, this is not a significant amount, the impact of such a decision can be assessed.

Where experience or vendor quotes are not possible, there are two major methods that can be used to obtain equipment costs from correlations. The first method is detailed by Guthrie (1969A), Ulrich (2009), Perry's, and Turton (2003), and others utilizes key equipment parameters (i.e. heat exchanger type, materials of construction, design pressure, and total tube surface area) to obtain a price from historical correlations.

A second method might be considered by some to be a variation of the Guthrie method. This method is built into some commercial software packages (i.e. those contained within Aspen/HYSYS or The Richardson Process Plant Standards) and uses process information similar to Guthrie. Some correlations are more complex and simulate those used in a fabrication shop to design and cost equipment. Simplified equipment prices based on the Aspentech program are presented by Loh (2002). A comparison of various costing programs is presented by Feng (2011).

It is worth noting that equipment costs are strongly affected by the selected materials of construction. Preparing a materials of construction flowsheet can aid in defining materials for equipment and piping. Merrow (1981) further discussed the risks to the process lifespan from contaminants.

Carbon steel is usually the starting point for material selection and, with the exception of polymers, is generally the least costly. The cost differential when using other materials, such as stainless steel, is not solely limited to the purchase cost of the materials. Welding practices, shop cleanliness, hydrotest water quality, quality assurance, and other fabrication requirements are significant to the final equipment cost. These factors make it difficult to confirm the material of construction factors in the correlations against currently known raw material costs without obtaining vendor quotes. To give a sense of the effects of materials of construction, the factors for different materials of construction for pressure vessels per Turton (2003) are in Table 9.2.

9.5.4.2 Cost Indexes – Timing Is Important

To address the issue of using costs from prior projects, most methods rely on an industry cost index such as the Chemical Engineering Plant Cost Index (CEPCI) to inflate the cost of equipment from

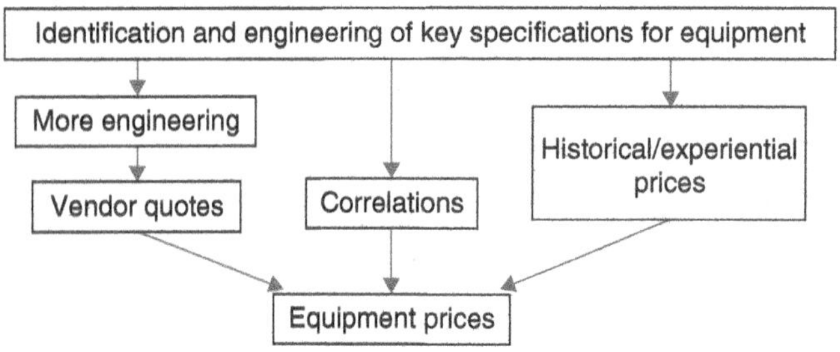

Figure 9.2 Routes to equipment pricing.

Table 9.2 Material of construction costs, adapted from Turton (2003).

Materials of construction	Factor
Carbon steel	1.0
Stainless steel clad	1.7
Stainless steel	3.1
Nickel alloy clad	3.6
Nickel alloy	7.1
Titanium alloy clad	4.7
Titanium alloy	9.4

a known date, or CEPCI index value, to the cost of the equipment now or at a future time. The estimator may find recent values for the CEPCI, but must forecast and specify the expected future CEPCI value for the project.

Clearly, all these methods rely on factors and assumptions to develop the total installed equipment cost. Regardless, the value for equipment costs can be good enough to make decisions, especially so by testing the sensitivity of the costs in the overall project decision and risk analysis.

After the equipment costs have been determined, the installed equipment and overall plant cost can be developed using the following methods.

9.5.4.3 Lang Factor Estimates

The Lang factor method is performed by multiplying the sum of the equipment costs by a single value to obtain the total installed plant cost (see Table 9.3).

The lower factors for solids processing plants do not imply that those plants are cheaper to construct, only that the equipment alone is a higher proportion (about 32%) of the overall cost versus a fluid processing plant where equipment is about 20% of the overall cost.

9.5.4.4 Guthrie Method

The Guthrie method has been modified, extended, and customized by numerous people and companies (Figure 9.3). It deviates from the Lang factor method by using separate factors for each

Table 9.3 Lang factors.

Type of plant	Lang factor
Solid processing plant	3.1
Solid–fluid process plant	3.63
Fluid process plants	4.74 (usually 5 or 5.1 is used now)

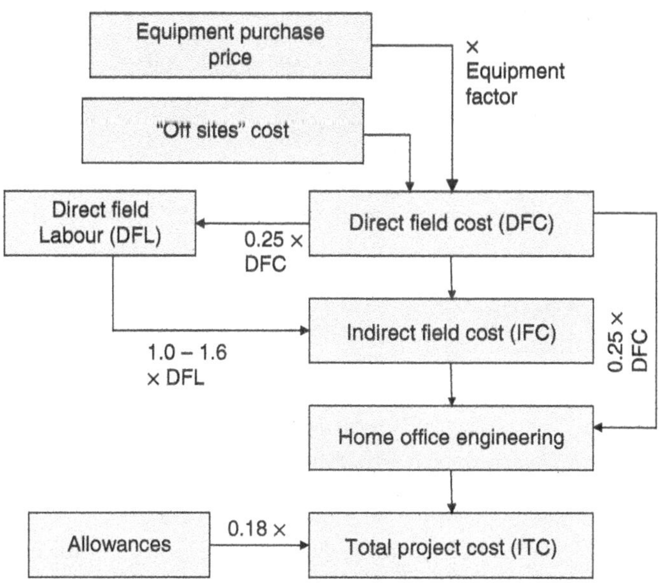

Figure 9.3 Modified Guthrie estimating method.

piece of equipment depending on the equipment type and materials of construction. This works well when providing different levels of accuracy or allowance for equipment costs that have been obtained with different methods.

A modified Guthrie method determines the plant cost as follows:

Where the equipment factors are from Table 9.4.

More expensive metallurgy utilizes lower equipment factors because the equipment costs are a higher proportion of the TIC, which includes items like wire, concrete, and structural steel, which do not change with process equipment metallurgy.

There are variations on Guthrie's method that calculates a "base cost" representing the cost of a piece of equipment in a standard material and pressure (usually carbon steel and 150 psig). The base cost is multiplied by a factor for installation to cover structural steel, concrete, electrical, instruments, etc. The multipliers on base cost will vary less. We prefer to use methods that work

Table 9.4 Equipment installation factors.

Equipment type	Direct field cost multiplier[a)]
Shop fabrication vessels	2.0–2.5
Compressors	1.5–2.5
Exchangers	1.7–2.5
Fired heaters	1.7–2.5
Pumps	2.5–3.5
Material handling equipment	1.5–3.0

a) More expensive metallurgy should use the lower numbers. Carbon steel equipment should use the higher multiplier values.

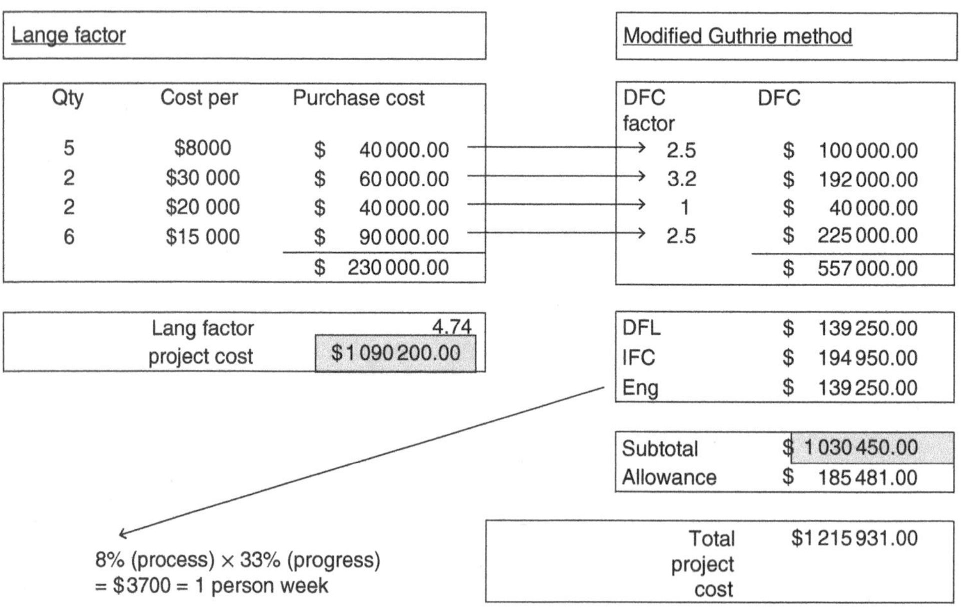

Figure 9.4 Comparison of factored equipment estimating methods.

Project: Costing example
Client:
Estimator:

Date: #########

Equipment tabulation

Item #		Base cost	Equipment cost	Quantity		Factor to DFC	DFL factor	Field labour	Bulk materials	Direct field costs	Labour units
Heat exchangers											
	e100	$ 35617	$ 101010	1	101010.45	2.344	50%	$ 23935	$ 23935	$ 148880	0.1
	Subtotal		$ 101010					$ 23935	$ 23935	$ 148880	0.1
Pumps											
	Subtotal		$ –					$ –	$ –	$ –	0
Vessels											
	Subtotal		$ –					$ –	$ –	$ –	0
	Totals		$ 101010					$ 23935	$ 23935	$ 148880	0.1
	Labour hours on-site (all labour)				Subcontract labour rate		$ 64.00	393	h		

Figure 9.5 Estimate equipment tabulation.

with a mixed method of obtaining equipment costs in actual materials being used, so they can be compared or more easily integrated with vendor quotes.

A comparison of the FEE methods follows (Figure 9.4), which illustrates that the two methods before allowances are added provided similar results. Figure 9.5 has a note that back calculates the approximate hours required to do the initial process engineering for such a project based on the engineering breakout for process hours of 8% of the engineering budget.

Engineering costs can range as a percentage of direct field costs (DFCs) from 10% to 30%, but 25% was the most common value used. Engineering costs are also often stated as a ratio of the total project cost (TIC) and historically ran 12% of the TIC. More recent values indicate a range of 12–30% of the TIC, with 18–21% being the average. Process engineering is about 8–12% of the total engineering budget.

It may be necessary to add additional line items or factors to these estimates to account for the following:

- Location factor: The cost correlations are generally based on plants built on the US Gulf Coast. Plants built in other areas can provide historical information on the location factor to build in those areas. Alternatively, there are published values for location factors. In most situations, the location factors affect the labor to construct and shipping, but not the equipment costs, so the modified Guthrie method approach above is a good one for utilizing a location factor. Uppal (1997) presented a table of labor productivities for American cities to provide location adjustments. Reemer (2002) provided a list of sources to determine location factors. Garrett (2012) had a list of construction cost location factors.
- Winterization: If the plant requires protection from winter, a 3% addition to the DFC can be added.
- Utility systems, site preparation, etc. may be required, and these factors vary from 1.2 to 2 of the IFC + DFC subtotal.
- DCSs: Typically, the instrumentation that was used when the equipment factors were developed was more primitive than what is used today. If P&IDs are available, the number of control loops can be counted, and a line item of $500/loop can be added for the DCS. A multiplier on older instrumentation costs is another approach.

A summary of the equipment costs could look as follows (Figure 9.5).

9.6 Detailed Capital Cost Estimates and Design/Build Projects

Detailed estimates use material takeoffs and per-unit costs that will include many line items. For example, with a single-pump installation, in a FEE, just the pump is costed and factors are used to account for the installation. In the case of a detailed estimate, the following would be considered along with the labor costs to install the pump:

- earthworks, foundations, steel, etc;
- pump and motor, labor to mount and align;
- motor control center (MCC);
- length and gauge of the wiring from the MCC to the motor;
- local switches for On/Off/Auto;
- pressure indicators;
- pressure switches, wiring back to the control system;

- vibration switches, wiring back to the control system;
- any other control hardware (flow meters);
- control valve;
- control system hardware;
- control system software and configuration;
- length and diameter of the piping, manual isolation valves, automatic valves, fire safety valves, strainers, filters, check valves, etc.;
- structural steel and hangers for piping, which may require piping stress analysis;
- relief valve installation and piping, which may also require a disposal system;
- diameter and length of pipe insulation;
- tracing, and controls.

Generating those numbers requires a high level of engineering design to be completed and an experienced estimator.

An owner's engineering team might complete smaller plant projects, but most expansion or new capital projects are completed by contractors to save time and costs, as well as bringing specialized knowledge into the design process. While the contractors often have experience in estimating and the database of costs, they are neither mind readers nor magicians, and the same rules of design development that apply to an owner also apply to them. To create an estimate in the face of uncertain information, a contractor might need to add substantial allowances or incorporate numerous assumptions and exclusion clauses. Alternatively, they could produce a lower-quality estimate, which typically leads to issues that can haunt the project later. Ensuring the engineering documentation is thoroughly prepared is crucial for producing a quality estimate. Failure to do so will leave a project that you, your operators, and your maintenance people may not like, or if you are unfortunate, a poor process with unexpected costs and lawsuits. Define the project, communicate the needs, and prepare a quality estimate to ensure the project has its best chance of success.

On small projects, a detailed estimate may be something that does not require a great deal of time and effort. Williams (2021) discussed matching the estimate type to the project size. Generally, large-project detailed estimates are prepared by specialists and require a lot of engineering support and are often done after the project has already been approved based on a less rigorous method (see Chapter 10, Project Management). If you are preparing a detailed estimate and it is more than a few pages in length, you will probably want to consult with an experienced project engineer to ensure you have not missed any considerations.

9.7 Hybrid Capital Cost Estimates

Estimates methods can be adapted to suit the situation. If a capacity factored estimate provides enough accuracy, apart from the environmental controls, then it can be combined with a FEE or a detailed estimate to provide the necessary level of accuracy.

In our view, an ideal estimating method allows you to obtain equipment prices via a mixed variety of methods (vendors, correlations, and manual estimates). Costs for more expensive, specialized equipment or those made from special materials that correlations do not cover can be determined by obtaining quotes from vendors. Less expensive or price-predictable equipment can be obtained with a faster less costly method. The estimate can combine the different price accuracies to determine an overall improved estimate accuracy.

9.8 Estimate Summaries and Additional Factors

A key deliverable in these estimates is usually a "priced equipment list", which could look like Figure 9.6. The estimate summary sheet (Figure 9.7) may add various line items for additional costs (vendor package, non-core process facilities, etc.), location factors, general project construction costs, and so forth.

9.8.1 Grass Roots Factor

The grass roots factor is sometimes applied at the base of FEEs. This factor would not appear as a separate line item in a detailed estimate since the equipment costs would be individually listed and included. The value, often ranging from 1.3 (according to Guthrie) to 1.5 times the process unit costs, accounts for general site facilities, such as loading/unloading facilities, maintenance, and utility facilities. Literature sources provide vague guidance on this factor because it is highly specific to the site and process. The significant impact of this factor on the total estimated cost introduces a considerable amount of uncertainty. While it might be adequate for early-stage estimates, it is advisable to attempt to estimate these costs directly and incorporate them into the capital estimate as early as possible.

9.8.2 Allowance and Contingency – Estimating the Unknowns

The reader should be advised that there is some contention in the definitions being used below. Communication between the estimator and stakeholders is essential to confirm definitions and the application of allowance, contingency, etc. A list of exclusions and inclusions written into the scope of work is equally as important.

This section also introduces projected risks, those possible issues that might have an impact on a project's completion. The topic of risk will be expanded upon below and in other chapters.

9.8.2.1 Escalation

This is the amount of money that needs to be allocated to address inflation over the period of the project. It might be added to the line items for the equipment, assuming a schedule is available, or just across the entire estimate.

9.8.2.2 Allowance – Probability 1, but Unknown Costs

An estimate can miss its stated value by items that are 100% needed and known about, but the actual cost value has some uncertainty (plus or minus). Examples are as follows:

- Depending on the type of estimate (feasibility or detailed), the piping was estimated to go from A to B, but later in the design, it is realized that it needs to go through C.
- In the feasibility estimate, you estimate that a pump will cost X based on experience, but quotes were not obtained. You are reasonably sure that your guess is close, so the solution is to add an **allowance** for line items that have some uncertainty at the point the engineering design is at.
- A compressor package is quoted from a supplier, but history would suggest as work on the design proceeds, there are additional costs.

These are sometimes called known unknowns, though it might be better to be called known issues. Note that quantity or amount errors on some items could also reduce the cost of the estimate. Some companies will not allocate very much for allowances, expecting they are overestimating

Project: Costing example
Client:
Estimator:

Date: March 31, 2024

Equipment list

Name	Type	#	Material	Size	Utility type	Utility cost (per year)	Pressure (barg)	Direct field cost	Additional	Additional
Heat exchangers							tube/shell		# Shells	
E-100	U-Tube	1	CS/SS	100 m^2	SteamMed	$ 551 661	10/10	$ 114 840	1	
Pumps										
P-100	Centrifugal	1	Carbon steel	25 kW		$ 12 045	10	$ 26 040		
Vessels									Demister type	
V-100	Horizontal	1	Carbon steel	1 m D × 3 m L			100	$ 92 261	None	

Figure 9.6 Equipment list.

Project: Costing example
Client:
Estimator: D. Mody

Date: September 1, 2019

Estimate summary

Factored equipment direct field costs							
	DFL	Labour factor		DFL	Bulk material	Equipment	DFC
Equipment	$ 23935	100%	of Gulf coast h	$ 23935	$ 23935	$ 101,010	$ 148,880
Winterization		0%		$ -	$ -	$ -	$ -
Design development		5%		$ 1197	$ 1197	$ 5051	$ 7444
Subtotal				$ 25,132	$ 25,132	$ 106,061	$ 156,324
Equiment/costs excluded from factors							
	User column	User column	User column	DFL	Bulk material	Equipment	DFC
Civil							
Site development				$ -	$ -	$ -	$ -
Concrete				$ -	$ -	$ -	$ -
Steelwork				$ -	$ -	$ -	$ -
Buildings				$ -	$ -	$ -	$ -
HVAC				$ -	$ -	$ -	$ -
Electrical							
Substations				$ -	$ -	$ -	$ -
Switchgear				$ -	$ -	$ -	$ -
Mechanical							
Equipment				$ -	$ -	$ -	$ -
Equipment				$ -	$ -	$ -	$ -
Piping							
Pipe				$ -	$ -	$ -	$ -
Pipe racks	# pieces of equipment:	15		$ -	$ -	$ -	$ -
	Estimated pipe rack cost:	$ 10,943.15					
Insulation & tracing	$12.30			$ -		$ -	$ -
Control systems							
DCS	$650			$ -	$ -	$ -	$ -
Analysers				$ -	$ -	$ -	$ -
Fire protection				$ -	$ -	$ -	$ -
Painting				$ -	$ -	$ -	$ -
Scaffolding				$ -	$ -	$ -	$ -
Subtotal				$ -	$ -	$ -	$ -
Freight, insurance, taxes	8%						$ 8081
Overtime premiums	0%						$ -
Total direct field costs							$ 160364
Indirect Costs							
Indirect field expenses		1.30	of DFL		(For project construction by subcontracts)		$ 32,671
Engineering		15%	of DFC's		8% of Total Proj		$ 24,055
IFC + DFC subtotal							$ 217,090
Allowances							
Escalation		0.0%		$ -	$ -	$ -	$ -
Sales tax							$ -
Duties							$ -
Contingency (IFC + DFC)		18%			12.9% of Total Project		$ 39,076
Catalyst and chemicals							$ -
Fee		2%					$ 4342
Demolition and removal							$ -
Grass roots factor	Include a grass roots factor?		Yes	50%			$ 41743.45
Total						USD	$ 302,252
					USD conversion		1.1
Total						CAD	$ 332,477
	Calculated Lang factor = 2.99		FEL 1 process engineering costs		$ 962	0.3 person wks	

Figure 9.7 Estimate summary.

as often as they are underestimating. Factors that can affect how much allowance to carry are as follows:

- How much design has been completed
- What type of equipment is included
- Typical experience in scope growth of this type of equipment or with that vendor
- Whether the project is a greenfield or brownfield (plant expansion), where the brownfield projects typically need a higher allowance.

A project might typically spend 10% extra on equipment not identified on the P&IDs at the time of the estimate (filters, small tanks, etc.). If the experience is that you will need this money on every project, placing it in the allowance category is sensible.

An owner may decide late in the design that instead of having one pump operating and one pump spare, they want two pumps operating and a third pump as a spare. The contractor will undoubtedly view this as a scope change (see Section 9.8.3), but an owner typically knows there will be small design changes they will want to do during a project and will need to allocate a pool of funds to these needs. Hence, while the contractor sees this as a scope change, the owner may see this as an issue dealt with by allowance funds (they may also categorize it as a contingency).

Communication between contractors and owners on these contentious definitions and what is included is critical. A good working relationship between the two parties will make these types of issues just part of the project process rather than a cause for concern.

9.8.2.3 Contingency – Probability <1

Another source of error in an estimate is for line items that do not have a 100% certainty of being needed, or unpredictable situations (e.g. weather). This is also contentious because an engineering contractor is not a mind reader, and they are encouraged not to pad an estimate for fear of losing the contract. However, there are risks, such as a project that is being built on the US Gulf Coast during hurricane season. You might have a hurricane, you might not, and the project might be stopped for a day, a week, or not at all. For these types of situations, either **contingency** or **insurance** is advisable. Contingency allocates a pool of funds to handle a few likely but uncertain situations, but not all of them. For instance, you might not allocate funds for both a snowstorm and a hurricane since both are unlikely to occur over the duration of the project. There may be a 50% chance that wet weather will cause a 10% increase in construction costs. There may be a 25% chance that it will be a dry construction period and save 5% of the cost. A risk-based approach can help determine how much contingency to include in the estimate.

9.8.2.4 Rand Contingency Analysis

In the 1980s, the Rand Corporation analyzed chemical process industry projects to understand factors contributing to project cost, schedule, and quality problems. Their original work was published as a series of four papers listed in the references (see Merrow 1979, 1981, 1983, 1986). While those papers are still relevant today, a company that provides updates and consulting services to an industry consortium was spun out of that work. For more information, refer to the company website Independent Project Analysis (www.ipaglobal.com).

The original papers provide further details for a person to understand and estimate some key factors that can predict the CAPEX uncertainty and start-up duration in projects. The factors are as follows:

1. New technology: The ratio of the cost of equipment used for new, commercially unproven technology to that used in proven technologies.
2. Impurities: The susceptibility of the process to problems that arise from the buildup of impurities.
3. Complexity: The number of process steps (not the pieces of equipment, number of steps, i.e. feed handling, gasification, reaction, and separation).
4. Project definition:
   a. Have pre-start-up personal costs been included?
   b. Have pre-start-up inventory costs been included?

c. Have land purchase costs been confirmed and included?
d. Is the design in the R&D stage, or (pre-)commercial?
e. To what degree has engineering been completed?
f. To what degree has site-specific information been obtained?

Utilizing those considerations, a cost growth ratio is determined (Figure 9.8) and used as a contingency in the project. The original paper states that projects using their method of contingency calculations will achieve an estimate accuracy of 10% regardless of engineering design development.

9.8.2.5 Insurance – Very Low Probability Factors

There are situations that have a very low probability, but if they occur, you need funds to recover and thus need to address them. A fire might be one example. These events might affect only a few projects in a basket of many, so it may make sense to use insurance, which you will see as one of the estimate line items. This is a risk you pass along to another party, and you obviously have a subsequent counterparty risk that they may not pay.

9.8.3 Scope Changes

There are times when external situations cause a design to not meet the requirements, and it needs to be changed, which has cost and schedule consequences. It is easy to imagine a regulatory change, or perhaps the process is to make toothpaste, and the market size has changed. These are unforeseen, "unknown unknowns", changes that cannot reasonably be listed, their effect cannot be predicted, and they cannot be given a probability. The only solution to these situations is to stop, decide the impact on the project, and decide on the soundness of the changes. In these situations, the budget for the project could be increased.

Depending on the contract, small design error changes or design omissions with relatively minor cost implications are passed to the owner (who must carry an allowance for these) or be dealt with by the contractor's inclusion of some allowance into their fee.

9.9 Economic Analysis

The primary objectives for constructing or modifying a process are to improve some, or all of:

- regulatory requirements,
- safety, environmental, societal impact,
- earnings of a company by saving money or increasing revenue.

In the case of meeting regulatory requirements or improving the societal impact, the objective from an economic point of view may be to minimize any negative economic impact. However, it is common to also look for opportunities that will improve the economic impact through better reliability, better quality, possibly high production rates, reduced waste and energy, lower environmental footprint, and so forth. Determining the economic impact of a process design is an important part of assessing the viability, and options available to a designer.

Following the preparation of the CAPEX estimate, the economic viability of the project is typically performed, followed by an analysis of opportunities and problems.

Project: Costing example				Date:	#########
Client:					
Estimator:					
		RAND cost growth			
Variable	**Value**	**Units**	**Coefficient**	**Notes**	
Installed cost of equipment representing new, commercially unproven technology:		$			
Total installed cost of all equipment:		$			
PCTNEW	0.00	%	0.002 97		
On a scale of 0 to 5, how susceptible is the process likely to be to problems that arise from the buildup of impurities?				0 is completely unaffected by impurity buildup, while 5 suggests that the process would be completely devastated should a buildup of impurities occur	
Impurities	0		0.021 25		
How many process steps are there?		Steps		This is the number of processing steps (e.g. feed handling, gasification, reaction, separation), not the number of pieces of equipment	
Complexity	0		0.011 37		
Have the following considerations been included in the cost estimate?					
Pre-startup personnel costs?	No				
Pre-startup inventory costs?	No				
Land purchase cost?	No				
Inclusiveness	0.00	%	0.001 11		
Is the design in the R&D stage, or the (pre-)commercial stage?	R&D				
To what degree has engineering been completed?				Value	
Limited engineering has been completed				3	
To what degree has site-specific information been obtained?				Value	
Site-specific information is assumed				3	
Project definition	6		0.063 61		
Cost growth ratio =	74.03%		=	Estimated cost / Actual cost	
The actual cost is estimated to be 1.35 times greater than the estimated cost					

Figure 9.8 Contingency calculation via rand methodology.

In situations where revenue is difficult to predict at the time of the estimate, Anderson (2009) mentions toothpaste as such an example, then the cost to manufacture is a metric that can be developed and further utilized by the organization. If revenue values can be estimated, then tools such as NPV, IRR, and a myriad of other metrics can be developed and utilized in the decision making.

The economic analysis allows the project to be analyzed from a variety of angles, can assist the technology development on how further research or engineering might be needed to improve the project. For instance, it might be recognized that a utility, a solvent, a waste stream, or other costs are disproportionally large, and the overall project would benefit from reducing their costs. Sensitivity analysis can be performed on the uptime of the process which can lead to reliability and redundancy guidance, thus feeding back to the engineering design and capital estimate.

This section begins with the determination of operating expenses and proceeds to the calculation of economics, risk analysis, and decision making.

9.9.1 Operating Costs, Profitability Risk Analysis, and Economic Optimization

9.9.1.1 Operating Expenses

Operating expenses (OPEX) for running the plant are the next step before the economic analysis can be completed. Similar to the CAPEX, there are direct costs and indirect manufacturing costs. Direct costs include: Cost of feed chemicals, waste disposal, and utilities should be gathered based on the throughput of the plant in $kg\,h^{-1}$ of product produced. Operating labor, supplies, maintenance, and repairs as well as laboratory charges and any relevant patent royalties should be included in the direct manufacturing costs. Indirect OPEX are those overhead costs associated with running the plant such as administration, human resources, accounting costs, some aspects of R&D, and so forth.

There are ways to estimate the number of operators a plant will require based on the equipment, but usually, an experienced engineer will have a good sense of how many DCS operators are needed and how many field operators may be required. Considerations for field operators should include the number of pieces of equipment, the physical distances in the plant, the likely need for multiple operators to help each other in the field, and the number and duration of manual operations such as rail/truck unloading or produce loading, and sample taking. Once the number of operators has been assessed per shift, the total number of employed operators will take into consideration: 12 hour shifts, days off, vacation and statutory holidays as follows:

$$\text{Total employed operators} = 4.5 \times \text{operators per shift}$$

Utility and Waste Disposal preliminary cost values Turton (2003) are presented in Figure 9.9. Ulrich (2006) presents a method to calculate utility costs based on fuel and electricity costs.

A series of equations and factors utilizing the capital and direct OPEX is presented below and mostly from Turton (2003) follows in Table 9.5.

Where typical values for the factors are in Table 9.6.

The Operating Income Before Tax (EBITA) may be calculated at this point. If there is a profit that year, the depreciation is determined from applicable corporate tax income laws, and is applied as a deduction against income, thus reducing the taxable income to a value no lower than 0.

Working Capital is the amount of money that must be spent on raw materials and the inventory of raw materials on an ongoing basis. For instance, the plant needs to be filled with raw materials before any product is produced and can be sold. Product usually must be inventoried, and typically 30 days are allowed for customers to pay. Working capital can be assumed to be 17.5% of Equipment

Utility costs

None	None	0	N/A
Steam: Low pressure (5 barg, 160 °C)	SteamLow	6.08	$ GJ^{-1}
Steam: Medium pressure (10 barg, 184 °C)	SteamMed	6.87	$ GJ^{-1}
Steam: High pressure (41 barg, 254 °C)	SteamHigh	9.83	$ GJ^{-1}
Dowtherm (300 °C)	Dowtherm	8.85	$ GJ^{-1}
Fuel oil	FuelOil	6	$ GJ^{-1}
Natural gas	Gas	6	$ GJ^{-1}
Custom heating	CustomHeat	6	$ GJ^{-1}
Electricity (GJ)	ElectGJ	16.8	$ GJ^{-1}
Electricity (KW)	ElectKWH	0.06	$ KWH^{-1}
Cooling water	CoolWater	0.354	$ GJ^{-1}
Refrigeration: Water (5=>15 °C)	RefWater	4.43	$ GJ^{-1}
Refrigeration: Low T (–20 °C)	RefLow	7.89	$ GJ^{-1}
Refrigeration: Very low (–50 °C)	RefVeryLow	13.11	$ GJ^{-1}
Refrigeration: Custom	RefCustom	6	$ GJ^{-1}

Figure 9.9 Typical utility costs. Source: Adapted from Turton (2003).

Capital if no better value can be determined (Turton 2003). In the final year of the plant, the working capital can be assumed to be removed and sold.

In Canada, the depreciation is dealt with using the term Capital Cost Allowance. Capital Cost Allowance has two accounts that must be tracked. The first account is the "Unused CCA" which begins at the total CAPEX. Each year a percentage of the Unused CCA, as determined by the CCA class, is transferred to the second account, the "CCA account". If the company has enough pretax income, it may claim the entire amount in the CCA account, thus zeroing it annually. However, the CCA cannot be used to create a loss. If a loss would result, the remaining CCA stays in the account and can be used against future year's income. One special situation is that the first year of claiming CCA, you may only transfer 50% of the allowable amount from the "unused CCA". This is to address the fact that, on average, the plant started up and began depreciating halfway through the year.

In the final year of the plant, the difference between any unused or unclaimed CCA and the scrap value (10% of the CAPEX) can be claimed as a loss or profit.

Taxable income is thus calculated as the before tax income minus the allowable CCA for that year. If a loss is incurred for that year, the loss is accumulated and can be used against future income. Large plants may have other tax implications that are specific to their location.

Positive taxable income is taxed at the corporate tax rate (assume 30%), and the remaining amount is the after-tax "operating income", or profit.

A summary sheet showing these calculations and the year by year DCFs follows (Figure 9.10):

9.9.2 Economic Comparison Methods

Measures of profitability include (see Table 9.7):

- Return On Investment
- Payback Period,
- Net Present Value
- Internal Rate of Return

Discounted methods rely on a "discount rate", also known as the "hurdle rate", or "minimum acceptable rate of return (MARR)". This value can be thought of as the interest rate that is expected

Table 9.5 Typical cost factors and equations.

Estimate line item		Calculation method	
Capital	Capital	=Direct field costs + indirect field costs + allowances	
Direct manufacturing costs			
Raw materials			
Utilities			
Additional costs		Waste disposal etc. any direct costs not included in raw materials and utilities	
Operating labor	Operating_ Labour	=Total_Operators × LabourRate × 4.5	
Other labour		=OtherLabour_factor × Operating_Labour	
Maintenance and repairs		=Maintenance_factor × Capital	
Operating supplies		=OperatingSupplies_factor × Capital	
Laboratory charges		=LabCharges_factor × Operating_Labour	
Patents and royalties		=Patents_factor × MfgCostEst	Iteration required to solve for this
Total direct manufacturing costs		Subtotal	
Fixed manufacturing costs			
Local taxes and insurance		=Insurance_factor × Capital	
Plant overhead costs		=labouroverhead_factor × Operating_Labour + capitaloverhead_factor × Capital	
Total fixed manufacturing costs		Subtotal	
General Mfg. expenses			
Administration		=adminlabour_factor × Operating_Labour + admincapital_factor × Capital	
Distribution and selling		=distribution_factor × MfgCostEst	Iteration required
Research and development		=randd_factor × MfgCostEst	Iteration required
Total general Mfg. expenses		Subtotal	
Total manufacturing costs		Total	

Source: Adapted from Turton (2003).

Table 9.6 Typical operating cost factors.

Additional labour factor	18%
Maintenance factor	6%
Operating supplies factor	0.9%
Laboratory charges factor	15%
Patents and royalties factor	3%
Insurance factor	3.2%
Labour overhead factor	70.8%
Capital overhead factor	3.6%
Administration factor (labour)	17.7%
Administration factor (capital)	0.9%
Distribution factor	11%
Research and development factor	5%

if an equivalent amount of CAPEX money is invested. An acceptable value for this varies based on the cost of obtaining the initial investment (cost of capital), the risk of the project not achieving the stated cash flow, and other factors. A project required for a regulatory requirement may only use the company's cost of capital, whereas an unproven new technology project may require a risk-adjusted cost of capital or hurdle rate that ensures a basket of such projects, in which some will fail, can still achieve an overall minimum investment return requirement.

At a high enough hurdle rate value, the profit becomes great enough that the project will pay for itself simply from the cash flow in the first year, and no extended external source of capital funds is required.

After-tax profit is carried forward to the enterprise's income statement, which would be reflected in higher corporate earnings and an improved Price-to-earnings ratio.

9.9.3 Determine the Cost of Manufacturing

A balance sheet of ongoing costs and revenues, along with a portion of the initial capital (assuming straight-line depreciation over a set number of years), can be used to calculate the cost of manufacturing.

The cost of manufacture is also the product selling price when the venture breaks even. There is likely to be some discussion as to what exactly is meant by "break-even", but it can be interpreted as when the NPV is calculated to be zero, when the discount rate is set to zero. An arguably better metric would be the cost of manufacture when the discount rate is set to the MARR value. When the Discount rate = MARR, the minimum selling price, or cost of manufacture, would be determined for the minimum viable project.

9.10 Risk Analysis of Project Economics

Risk, regardless of the context in which it occurs, is the multiplication of probability and consequences. If the issue is 100% going to happen, it's not a risk; it's an issue, so you start making plans for its occurrence.

$$\text{RISK} = \text{Probability} \times \text{consequences}$$

OPEX and profitablity worksheet			Net present value		$ 12 355.24 K$		Internal rate of return		42.75%				
Discount rate	18	%	Payback period			Years (from year 2)		Discounted payback period		4	Years (from year 2)		

Year	1	2	3	4	5	6	7	8	9	10	11	12	13
Revenue													
Sales			3011	24090	24090	24090	24090	24090	24090	24090	24090	24090	24090
Additional revenue			0	0	0	0	0	0	0	0	0	0	0
Total revenue			3011	24090	24090	24090	24090	24090	24090	24090	24090	24090	24090
Direct manufacturing costs													
Raw materials				8030	8030	8030	8030	8030	8030	8030	8030	8030	8030
Utilities				803	803	803	803	803	803	803	803	803	803
Additional costs				1000	1000	1000	1000	1000	1000	1000	1000	1000	1000
Operating labour		# Shift Op's	4	1440	1440	1440	1440	1440	1440	1440	1440	1440	1440
Other labour		# Total Op's	16	259	259	259	259	259	259	259	259	259	259
Maintenance and repairs				600	600	600	600	600	600	600	600	600	600
Operating supplies				90	90	90	90	90	90	90	90	90	90
Laboratory charges				216	216	216	216	216	216	216	216	216	216
Patents and royalties				76	76	76	76	76	76	76	76	76	76
Total direct manufacturing costs			3129	12514	12514	12514	12514	12514	12514	12514	12514	12514	12514
Fixed manufacturing costs													
Local taxes and insurance				320	320	320	320	320	320	320	320	320	320
Plant overhead costs				1380	1380	1380	1380	1380	1380	1380	1380	1380	1380
Total Fixed Manufacturing costs			425	1700	1700	1700	1700	1700	1700	1700	1700	1700	1700
General Mfg. expenses													
Administration				345	345	345	345	345	345	345	345	345	345
Distribution and selling				278	278	278	278	278	278	278	278	278	278
Research and development				126	126	126	126	126	126	126	126	126	126
Total general Mfg. expenses			187	749	749	749	749	749	749	749	749	749	749
Total manufacturing costs			3741	14963	14963	14963	14963	14963	14963	14963	14963	14963	14963
Operating income before tax (EBITA)	0	0	−729	9127	9127	9127	9127	9127	9127	9127	9127	9127	9127
Capital				Total capital cost =		10000		Scrap value =	1000				
Fixed capital investment - K$	1000	6000	3000										
Working capital			1750										−1750
Unused CCA			10000	8500	5950	4165	2916	2041	1429	1000	700	490	343
Capital cost allowance			1500	2550	1785	1250	875	612	429	300	210	147	−657
Accumulated CCA			1500	0	0	0	0	0	0	0	0	0	0
Taxable income	0	0	−729	5077	7342	7878	8252	8515	8699	8827	8917	8980	9784
Accumulated tax loss	0	0	−729	0	0	0	0	0	0	0	0	0	0
Taxes	0	0	0	1304	2203	2363	2476	2554	2610	2648	2675	2694	2935
Operating income after tax	0	0	−729	7823	6924	6764	6651	6573	6518	6479	6452	6433	6192
Cash flow	−1000	−6000	−5479	7823	6924	6764	6651	6573	6518	6479	6452	6433	7942
Discount factor	1.0000	0.8475	0.7182	0.6086	0.5158	0.4371	0.3704	0.3139	0.2660	0.2255	0.1911	0.1619	0.1372
Discounted cash flow	−1000	−5085	−3935	4761	3572	2957	2464	2063	1734	1461	1233	1042	1090
Accumulated Disc. cash flow	−1000	−6085	−10020	−5259	−1687	1269	3733	5796	7530	8991	10224	11265	12355

Figure 9.10 Discounted cash flow analysis worksheet.

Table 9.7 Typical economic analysis methods.

	Considers time value of money	Formula	Notes
Annual or net return on investment	No	% ROI = Net or annual profit/total investment	
Payback period	No	Point in time that sum of after tax profit offsets initial capital	Discounted Payback Period may also be calculated by discounting the after tax profit
Net present value	Yes	The equation in excel is: =Cash flow year0 + NPV(Cashflow year_1:Cashflowyear_n,discount_rate)	A positive NPV is not a breakeven point. A zero NPV is telling you that the project has just barely met the desired hurdle rate (discount rate). Depending on what hurdle rate is chosen, the venture is (extremely) profitable at a zero NPV
Internal rate of return	Yes	Vary the discount rate across a range of values. Where the NPV = 0, the discount rate at that point is the IRR. IRR is root finding method. If there are alternating loss and gain periods (therefore possibly multiple roots), you should examine the use of the external rate of return method Excel provides an equation for IRR, but requires a guess value and thus will not find multiple roots to the equation. It's worth knowing if there are multiple roots	IRR represents the interest rate on the initial investment amount

Risks are composed of:

- an existing condition(s) or cause(s) which is something that is true today,
- an uncertainty or risk factor of a triggering event, and
- an effect if that risk factor becomes true.

Risk Analysis enters many aspects of engineering design and business. Projects in the chemical process industry are characterized by their complexity, significant scale, high costs, extensive reach, and long timelines. The risks have implications in various domains:

- Environmental and public health impact: These projects can affect the environment and the health and safety of extensive populations across broad geographical regions for extended periods.
- Raw materials and waste management: The handling of raw materials, management of waste streams, and the risk of accidental chemical releases or energy discharges pose localized yet substantial risks.
- Financial investment: The construction of facilities often demands substantial financial investment, with costs frequently reaching into the billions of USD, making economic losses a major concern.

- External uncertainty: Factors such as fluctuating customer demands, competitive dynamics, regulatory changes, societal acceptance, technological shifts, and business cycles critically influence project necessity and viability.
  - A facility initially deemed economically viable may face budget constraints and cost-cutting measures in subsequent years, affecting safety of operations and community relations.
- Internal uncertainty: The reliability of underlying technology and the effectiveness of project execution are pivotal to a project's success, with uncertainties in these areas posing significant risks to its feasibility.

This underscores the varied risks and considerations inherent in chemical process industry projects, highlighting the need for meticulous planning, risk management, and adaptability in response to both internal and external uncertainties.

From a project execution perspective, the effects or consequences of risks within a project can come in the form of missed objectives in cost, schedule, or quality. These can, in turn, be risk factors, along with others, in the long-term viability of the project. Issues such as safety, environmental impact, and so forth also need to be considered, but in this chapter, we will limit our risk analysis efforts to economic ones.

Risks can be opportunities or threats, and both could be examined for their potential benefits and harm. The general process of Risk Analysis includes these steps:

1. Define the objectives of the risk analysis: It may be as simple as achieving a certain economic return value, but there could be others to document.
2. Risk identification: Utilizing a method such as a checklist, brainstorming, or other tools, define the factors that lead to undesirable (or desirable) outcomes.
   a. This may include reviewing factors used in the estimate and their certainty. To save time, break the risk analysis into just a few major sections of the economic analysis. If one section shows a high sensitivity, that section can be broken down further.
3. Risk evaluation: Prioritize the risk impact by determining the effect (possibly through economic modeling).
4. Define responses: Determine what should be done to mitigate the risks.
5. Execute the responses as required.
6. Review the responses for their desired effect and change as needed.

Some Tools for performing a risk analysis overlap with process hazard analysis risk tools and include:

- Risk registers: Are a useful tool for documenting and managing the risk management process. See Table 9.8.
- SWOT (Strengths, Weaknesses, Opportunities and Threats)
- Checklists: provide flexibility to tailor to any situation, but require expertise to generate
- Experts (Delphi Technique)
- FMEA (Failure mode and Effects Analysis)
- FTA (Fault Tree Analysis)
- Risk Matrices (Risk Impact/Probability Charts)
- Pert Analysis (Program Evaluation and Review Technique): for schedule risk
- Sensitivity Analysis
- Scenario Analysis
- Monte Carlo Simulations
- Rand/IPA analysis

Table 9.8 Risk register example.

Risk ID	Risk description	Probability	Impact	Risk level	Mitigation strategies	Responsible	Status
R1	Delay in obtaining environmental permits	High	High	Critical	• Engage with local authorities early • Hire a specialist consultant	Project manager	Monitoring

The economic model is a useful tool for determining the consequences and ranking of risks, either individually in a sensitivity analysis or collectively for scenarios and Monte Carlo analysis.

Sensitivity of metrics to individual inputs:

- The economic model can be tested for positive NPV using various input values, such as high and low CAPEX estimates or high and low revenue values.
- By providing individual equipment accuracies, it's possible to develop an estimate of the CAPEX range.
- A worst-case value that would make the NPV = 0 can be determined for input variables.
- Test for the effects on NPV from a longer-than-expected start-up or production rates that do not meet targets, either due to technical problems or lack of customers.
- Scenario development: Scenarios are an effective tool for developing an economic assessment given multiple factors that have all changed in some way. This method is popular with Royal Dutch Shell, and several online articles discuss their use of it.
- Monte Carlo analysis: This method provides the probabilities of achieving a range of economic metrics given ranges in multiple input assumptions. A simple Monte Carlo Analysis is surprisingly quick to add to a spreadsheet using a macro. More sophisticated tools are available commercially for that purpose.

9.10.1 Managing Risk in the Project CAPEX

By definition, an estimate is never completely accurate. You hope the final project will fall within the range of accuracy you're aiming for, but thought and care must be applied to make that happen.

Managing the risk in the CAPEX is done by incorporating the previously discussed estimate line items of Allowance, Contingency, Insurance, Escalation, and Scope Change.

9.10.2 Decision Making: The "So What?" of Risk Analysis

Risk analysis identifies the factors and consequences. Decisions can then be made about how to address the possibility of those factors becoming true. A risk register provides an excellent way to document and control risks that are ongoing in a project. However, there is a strategic go/no go decision-making process also.

Decisions to continue a project are both economic and strategic. For instance:

- The company may be limited in the amount of capital available at the time of the evaluation. A project may have to be halted due to lack of capital.
- Other projects competing for capital may be ranked higher in return, importance, need, or strategic fit.
- The payback period, if less than a year, implies the capital can be paid through the project's own cash flow. That's usually a very attractive situation.

- How long is the project cash flow required to pay the project back, and how certain is the cash flow? Longer-period projects are subject to greater risk in the business environment; thus, cash flow predictability is a risk.
- Projects, ones with long-term contracts for raw material purchases and/or product sales, are potentially less risky. Counter-party risk becomes an aspect that may require examination.
- If the project is strategic and allows the company to increase market share or leverage other corporate assets or competencies, it may be more attractive.

The information from a project's economic and decision risk analysis can help engineers or business professionals focus efforts on areas needing improvement or uncertainty reduction.

A range of values can be trialed in the economic model, and the results could indicate the plant should definitely be built and operated. More likely, as in the case of a Monte Carlo analysis, the outcome may show a positive NPV across a percentage of scenarios. That leaves the analysis possibly creating as many questions as it started with. However, the analysis can point to specific areas of uncertainty in the technology or business situation that need further effort to be studied and accuracy reduced. It may point to where variability needs to be reduced through contracts or other risk management techniques. The solutions will be very project-specific, so creative problem-solving and returning to economic modelling may be advisable. Christensen (2008) discusses how a reverse balance sheet can help direct business, research, and engineering efforts more effectively and avoid financial tools such as DCFs from killing innovation.

References

Anderson, J. (2009). Determining manufacturing costs. *Chem. Eng. Prog.* 1: 27–31.

Chilton, C.H. (1949). Cost data correlated. *Chem. Eng.* 56 Jan: 97–106.

Christensen C., Innovation killers how financial tools destroy your capacity to do new things, *Harv. Bus. Rev.* 2008 86 Jan, 98–137.

Douglas, J.M. (1988). *Conceptual Design of Chemical Processes*. McGraw Hill.

Dysert L., Capital cost estimating, *Chem. Eng.* 2003A 45 Jun, 22–30.

Dysert, L. (2003B). Sharpen your cost estimating skills. *Chem. Eng.* 108 Jun: 70–81.

Feng, Y. (2011 Aug). Evaluating capital cost estimation programs. *Chem. Eng.* 118: 22–29.

Garrett, D. (2012). *Chemical Engineering Economics*. Springer.

Guthrie, K.M. (1969A Mar). Capital costing. *Chem. Eng.* 24: 114–142.

Guthrie, K.M. (1969B Mar 24). Data and techniques for preliminary capital cost estimating. *Chem. Eng.* 114–142.

Hand, W.E. (1958). From flow sheet to cost estimate. *Petrol. Refiner.* 37 (9): 331.

Lagace, J.C. (2006 Aug). Making sense of our project cost estimate. *Chem. Eng.* 113: 54–58.

Lang, H.J. (1948). Simplified approach to preliminary cost estimates. *Chem. Eng.* 55 (6): 112.

Loh H., Process equipment cost estimation final report, 2002, DOE/netl-2002/1169.

Merrow E., A review of cost estimation in new technology implications for energy process plants, Rand, 1979, R2481-DOE.

Merrow E., Understanding cost growth and performance shortfalls in pioneer process plants, Rand, 1981, R2569-DOE.

Merrow E., Cost growth in new process facilities, Rand, 1983, P6869-DOE.

Merrow E., Understanding process plant schedule slippage and startup costs, Rand, 1986, R3215-DOE.

Peters, M.S. and Timmerhaus, K.D. 1991 *Plant Design and Economics for Chemical Engineers* 4th ed. McGraw Hill.

Remer, D. (2002). Cost and scale-up factors, international inflation indexes and location factors. *Int. J. Prod. Econ.* 84: 1–16.

Silla, H. (2003). *Production and Capital Cost Estimation*. Taylor & Francis Group LLC.

Turton, R. (2003). *Analysis, Synthesis, and Design of Chemical Processes*. Pearson.

Ulrich, G.D. and Vasudevan, P.T. (2004). *Process Design and Economics*, 2nde. ISBN0-9708768-2-3.

Uppal, K.B. (1997). *Cost Estimating Made Simple*. Hydrocarbon Processing.

Ulrich G., How to estimate utility costs, *Chem. Eng.* 2006 113 Apr, 66–69

Ulrich, G. (2009 Apr). Capital costs quickly calculated. *Chem. Eng.* 116: 46–52.

Williams, T.J. (2021 Apr). Early-stage capital cost estimation. *Chem. Eng. Prog.* 117: 42–47.

Williams, M.R., Williams, T.J., and Williams, J.A. (2021 Aug). Detailed capital cost estimation – part II. *Chem. Eng. Prog.* 52–55.

10

Project Management

10.1 Introduction

The scope of what constitutes a "project" is as varied as the challenges they present. A robust, tailored, project management (PM) process can handle any of these scenarios. This chapter discusses the application of PM principles and tools across a range of projects, from the simple writing of a report to the complexity of designing and constructing large-budget chemical facilities. Beginning with outlining the general steps and tools of PM planning, then narrowing the focus to the chemical engineering discipline, and finally providing some guidance for process engineering tasks.

Recognizing that chemical engineers engage in projects outside their core field, this chapter introduces the general PM tools and methodologies for any project and subsequently integrates them with the details tailored for chemical engineering projects. PM terminology traditionally used in the chemical industry does not align perfectly with that of the Project Management Institute (PMI), which we choose to use in this text. The PMI framework is designed to be universally applicable, but must be customized for specific industries. Portny's (2022) text is a valuable resource for further application of the PMI framework. We suggest reading this chapter in combination with Chapter 2 (Documentation) and Chapter 9 (Capital Costing).

10.2 Comparison of Academic Versus Industry Project Environments

We felt this chapter was necessary because PM, as Peachey (2007) and this author's experience suggest, is a challenging subject to address in university due to the lack of complex multi-stakeholder situations to apply the theory. PM systems are intended to facilitate the generation of products efficiently, inexpensively, and with high quality in an environment that has many internal and external variables affecting the outcome. University projects aim to teach a particular set of skills or thought approaches, but the projects are typically highly defined or constrained, short-term, and (hopefully) insulated from third-party interactions that could sidestep the learning objectives. Highly defined projects exist in industry too, as tasks and often as part of a larger project; like their counterparts in the educational system, they are manageable with tools like to-do lists. Projects that involve longer durations, complex interactions between stakeholders, and third parties without direct benefit from the project's success will benefit from PM's more in-depth project initiation steps and tools, such as network charts and Gantt charts, for effective planning and management.

Practical Process Design for Chemical Engineers, First Edition. Keith Marchildon and David Mody.

10.3 The Core Principles of Project Management

No PM discussion would be complete without beginning with the three constraints: cost, schedule, and scope; and the joke, "choose two" (Figure 10.1). The reality is that a project manager is challenged to fulfill the needs of all three while maintaining the necessary quality.

Engineers are experienced in optimizing multiple objectives and constraints in their field. Successful projects require identifying the objectives and constraints as early as possible by knowing who to consult and inform, achieving a design that best accommodates them, and informing the stakeholders of their status.

The objectives and constraints can be financial, schedule, and performance (anything else). The planning stages of PM involve revealing an onion of layers composed of stakeholders, their requirements, and ideally an understanding of why. As much as possible, these layers are identified during the project planning phase, but inevitably some slide into later phases of the project and must be adequately accommodated through a change management system.

Objectives and requirements of a project can be non-tangible, for instance, an **objective** could be for "the project to position the company as a world leader in the production of ABC." Objectives need to be converted into tangible **deliverables**, like a report or a chemical plant with a defined production rate and quality. The section on project planning will delve into this in more detail.

10.4 Phases of a Project

A good general PM system will work for any type of project. Whether it is for a vacation, renovating an office, writing a report, designing a product, or building a chemical plant, the steps and their management are very similar. Portny (2022) described the PMI process and deliverables. Rosentrator (2013) related PMI practices to the chemical process industry discussed later in this chapter.

There are four phases to every project:

1. initiation,
2. planning or front-end engineering with design iterations ending with project definition and approval,
3. execution, detailed design, construction, and start-up, and
4. project closing.

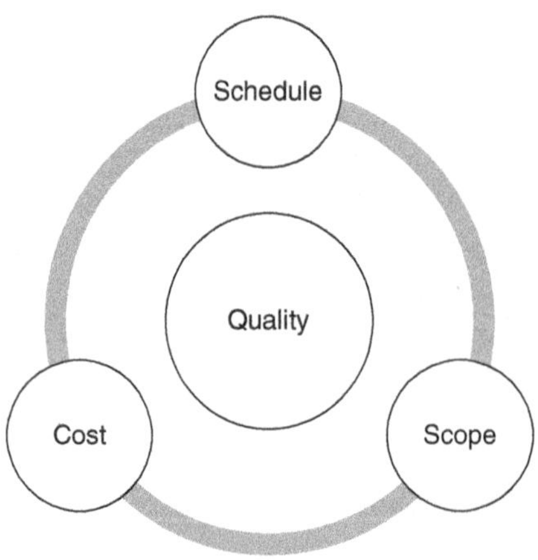

Figure 10.1 Project management triple constraint.

10.4.1 Initiation

In the "100 Rules for Project Managers," Madden and Stewart (1996) says, "The seeds of problems are laid down early. Initial planning is the most vital part of a project. The review of most failed projects or project problems indicates the disasters were well planned to happen from the start."

Initiation involves collecting the objectives, known inputs, and constraints to the problem and usually presenting them in some communicable format. If someone, the primary stakeholder (you), were planning a vacation trip, you might say your objective is to go on a relaxing beach vacation with the following constraints: You have two weeks in December and USD $5000, and you would like to have a spot that has plenty of nice restaurants. In PMI terminology, this would be called the project charter; in the engineering field, some people would call such an input a "business objectives letter." It will be short and contain key information about the project's objectives and rationale, but not a lot of background or details.

A second document could accompany the project charter or business objectives and is sometimes called the project briefing document or the first version of the "Basic Data." The second, more detailed document is a place to communicate everything you know about the problem and any prior work that might have been done on the project. Basic data is going to be specific to the industry and specific to the project.

In the travel example above, information such as you have been to three locations already, you want to try somewhere different, you really like Italian food, it is the rainy season in Bali during your time off, you have a friend's itinerary from two years ago that you like, and a magazine article on the best snorkeling trips in the world could all be relevant information and included in that background project briefing document. The chemical process industry project will have background information such as laboratory testing, reaction kinetic data, prior design work, and people who should be consulted. A product design project would have relevant market and customer background and needs information.

There are only three times when a project can be derailed: the beginning, the middle, and the end. With the right tools and templates, the project manager can help avoid the project problems from the beginning. Using a tool such as the "five why's" (root cause analysis) during the project charter will help to uncover and understand the nature and drivers of why the project is being undertaken in a way that will assist future design decisions [Read (2015)].

Larsen (2005) and others described the initiation phase as one with extensive communication between the sponsor, the project manager, and others.

Schatz (2018) discussed the metrics for project success that are different from the historically used metrics. He also applies those metrics to the results of several projects in the chemical industry. In addition, he emphasized prior project closeout reports should be consulted.

No doubt, the reader of this chapter has worked on projects that have had changes occur because initial objectives were not clearly defined early on or opportunities for improvement were not followed because of a lack of project understanding. The initiation of the project is well documented by Portny to have several steps that seek to document and communicate the project in a way that will help reduce these situations.

Objectives can be murky sometimes, and it is not uncommon in development projects for the objectives to gain clarity once more engineering or business effort is expended. Getting a clear understanding of the purpose early on will make a project manager more likely to be successful.

Portny (2022) described the following tasks in the initiation phase:

1. The initiation and planning phases involve gathering information to decide whether a project can be done and whether it should be done. That simple phrase reveals quite a bit of insight, and when unwrapped, it reveals a great deal about the PM steps.

2. Look at the big picture: How your project fits in and why you are doing it by identifying the initiator (not necessarily the person who asked to do the project – dig deeper to find that person).
3. Perform a cost-benefit analysis: The idea of doing a cost-benefit of the project early on is an effective tool for understanding what is important about the project, uncovering metrics that should be included in the project objectives, and deciding if the project should be done.
4. Creating an audience list that identifies:
   a. Project champion: A person in the organization who has significant influence to help the project be successful.
   b. Project drivers: Those people who have a say in what the project should accomplish or affect its scope. Drivers may be internal (sale reps, business people, operations people, maintenance people, etc.) or external (ultimate customers, regulatory and government officials, etc.).
   c. Project supporters: Those people who will help you with the project.
   d. Others: People who are affected by, or will affect, the project in some manner.
5. Determine the project drivers' real expectations and needs. There may be public relations or other less obvious goals that are not stated in the project charter.
6. Uncover other activities that are occurring and relate to your project. Coordinate with those efforts as needed.
7. Use the SMART (Specific, Measurable, Aggressive, Realistic, Time sensitive) technique to describe the desired project objectives.
8. Document limitations, constraints, and assumptions.
9. Review, review, review.

10.4.2 Planning Phase: Project Plan Development and Scope Finalization

Once the project is initiated, it proceeds to the planning, preliminary, or front-end engineering phase to explore options and add necessary detail to create the core project plan elements as well as other optional project plan elements listed by Rosentrator (2013) in Figure 10.2.

In the chemical process industry, the planning phase is often divided into three phases [Front-End Loading (FEL) 1, 2, and 3 explained in detail below] that are in reality, progressively

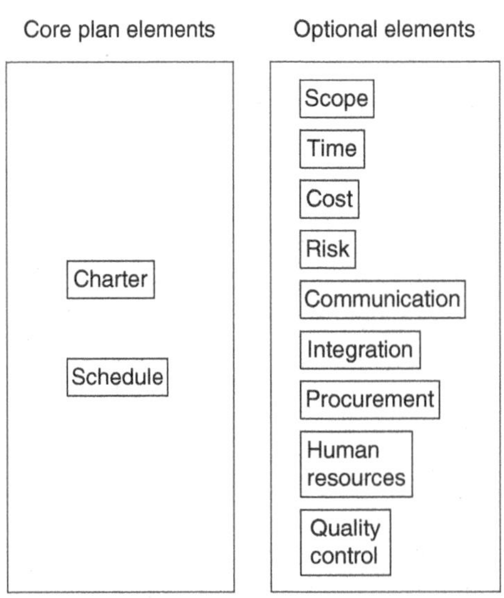

Figure 10.2 Project plan elements.

more detailed feasibility studies. By adding engineering definition to the project and certainty to the cost plan in phases, the project can minimize costs and focus on issues in a timelier manner.

A key step in the planning phase is to turn the project objectives into a list of deliverables which is documented in the Work Breakdown Structure (WBS).

The WBS should be a list of tangible deliverables, not activities or metrics. The following is debatable in the PM community, but we believe it is better to stick to deliverables being physical objects you could give to someone, and leave the non-tangibles in the project objectives (e.g. an objective is to achieve a 90% performance metric, and a deliverable is a printed report documenting the plant performance metric). Portny (2022) presented the example of a WBS for building a house and showed how it progressively decomposes the project's intent into physical details of the house that describe what will be delivered to the owner. A different example could be a project that delivers a report. The WBS for that project could be the table of contents describing what the report will contain to the level of detail necessary for the stakeholders to agree upon. The WBS should break a project down to the level that the deliverables are clear to everyone who needs to create that deliverable and can estimate how long it will take to produce it, and everyone who will be receiving the deliverable agrees they will have what they want. Depending on how familiar people are with a particular deliverable, there may need to be more detail. Sometimes the deliverable can be difficult to describe if the team or project sponsor has never executed a similar project. In such situations, try to find an example project that can provide a template of deliverables, plan a little more than you think you need, and be prepared for the WBS and all its subsequent documents to evolve during the life of the project. Upon completing the WBS, it logically leads to creating the list of activities, the network diagram, and the Gantt chart.

There are numerous tutorials for creating Gantt charts, starting with a spreadsheet, which is a useful first-time exercise, or specific scheduling tools such as MS Project. The discussion below will focus on the preparation needed in creating the inputs for a Gantt chart, rather than the mechanics of building the chart using the software of your choice.

Once the WBS is generated, with the help of the project team, the next steps (Figure 10.3) move this information to a schedule that will be used to monitor and manage the project. Since the WBS is a hierarchy of deliverables, not tasks, the next step is to create a list of work packages or activities that produce the deliverables. The activity/work packages can include one or more deliverables that will be assigned to a single person for accountability purposes. The list of activities is then numbered for ease of the next step. Activities are given a calendar duration to complete them and a list of the prior activity names or numbers that must be performed before this activity can begin (or finished), see Figure 10.4. A word of advice, in some situations, it may help in the communication

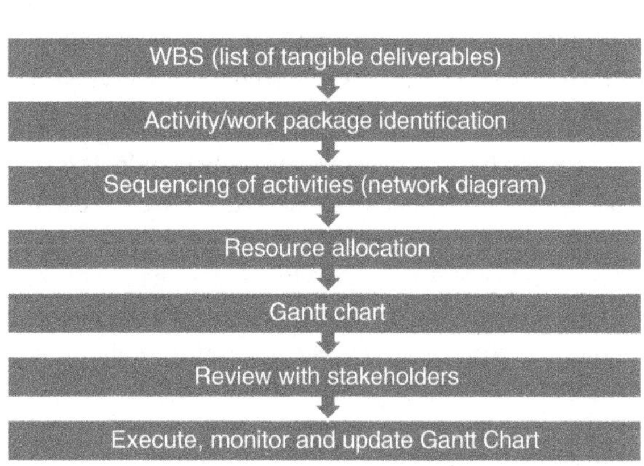

Figure 10.3 Work breakdown structure to schedule.

#2.8 Draw P&IDs

(ready for intial review and factored estimating)

- Duration: 1 wk
- Predecessors: PFD, Simulation Rationalization

Figure 10.4 Brown paper task definition.

with others if you clarify the state of the deliverable in the activity description. In a sense, this is clarifying what the successors are, which is not strictly required, but it can help with communication with team members.

For example, in a chemical engineering design, P&IDs are a key document that both initiate many other discipline's work and undergo continuous development; they are a living document. The implication is that other people will begin using a P&ID at various stages of its completion and will, in turn, contribute to the P&ID. For instance, the electrical department might use the P&IDs at an early point in the project to count motors and determine electrical loads, the instrumentation department may use them to estimate the number of control loops, and the piping department may use them to build a line list (list of pipes). The mechanical department will use them to add to the equipment list. Thus, the activity to draw the P&IDs needs to be broken into phases and qualified with a state that others will understand and decide when their activities can begin. Typical milestone states are IFR – issued for (internal, client, and hazard) review, IFE – issued for estimate, IFD – issued for design, IFC – issued for construction, and As-Built. Another way to accomplish this is to qualify the states as various %'s of the completed activity.

There are two approaches to creating the Gantt chart that are quite successful.

One approach is to gather the people responsible for creating the deliverables, give them each a different colored pad of sticky notes, and ask them to write down the activities they need to perform to create the deliverables (in most cases, each deliverable will convert to one activity), one activity per sticky note (Figure 10.4). Along with the activity description is the calendar duration (not the number of work hours) that the activity will take to complete, and a list of any predecessors. Planning in units of weeks or days usually makes sense, but any duration can be accommodated. Any task that is longer than weeks is probably one that would benefit from being broken into stages for easier tracking.

A little aside, choosing a calendar duration is less straightforward than it appears because you can be constrained to hit certain dates or constrained by your available resources. The duration can be determined using either a top-down or a bottom-up approach. A top-down approach looks at the overall deadline for the project and estimates the activity needs to occur over a certain period of time, and resources are allocated to achieve that date. A bottom-up approach looks at the "effort" hours it takes to complete a task and divides it by say eight hours per day to determine the calendar duration, but it does not consider the following:

- A task can be completed by multiple people. It is worth noting that there is a limit to how many people can be applied to a task to decrease the calendar duration. More people may slow the task down.
- The number of people working on the task may change during the calendar duration.
- The hours per day people spend on the task may change from day to day. You may spend a few hours a day on the task at the beginning, eight hours a day in the middle, and a few hours toward the end.
- There may be gaps in when the work is being completed, and this could be handled by breaking the task into subtasks.

You can assume some average work hours per day are spent on an activity, or there are advanced scheduling tools that accommodate nonconstant resource loadings, but it is probably best to decouple the two concepts and use your experience to monitor calendar progress and effort hour progress separately. To determine the duration using a bottom-up approach, look at the activity and the work hours needed, make an assumption about the number of people available to work on the activity and how many hours per day they will be able to work, and use a safe percentage of that. Try to check with other projects to see if that duration makes sense.

It is entirely possible a top-down approach is being asked for by the management to achieve particular dates, and you must use a bottom-up approach to determine the reasonableness of the resources that are needed to achieve this. Remember, there is often a limit to how many people can simultaneously work on a particular task.

The sticky notes are placed on a wall, or long roll of paper (the name of the meeting has been called a "brown paper exercise"), in the order in which they can be completed. Where one activity requires another to be completed (has a predecessor), an arrow is drawn from the "finish" of the predecessor to the "start" of the next activity. This is called a Finish to Start (FS) relationship. More complicated relationships can be defined (see Portny 2022). You will have a Network Diagram at the end of this. Add in the start date, use some software, and the Gantt chart is the result. At that point, you can check to see when the project finishes and compare it with the needed date for completion. The Gantt chart can be used to further understand the critical path and modify or optimize it accordingly.

A second way to create a Gantt chart that works well for small teams is to gather the deliverables and create a list of activities and calendar durations in a spreadsheet. You can copy those two columns into project planning software (e.g. MS Project) and then add the predecessors. The software will create a network diagram, and the Gantt chart and find the critical path, and you can proceed to optimize the schedule as needed.

You can imagine your vacation trip itinerary is the schedule developed from the list of the locations and hotels you wish to visit, your WBS. A budget can then be developed with a list of quoted and estimated costs (airfare ticket quotes, hotel costs, rental car costs, and possibly you have peeked at the restaurant menus and the prices). There will still be some known unknown estimated costs, like additional taxi costs or snorkel trips, so you add some design allowance, and there may be some unknown unknowns, so you add some contingency to both the budget and the schedule (plan to leave the airport early to accommodate unknown delays). Larsen (2005) discussed known unknowns, and unknown unknowns in the context of chemical engineering projects, as does Chapter 9 of this book. Unknowns and assumptions logically lead to the risk plan.

There are several possible components in a project plan (Figure 10.2), such as budget and schedule, but the **Risk Plan** is a source of confusion with engineers and especially chemical engineers who are familiar with process hazard analysis tools. While some authors indicate that a process hazard analysis is a risk plan, that is not completely correct. Portny (2022) described the project planning phase as answering two questions: **"Can the project be done? and should the project be done?"**

The first question, "Can the project be done?", is answered by an analysis of available versus required resources, budget, timeline, assumptions, and other risk factors using a risk analysis and management tool like a risk registry. In the travel example, if the trip involved a small boat trip to a remote island in a location that was prone to hurricanes, there is a very real possibility that you might not get to the destination (weather is a risk factor). Similarly, for an engineering project, there are numerous risk factors that you may need to understand to ensure you can complete the project as required in the project plan. For instance, if the project is to deliver a written report, you may not ever find information that is needed because it does not exist, or you do not have the budget to obtain the information. If it is a chemical plant that will need permits or other elements to complete it, they may not be granted (politics is a risk factor). These are risks that you cannot deliver what is required (the report, the plant, the product, etc.) to meet the project objectives. To answer the question, "Can the project be done?", conducting a risk analysis of the project's execution, via brainstorming or a checklist using a risk register, is advisable.

The second question is about whether you **should** deliver the project? This is the more familiar risk analysis chemical engineers deal with. If the project creates an unacceptable societal or safety hazard, then the answer is no, or at least the design needs to change so it is acceptable. This second question is answered by a process hazard analysis, an environmental review, in the case of a chemical project, or other tools such as FMEA and FTA for product development projects, and strategic review of the project as a whole.

Hence, while the first question answers to the execution of the project and thus should be part of the project plan, the second question is both integral to the design development and the project plan. You need to be able to say, "Yes, we can" and "Yes, we should" deliver this project before continuing to the execution phase.

A third type of review(s), a design review, answers the question, are we making the best plant or product possible? That question would be answered by reviews that include the following:

- design reviews,
- construction reviews,
- design for manufacturing analysis,
- Quality Function Deployment (QFD) analysis, and so forth.

While these last reviews are a normal part of the optimization of the design, they are a significant element of planning by deciding what the final design will be.

There are more documents that PMI advises are part of the project planning stages, which we will discuss below, but as you can see, the essence of project planning is easy to relate to any type of project. Portny (2022) discussed additional elements for effective project execution.

The project plan is finalized at the end of the planning phase and involves a detailed outlining of the project's scope, objectives, timelines, cost estimates, resources, risk management plans, and communication strategies (Figure 10.2). The objective of the planning phase is to establish a roadmap for the project team to follow, ensuring all project activities are aligned to achieve the project goals within the defined constraints.

However, it is important to note that while the project plan is "finalized" at the end of the planning phase, it is not set in stone. In reality, the project plan is a living document that may be revisited and revised throughout the project lifecycle. Changes in project scope, unexpected risks, or shifts in stakeholder requirements can necessitate updates to the plan. Effective PM requires flexibility and responsiveness to such changes, with appropriate adjustments made to the project plan as needed, following established change management processes.

Therefore, while the project plan is initially finalized at the end of the planning phase, it remains subject to review and revision to accommodate changes and ensure the project stays on track toward its objectives. This process is often called management of change.

In the chemical industry, there are specific "deliverables" at the end of the front-end engineering that are discussed in Chapters 2 and 9, throughout this book, and below.

At logical points in the project, you have an ever-increasing knowledge and confidence in what you plan to do, what the final deliverables will be (**Scope**), what the cost of the project will be (**Budget**), what the timing/schedule will be (**Schedule**), and whether you are likely to meet the project's objectives. At key points in time, projects pause to make a decision. Does this project meet the project objectives? For the vacation trip, will it be relaxing, will there be some good restaurants, will you be on a beach, and will it likely cost less than $5000? This is called a **Stage Gate Review**. If the objectives are reasonably sure to be met, and there are no other issues, you proceed to the next stage gate, or the **Execution Phase**. If the objectives are not met, then you have a choice to stop the project (not go on vacation) or recycle the project by looking at different options and possibly alter some objectives and constraints.

At the end of the planning phase, the decision is made to move forward with the project or not.

10.4.3 Execution Phase

Execution implies you are proceeding with the project. Money is being spent, commitments are being made, and details are being finalized that could not or did not need to be done in the planning phase, and you are following your plan to completion.

For the trip, you start booking airline tickets, reserving hotels and restaurants, and monitoring that the actual expenses are what you expected them to be in the estimate. Last-minute design details are performed, such as getting your bags packed now that you know where you are going, for how long, and that you will be snorkeling and sitting on the beach, and a last-minute check of the weather to assess your clothes for the duration. You hop in a taxi and off you go according to your plan with some minor changes as might be necessary. When you come back, you should complete the project closing phase.

In theory, the project would be formally approved at the end of FEL III and thus would proceed to detailed design. However, quite often, there is enough confidence in the FEL II estimate, the economics, and the design that some long-lead items will be ordered, and some construction items, such as site preparation, will begin.

10.4.4 Project Closing Phase

In the travel analogy, you might review your initiating process and see if there were any objectives that you missed early on. Perhaps you forgot that in the back of your mind, you wanted to make sure the beach location had calm water for swimming. You look at your estimate and see if it matches your actual expenditures and review your process for making an estimate. You might review what went well during the trip, jot down some notes and create a checklist for a future experience, and even pass that information along to others who are doing something similar.

10.5 Business Phases of a Project

Before getting into specific details of executing chemical engineering design projects, we will examine how chemical plant capital projects fit within the context of bringing a new chemical product to market.

The steps from product discovery to production are discussed by Berg (2011) and in Figure 10.5, to which the FEL phases of engineering asset development have been added. The diagram illustrates the parallel path that often happens in product development to accelerate an idea to market and certainly applies to other products, with prototyping and customer verification as key elements of the development. You can imagine the diagram below broken into many interconnected business, research and development, and engineering (sub)projects to reach the end goal.

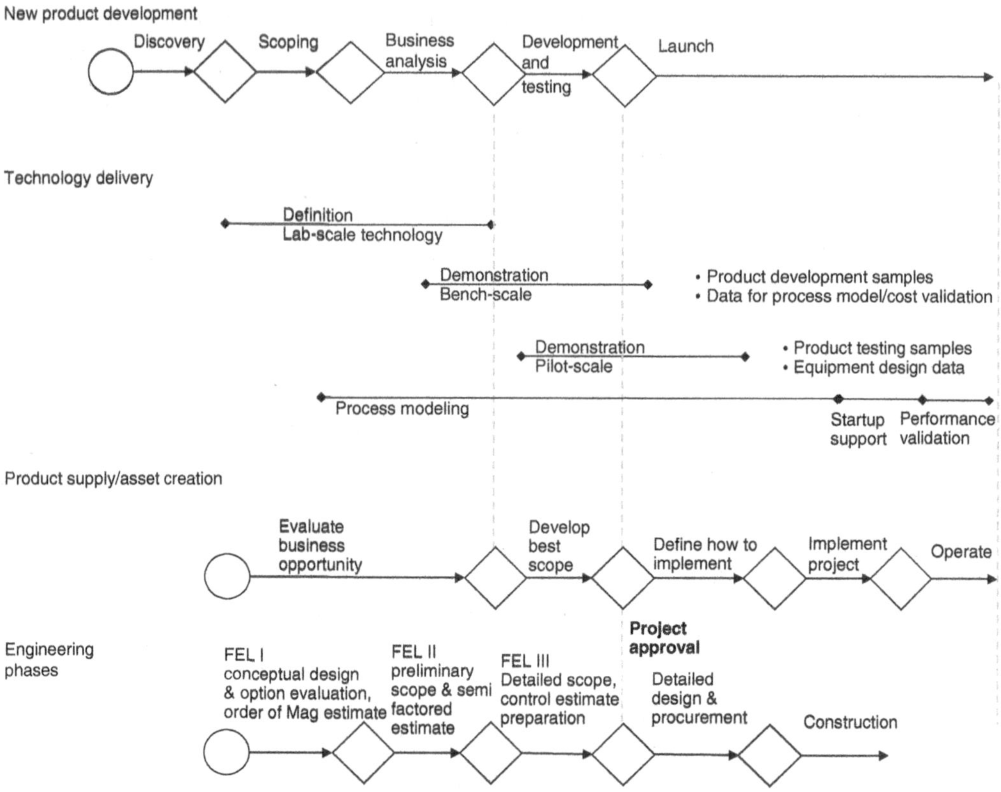

Figure 10.5 Product and process development. Source: Adapted from Berg, 2011.

There is a natural conflict between the goal to define a project so well that changes will not affect cost and schedule, and a project that is able to, or must, innovate in some way as it proceeds (Ben Mahmoud-Jouini 2016). This conflict is especially prevalent in design projects where the customer, market, and thus the product have not been established. Recognize, though, that change occurs in every project, and PMI practices provide systems to manage the changes. Larsen (2005), Berg (2011), Karol (2002A&B), and Eldon (2005) discussed the challenges of executing Research and Develoment (R&D) projects. The Independent Project Analysis (IPA) company provides a consulting service to improve the PM practices of companies in the chemical industry. An excellent introduction to their methods is presented in Understanding Cost Growth and Performance Shortfalls in Pioneer Process Plants (Merrow 1981). Three other articles in that series complete the background and are relevant today for understanding cost growth issues in plant designs. The IPA methods can help with risk management and can provide contingency planning for chemical engineering projects (see Chapter 9).

10.6 Engineering Project Phases

Chemical engineering projects that are executed in phases or stage gates (Front End I, FE II, FE III, or similar name) enable efficient classification of viable and nonviable projects with a minimum of engineering costs and time. The tools of capital costing discussed in the previous chapter allow decisions to be made earlier and can help focus the project on the factors that make the most

Figure 10.6 Cost to make changes to a design.

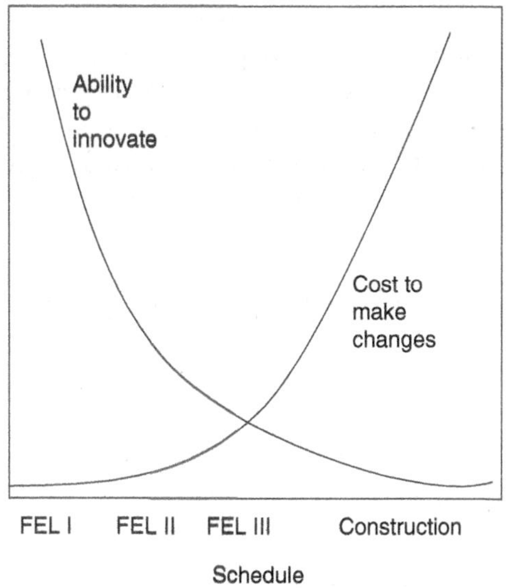

difference to its viability. If technology and financial issues can be identified early on, changes can be made to the design that will have a minimum effect on cost and schedule (Figure 10.6), thus allowing for greater innovation and more efficient use of available funds.

The stage-gate process allows for innovation while minimizing engineering costs by deciding at each phase whether to pause, recycle, or stop in situations when the cost, schedule, or performance objectives cannot be met. The process also helps to communicate issues to the management level more quickly. Many projects are shelved during the review process, thus incurring only the engineering design cost to that point. The deliverables that typically accompany these stage gates are discussed below.

10.7 Project Initiation: Project Charter and Project Business Objectives

The Project Charter or business objectives is written by the primary stakeholder (the business person) explaining the need for the project. What is the problem to be solved, or what opportunity is to be capitalized upon? An "objective" needs to be defined, along with any constraints (such as time, money, performance, etc.). The best projects clearly articulate the challenge or "customer pain point," enabling designers to explore a wide range of solutions. However, considering other projects might explore alternative options, setting exclusions could be beneficial to narrow the scope and expedite the project's progress. If there is a great deal of information that must be communicated in an engineering project, then typically a more detailed document (basic data or project briefing document) will accompany the shorter project charter. Read (2015) and Pavone (2006) discussed basic data for chemical projects in more detail.

Objectives of the project should be formulated to be "SMART," which is to say they are specific, measurable, achievable, relevant, and time bound. Objectives can be tangible (a report or a chemical plant), but they can also be non-tangible, like increase production by 10% of the current production rate ($\times$ kg h^{-1}) within one month of start-up. Deliverables should only be tangible (the facility, the report, etc.).

Taking the authoring of a report as an example, approaching it as a project employs a similar process. Understanding the report's purpose, its audience, and usage provides crucial insights for the authors as they consider the necessary content. Similarly, engineers designing products and processes must identify the end users and their requirements. The project initiator might also need to specify needs and opportunities during project initiation, so be aware that the initiation phase is often iterative with all the stakeholders.

10.8 Project Initiation for Chemical Engineering Projects

The Initiation phase for a chemical engineering project should include a basic data document. Pavone (2006) discussed the general topic of project initiation. Basic data begins at initiation and ends at the planning stage and will contain constraints and objectives that are driven by the following:

- Internal:
  - sales needs, plant reliability, labor availability and skills, etc.,
  - financial goals,
  - the physical plant location, and
  - utilities, existing infrastructure, or lack thereof, transportation methods, raw material sources, and weather.
- External:
  - customer requirements,
  - legal and intellectual property,
  - environmental,
  - government regulatory,
  - goodwill with neighbors and other societal needs, or ethical factors, and
  - sustainability.

The project initiation documents should include the objectives and constraints that are known. This might include known customer quality requirements for the product, and growth plans for its production. Why? A designer can decide to size pipes a little bit larger for an easier plant expansion at minimal expense or to keep the cost of maintenance spares to a minimum if initial capital cost is more important than plant uptime and reliability. This stage of the project will likely involve many unanswered questions, but tabling as many of those questions as early as possible will provide a facility that achieves objectives earlier and more cheaply.

The project initiation documents should include an evolving list of stakeholders, those categories or groups of people who will have input into the project and have needs or constraints. In the chemical engineering domain, this might include identifying legal and regulatory groups, operations and maintenance groups, community groups, and downstream users.

Glavic (2021) described the future "Circular Economy" involving product manufactures (including chemical producers) taking responsibility for the end-of-life destruction of their products. This possible future requirement affects the overall economics of producing a product and, ultimately, the design of a process that economically and socially prioritizes reuse or recycling options ahead of destruction. Objectives, such as zero discharge plants, are key to the design planning process.

It is normal to create an initial list of people (or functions, e.g. environmental group, utilities group, maintenance group, and the general public) who need to be consulted or informed during the project definition phase. This list will grow and change over time and into the execution or construction phase of the project, but certainly it is important to include other people in the definition of most projects.

A RACI chart (Responsible, Accountable, Consulted, Informed), and sometimes linear responsibility chart, is a useful tool for defining what team members are responsible for on the project throughout its execution.

10.9 Chemical Engineering Project Planning

The project plan for a chemical engineering project will need to contain the elements described by Pavone (2006) in his article about the "How to Prepare a Process Design Basis," which is not a PMI term, but the elements overlap with a PMI project plan requirement. Some of those elements, as indicated in the article, are necessary at the initiation phase, and some may be completed as part of the front-end engineering. Review the list and upgrade the project plan during the planning phase as necessary.

A flow diagram for completing the FEL I process engineering design steps is shown in Figure 10.7. The primary objective of these steps is to create a technical, economic, and safety risk analysis of the process idea. Should there be options, alternatives, or optimizations that need to be considered, and they can be assessed as a part of these steps or complete cycles. FEL II and III would add additional deliverables and definition based on these, such as line lists, line sizing, instrument data sheets, relief valve sizing, further hazard and environmental analysis, and layouts, providing progressively more information for other disciplines to perform their estimates and thus accuracy to the capital and operating estimates of the business analysis.

At the beginning, the project is initiated with the business objectives, basic data, and sometimes a block flow diagram or PFD. For a heat and material balance to be prepared, usually with a simulation or model, several actions need to be performed:

- If there is a reaction occurring, the reaction equations and kinetic data must be determined.
- The molecular species that need to be tracked in the simulation need to be identified, and an appropriate physical property model/library/methods must be selected and possibly checked.
- Identifying the information you wish to obtain from the completed simulation may guide what species you model and to what level of rigor. Species that do not already exist in the simulator can usually be added, but it may take some time to obtain the relevant physical property data, and some expertise to fit the data to the physical property equations.
- Determining the property package to be used can be challenging. Guidelines provided by the simulator vendor will help. Checking the predictions of the property package may be necessary. Ultimately, it is worth recognizing that the materials are virtual, and thus, only the properties that matter to the simulation you are preparing need to be modeled accurately.
- It might not be possible or desirable to model all that is within the process. Small contaminants that do not significantly contribute to the heat and material balance, but may cause corrosion or catalyst issues could be dealt with outside of the simulation. Unusual unit operations or equipment may need to be modeled in a simpler fashion in the simulation software, leaving the details for manual material balance calculations, or the equipment sizing. However, others need to be made aware of the simplifications that may exist in the simulation model, so documentation is key.
- While pressure drops can be modeled in process simulations, most people choose to only model the major groupings of pressures and leave the piping and gravitational pressure changes for later analysis (see process rationalization) and refined further once equipment sizes are available.

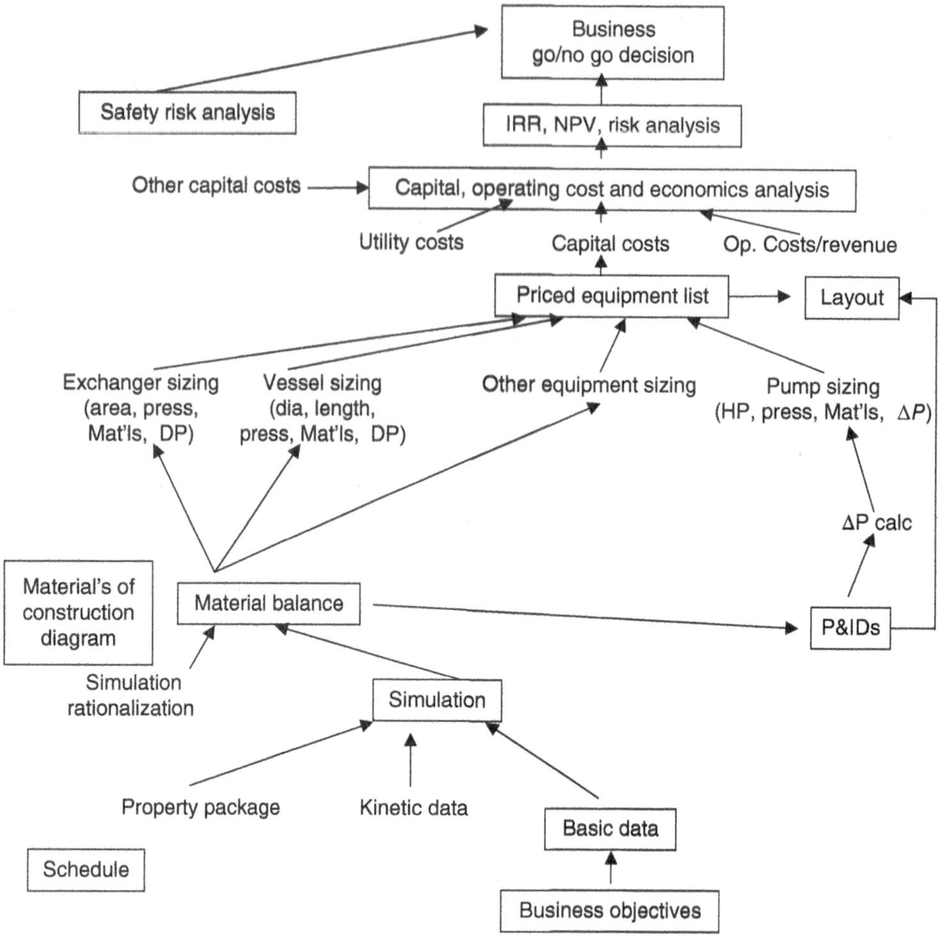

Figure 10.7 Chemical engineering front-end engineering I process design steps.

- The process of finalizing the simulation may undergo several iterations that attempt to optimize it (perhaps by comparing the capital costs to the operating costs). Therefore, there may need to be a route to the economic analysis that bypasses some steps (detailed equipment sizing, P&IDs, or hazard analysis) only to iterate back and complete these steps in a later design cycle.
- The simulation can be used later in the design process (once vessel volumes and so on are known) to assess control system ideas and tune controllers.

Owing to the simulation having some simplifications (e.g. not including pressure drops, or missing some chemical components, and the need to check the simulation for errors), a step we call "simulation rationalization" produces a final material balance/PFD should be completed to identify and deal with issues that matter in equipment selection, sizing, and further process design. This may be documented with a final PFD and material balance in a spreadsheet. A reliability strategy will guide you toward what equipment (rotating equipment, holdup tanks, etc.) must be spared and must be shown on the P&IDs.

Arguably, copying the simulation material balance information into a spreadsheet by hand is superfluous and time-consuming, but adding the equipment needed in the simulation rationalization into a final PFD is very much necessary. Both are usually required to promote effective communication with others, especially later. An idea of equipment sizes and a rough elevation from the layout of equipment can provide an excellent basis for preliminary pump sizing.

The selection of materials of construction should be completed before equipment is costed and should consider the possibility of corrosive chemicals in the process that may not have been modeled in the simulation, especially in situations where those chemicals can build up in concentration.

Equipment is usually sized using a combination of:

- Manual methods using a spreadsheet or, preferably, Mathcad. Mathcad is adept at catching errors due to its handling of units and its paper calculation look that shows the equations being used.
- Software programs, such as for sizing heat exchangers, tower trays, fans, etc., provided by a simulation software provider, vendors, or other third parties.
- Vendor or in-house sizing software and methods for specialized equipment such as tower internals, compressor sizing, and so forth.

Although not strictly called for during the early FEL stages, P&IDs are invaluable for instrument engineers and other disciplines to prepare their estimates. They enable the preparation of more accurate cost estimates for control hardware, process plant computer hardware, and software, as well as better estimates for the piping, electrical, and civil disciplines. P&IDs can help identify items that might be overlooked or excluded from an equipment factored estimate. Depending on the size of the project, the need for other disciplines to contribute to the FEL II and FEL III, and the need for vendor information, some aspects of the P&IDs are completed at different stages of a project, but in general, the P&IDs will be completed according to the following steps:

First draft P&IDs include the following:

- all known equipment and controls,
- the basic control scheme, including interlocks,
- known standard piping details, such as bypasses on pumps, and control valves,
- known critical dimensions as determined by the process engineer (e.g. elevations for NPSH, mixing or separation that are inline, etc.),
- known drains and vents around equipment,
- secondary containment,
- relief valve locations,
- process lines, drain and fill lines, and utility lines, and
- vendor package scope demarcation and known vendor requirements for utilities.

Issued for Engineering Internal/Engineering Review:

- Updates from the internal review are made and issued for client review.

Issued for Client Review (including operations and maintenance input):

- Updates from the client review and/or response to questions from the review,
- prehazard review notes and operation notes are added, and
- in the event, the project is performed solely as an owner arrangement, not a contractor/owner situation, and the P&IDs are issued and reviewed with the stakeholders, maintenance, and operation people.

Prehazard Review Activities (line numbers and instrument numbers are required for the HAZOP method):

- Line numbers,
- instrument numbers and details, and
- off-page connectors match.

Issued for Hazard Review (HAZOP):

- Updates from the Hazard Review are addressed, and drawings may be issued.
- In situations where there are unresolved design aspects either from the PHA, from vendors, or otherwise, the affected areas are clouded and put "on-hold for …" to denote the design details will be completed at a later date.

Estimate Preparation (supporting discipline factored estimates with some takeoff type estimates):

- Vendor data as available is added.
- Line sizes, pipe specifications, and insulation requirements suitable for the line list and factored estimate by piping, as well as input for piping stress analysis. In some situations, piping stress analysis may be completed for some lines in FEL III, with the remaining pipe analysis being done during detailed design.
- Motor sizes (may be documented on the equipment list instead of the P&ID), but the electrical discipline usually maintains their own motor list that coincides with their single line diagrams.
- Equipment sizes and weights (may be documented on the equipment list instead of the P&ID). Different companies have differing philosophies with respect to how much information will be placed on the P&IDs about equipment for intellectual property security purposes.
- Relief valve disposal system.
- Utility P&IDs will likely be started at this stage and completed during detailed design.

P&IDs issued for estimate (FEL II, III, or Design)*:

Depending on the owner/engineering firm culture, the end of the FEL II stage may approve the project to purchase long lead equipment and do site preparation. Thus, engineering disciplines such as instrumentation, piping, electrical, and civil (pipe bridges) will ask for the P&IDs to have a reasonable level of completion for their (factored) estimates. For instance, instrumentation's early estimate method may count control loops from the P&IDs and assume some average cost per control loop for hardware and distributed control system (DCS) configuration.

The FEL III estimate will utilize more material takeoffs, possibly with some factored estimating, and thus requires better P&ID definition than the previous design phase. During the detailed design phase, a "Construction Control Estimate" is produced based on a complete bill of materials and material takeoffs; however, the expectation is that the estimate will not change significantly from the FEL III, or even the FEL II values. Thus, disciplines will push to have as much design definition to base their estimates on as early as possible.

In other situations, especially for smaller projects, the project approval may happen at the "issued for design" phase, in which case some P&ID details above may move out to coincide with that stage.

P&IDs issued for construction:

- These are issued to the construction contractor as the approved final design.

P&IDs issued as-built:

- Inevitably, there are changes that occur during construction, start-up, and commissioning, and these changes need to be updated on the P&IDs, so the plant records are up to date. Some changes occur because the design will leave small bore piping, heating, ventilation, and air conditioning (HVAC) ducts, electrical wiring conduit, etc., to be field run by the contractor. This can sometimes affect the overall design in minor ways. Alternatively, the construction contractor may see opportunities to reduce costs either by constructing the project in a different way than was imagined, or through design changes. These construction changes can be opportunities for reducing costs as long as the integrity of the design is not compromised. A thorough review of the final project is required, and the as-built P&IDs are a part of that final review and sign-off.

During FEL 1 and II, a tool such as SketchUp and a prebuilt library of process equipment (even just simple cubes) can be used by a process engineer to quickly generate a rough 3D equipment layout model of the plant (without including piping) in a surprisingly short period of time. That model can guide the proper plant layout and provide significant insight during early process hazard analysis studies. A model is especially helpful if placed in the plant location via Google Earth for assessing off-site hazard consequences.

Process hazard analysis studies should begin at the FEL I/PFD stage. Knowledge of the chemicals and reactions that are present, approximate operating conditions, and possibly a heat and material balance will provide an adequate amount of information for this review. Substituting hazardous chemicals and operating conditions to make the process less hazardous via an inherent safety analysis during this early phase of design can save a great deal of engineering and capital costs later. A checklist hazard review is also appropriate at this stage. The process hazard analysis (PHA) may develop a list of questions that need to be answered as part of further design development.

The next phase of process hazard analysis should take place when P&IDs are available. Owing to the ease of creating a 3D model with simple tools (e.g. SketchUp), it is advisable to have a draft of the 3D model for reference. PHA methods such as checklists, HAZOPs, and so forth make sense at this point. Major safety issues identified and dealt with early, can be done so without causing significant design changes to other disciplines, thus keeping the cost of changes to a minimum.

Using the material balance and PFD, the equipment list can then be developed in combination with sizing calculations, discussions with vendors as necessary, and a materials of construction review. Subsequently, equipment data sheets and the equipment list evolve together. Pricing the equipment list via vendor quotes is preferable but not always practical during the early stages of the design. Where vendor quotes are not practical, cost equations or in-house costing methods can be used.

With the deliverables described above, an overall economic model including utilities, raw materials, product sales, and capital costs can be developed and analyzed. Where there is uncertainty in some variables, a Monte Carlo analysis can be performed, and a range of economic values (net present value (NPV) and internal rate of return (IRR)) can be developed along with probabilities (see Chapter 9).

10.9.1 Chemical Engineering Deliverable Time Requirements

It is difficult to determine how many hours it will take to prepare the chemical engineer's deliverables in a design project. The hours can vary significantly if the project is a modification of an existing design or a novel design, which is likely to have many unforeseen challenges. Other disciplines often rely on the capital or total engineering costs to determine their engineering estimate. Often, the chemical engineer's work is significantly complete before a capital cost estimate is complete for a project. The timing makes using historical budgets or hours difficult based on the project's overall cost. Below we provide some guidance for estimating the hours for different design situations.

If there is an estimate for the project, either a capital estimate or an engineering estimate, the following guidance can be used. A process department budget is commonly 2.5% to 8% (average

5%) of the engineering costs, which is, in turn, about 12% to 30%, with an average of around 15% to 17% (12% in past years). Lower percentages (12%) for larger projects that are proven technologies, and higher numbers (20% to 30%) for small projects and/or unproven technologies of the total installed project costs. About 50% to 80% of the process engineering budget costs would be spent by the time stage gate III is complete.

If the design development is to the stage where a known number of P&IDs for the project is available, the total process department budget can be estimated to be 150 hours/P&ID. This includes all process engineering tasks such as data sheets, equipment lists, preparing P&IDs, hazard reviews, etc. Bear in mind, there can be exceptions. A revamp job may only require a single P&ID, but the fact that it interfaces with X other existing P&IDs will make the process hazard analysis X + 1 times more involved than it might appear.

Another method is to draw a rough process flow diagram and estimate the hours based on the number and type of unit operations present. You can assume one major piece of equipment per P&ID, and 150 hours/P&ID. The PFDs at an early stage in design tend not to have included all the equipment so add some margin (say 20%) to this and consider the need for utility P&IDs. One project manager has observed that typically, about 10% of additional equipment items are added to the P&IDs after stage gate III (during detailed design).

Alternatively, you can build the estimate from the ground up using the guidelines below. While we have tried to use our experience to give you some idea of the time required to develop the deliverables in the table and notes below, the specific situation at hand must be considered.

Table 10.1 and the following notes provide some experiential guidance for estimating process design engineering hours. In general, effort hours are described rather than the required durations for a schedule. To convert effort hours to durations, it is suggested that you sum the various activity effort hours, round up to the nearest day and add 15% to 20% overhead to the task.

Notes: If a process simulation or other computer-based heat and material balance is utilized, the following primary steps of producing that deliverable will require varying amounts of work:

Preparation – Step 0: Preparation includes determining objectives of simulation work, determining components, process topography hand sketch, and listing constraints and objectives.

Decide what information the model needs to provide in its finished state and what questions need to be answered by the modeling. This can be a little more insidious than it might appear at first glance. Give it some thought, and if necessary, ask yourself, "What if I had that information today? What would I do with it?" Often further questions are uncovered in this way.

Physical properties – Step 1: Decide what chemical components will be included in the model. Decide if one property package in the simulation will suffice, or if there will need to be a break in the simulation with a second property package being used (that often adds significant complexity to the model when recycles are present). The location of recycles can make or break the easy convergence of a simulation. Check to see if the components you want to model have the needed data for the property package you want to use, or if you will need to obtain and input that data into the program. This may require library and possibly laboratory research, followed by property package curve fitting and integration.

A duration of longer than 1 to 2 weeks indicates there is significant missing thermodynamic information, and a separate R&D subproject might be required.

The shorter duration occurs when you have significant experience and confidence with the components, operating conditions, and type of system that will be modeled.

Longer durations occur when the chemicals and operating conditions are not familiar and/or there is the possibility of azeotropes, or other situations that may require that you generate binary coefficients and check them. The assumption is that literature values are available, accurate, and reliable.

Table 10.1 Process engineering hour estimation.

		Process hours estimate					
					Effort hours		
FEL phase	**Activity description**	**Low**	**High**	**Qty of units**	**h per Qty low**	**h per Qty high**	
1	Basic data/project initiation activities	8	100				
1	Simulation preparation	8	40				
1	Property package Sel	2	80				
1	Reaction selection	2	80				
1	Process simulation modeling	16	48	2	8	24	Distillation units
1		5	10	5	1	2	Heat exchangers
1		5	5	5	1	1	Pumps
1		3	9	3	1	3	Separators
1		4	20	1	4	20	Reactors
1		0	2	2	0	1	Storage tanks
1		16	48	2	8	24	Recycles
1	Process model basic checks	8	24	1	8	24	Basic checks
1	Process simulation studies and optimizations	80	160	2			
1	Simulation rationalization	5.7	35.5		10% of sim h	25% of sim h	
1	Final PFD	5	25				
1	Final material balance	4	80				
				18	**P&IDs estimate assuming 1 sim unit OP per P&ID**		
	P&IDs				h low	h high	
2	Initial draft	72	144		4	8	h per frame
2	CADD checking	18	18		1	1	h per frame
2	Issued for client	450	720		25	40	h per frame
3, design	Vendor markups	216	216		12	12	h per frame
2,3, design	P&ID markups based on line list	18	72		1	4	h per frame
	P&ID overhead	154.8	234	1	20%	20%	

(Continued)

Table 10.1 (Continued)

		Process hours estimate					
		Effort hours					
FEL phase	**Activity description**	**Low**	**High**	**Qty of units**	**h per Qty low**	**h per Qty high**	
1	Client review meeting	72	144		4	8	h per frame
2	Client review 2	72	72		4	4	h per frame
3	client review 3	0	0				h per frame
1	PHA – Pre-simulation, hazard identification, and inherent safety analysis	24	240				h per PFD
				Assume	4	Findings per P&ID	
2,3	PHA detailed (HAZOP) review meeting	36	72		2	4	per P&ID frame
2,3	PHA report writing	36	144		2	8	per P&ID frame
2,3	PHA resolving findings	144	288		2	4	per finding
1,2	materials of construction input/review	4	40				
1,2,3	Equipment calculations & data sheets			Qty	h low	h high	
	Pumps	75	125	5	15	25	
	Heat exchangers	125	200	5	25	40	
	Separators	60	150	3	20	50	
	Distillation towers and internals	50	80	2	25	40	
	Relief valve calcls	75	240	3	25	80	
	Reactors	8	80	1	8	80	
	Storage tanks	16	80	2	8	40	
	Air coolers	0	0	0	20	30	
	Contactors	0	0	0	20	50	
	Compressors	0	0	0	30	60	
2,3,DD	Vendor equipment (boilers, evaporators, fired heaters, absorbers, and gas dryers)	0	0	0	20	50	

Table 10.1 (Continued)

Process hours estimate							
				Effort hours			
FEL phase	**Activity description**	**Low**	**High**	**Qty of units**	**h per Qty low**	**h per Qty high**	
2,3,DD	3D plant model review	10	50	h per system			
				Lines per P&ID			
1,2,3	Line sizing	5	40	10	0.5	4	h per line
2,3,DD	Line list	5	20	10	0.5	2	h per line
				Instrument loops per P&ID	h high	h low	
3	Instrument data sheets	360	720	5	4	8	h per loop
2,3	Control narrative	36	72		2	4	
2,3	Cause and effect [range, alarms, trips, shutdowns (RATS list)]		36	72		2	4
2,3	Process description	20	40				
1,2,3	Capital estimate/economics support	8	8				
	Subtotal	2377	4955				
	Overhead	356	743	15%			
	Total	2734	5699				
	Man months	16.4	34.2				
	Total/P&ID	152	317				

Table 10.1 illustrates an example of of estimating the hours from an initial PFD that is expected to contain 2 distillation units, 5 heat exchangers, 5 pumps, 9 separators, 1 reactor, 2 storage tanks, and 2 recycles in the process.

Reactions – Step 2: Identify what chemical reactions you will, and will not, model (remember this for the simulation rationalization step). Heterogeneous catalysts might be reported in the literature in ways that may not be consistent with the equations used by the simulator (free volume vs reactor volume vs mass basis). Determining reaction rates, reaction equilibrium coefficients, inhibiting or poisoning conditions, and checking the simulator output can take substantial amounts of time in our experience (a week to a month duration). If it is more than a couple of weeks in time, then you may need an alternative approach because the data simply is not there for you to proceed. You may need to model the system using a conversion reactor with an assumed conversion and come back to the reactor design later.

A duration of more than 1 to 2 weeks indicates significant missing reaction information, and a separate R&D subproject is likely required.

Consider using conversion or equilibrium reaction schemes to speed up this task while simultaneously developing necessary kinetic-type reaction modeling.

Heterogeneous catalyst modeling requires significant effort hours prior to modeling and checking the reaction kinetics against laboratory data.

Build the simulation – Step 3: The time required to build a simulation and converge it is affected by variables such as the number and type of unit operations in combination with the property package. Whether or how many distillation towers are present (one to three days per tower). How many recycles are in the simulation (one day per recycle), and how many other unit operations are present (one hour per exchanger, pump, compressor, and vap/liq separator). The property package can affect how stable the solutions are in complex operations like towers. Typically, once the simulation is first converged, there are numerous days spent tweaking inputs to check and obtain the desired product flows and compositions. Optimization studies typically take a week or more per study to run the various cases, perform calculations on the output variables, and assess the economics.

Hours expended as part of doing equipment data sheets are considered part of the data sheet hours.

10.9.2 Process Modeling Using a Spreadsheet

In circumstances that either do not lend themselves to using a commercial simulator or that are straightforward, engineers sometimes elect to use a spreadsheet to create a process model. Typically, this is a week or so per major piece of equipment. Tweaking, checking, improvements, and automation will probably consume a month.

Spreadsheet models are often based on an existing process from which you can retrieve operating data. Physical property calculations are complicated to include, so there is a tendency not to use a spreadsheet if extensive physical property data is required in the modeling.

Batch processes are often modeled using a spreadsheet, although there are commercial batch process simulators that provide scheduling.

10.9.3 Process Modeling via Custom Programming Language

These models are used when commercial simulators do not provide adequate modeling features. Specialized property data or special equipment are examples where a custom model might be preferred. Refer to the chapter on modeling.

Sometimes, a custom model can be combined with a commercial simulator to achieve the objectives.

10.9.4 Simulation Rationalization

This deliverable is typically not discussed by other design texts, or it is inferred in the preparation of a PFD. Our experience is that it helps to separate it into a definite step. The purpose is to convert the simulator output to a document(s) that further design work can proceed with, such as equipment sizing and P&IDs. It typically involves a hydraulics analysis to add pumps or compressors due to simplifications or lack of elevation or piping pressure drop information in the simulation. Storage tanks and other outside battery limits equipment should be considered for inclusion. Temporary process holdup should also be considered at this time. It should include at least a brief check of the simulation heat and material balance and document any further simplifications in the simulation that need to be addressed "manually" by the process engineer. Reactions that were ignored should be addressed. This usually takes a small percentage of the time to prepare the simulation (a few days or 10 to 25% of the time required to prepare the simulation), but situations vary. The result of the simulation rationalization is a PFD drawn with all the necessary equipment to continue with the design, and the simulation material balance with notes, or a final material balance prepared by hand.

10.9.5 Final PFD and Material Balance

If deemed necessary, and it usually is, the simulation rationalization and the simulation model may need to be combined and transcribed into a final spreadsheet material balance and PFD. Hours spent transcribing the simulation output can sometimes be reduced with export functions.

10.9.6 P&IDs

P&IDs are the gateway for other disciplines to get started on the facility design. They become authored by numerous groups, but the process engineer is ultimately responsible for signing off on them to ensure safety and adherence to the proper design. They are a living document, so the hours spent on them extend throughout the project, and it can be challenging to estimate how many hours are needed over the project's life. In addition to the core process P&IDs, utility P&IDs are usually needed. Also, developing P&ID lead sheets keep everyone speaking the same language because there is often some latitude in interpreting ISA P&ID symbology standards. We have also found that creating a P&ID map, which is just a block flow diagram of P&IDs, has helped to organize and find drawings in a large project with many P&IDs.

Preparatory work includes a marked-up PFD with controls and usually some rough sketches.

Assuming you have a PFD with the basic control scheme sketched out, the initial drawing of a P&ID is about 4 hours per frame; the quickest method is to hand draw them on 17 × 22" paper, with usually one piece of equipment per drawing. The remaining effort is spent reviewing and updating them with items such as details of the instrumentation (seals, isolation valves, interlocks, alarms, and controller interactions), piping line numbers, pipe codes, manual valves, specification breaks, integrating operation maintenance, and safety review comments, and vendor data as it becomes available. More time will be taken during detailed design in reviewing vendor equipment and making the P&IDs consistent with vessel nozzle schedules, specifying critical locations or dimensions on the drawings and other details.

Materials of construction analysis is best done in combination with a process metallurgist who is familiar with the site and the utilities present. The operating conditions and chemical species, along with the PFD, are passed to them along with a process description that should be adequate for them to understand any unusual situations that might be present. Small component concentrations and less common operating conditions (start-up, shutdown, cleaning, hydrostatic testing with water, and relief conditions), or cycling can cause corrosion, fatigue, and premature failure. The RAND report mentioned above will convince you of the importance of spending adequate time discussing this analysis with operations, maintenance, and the process metallurgist.

10.9.7 Equipment Sizing Calculations

The hours for these can also vary significantly. Vendors can be utilized to help size heat exchangers, but a check of their calculations is prudent. Sometimes more process simulation work will be necessary. Distillation column internals are typically the task of the vendors, but some checking calculations are a wise thing to do. Pumps, knockout vessels, and simple holdup tanks might take about 4–8 hours to complete the data sheet for vendor costing, but there are a significant number of hours spent checking vendor-submitted drawings against the P&IDs and other drawings, so allow for 20–50 hours. Reactors, solid flow equipment, and other equipment can take 20–60 hours, or more. If you have never sized a pump, it can take you 40 hours to do one, but with practice, you can get the calculations and a data sheet done in 3 or 4 hours, with the rest of the time required (15–25) for vendor drawings, verifying the final piping design against your assumptions around piping fittings and checking the NPSH and so forth.

10.9.8 Line List

A process engineer is necessary to create and maintain their portion of the line list. This is a list of all pipes, their operating conditions, their maximum operating conditions, the pipe specs, insulation requirements, and so forth. Budget 5–15 minutes per line, assuming you have all the information at your fingertips. In revamp jobs in which owners do not have up-to-date pipe codes or P&IDs, you can expect to spend time in the plant reading valve model numbers and working with mechanical integrity people.

10.9.9 Relief Valves

Relief valve calculations can be 8 hours for a preliminary sizing in which you have done the system design and the valve vents to the atmosphere, but more typically, about 40 hours is needed to provide thorough documentation. Two hundred hours is not uncommon when a disposal system is present, and relief valve variable back pressures need to be determined. The hours to complete a set of relief valve calculations depend on the availability of information for existing equipment (especially design pressures and temperatures), piping (especially pipe specs), instrumentation documentation, mechanical integrity information, and whether there is a common disposal header system. Usually, a field visit is required to acquire the necessary data.

The process engineer's job is to communicate the design intent to others. As such, a good process description is necessary (8–20 hours), and time allowance for informal and formal reviews with other engineers, environmental, operations, maintenance, management, and so forth is essential (15% of your time).

References

Ainsworth, D. and Brocklebank, M. (2003 Jul). Multiproduct plant design. *Chem. Eng.* 110 (7): 42–49.

Ben Mahmoud-Jouini, S. (2016 Apr/May). Contributions of design thinking to project management in an innovation context. *Proj. Manag. J.* 47 (2): 144–155. https://doi.org/10.1002/pmj.

Berg, D. (2011 Jul). Reduce piloting time and cost. *Chem. Eng. Prog.* 107: 34–38.

Eldon, L. (2005 Jan). Apply project management concepts to R&D. *Chem. Eng. Prog.* 101 (1): 47–50.

Glavic, P. (2021). Process design and sustainable development – a European perspective. *Processes* 9: 148.

Karol, R. (2002A Jan/Feb). Better new business development at DuPont – I. *Res. Technol. Manag.* 45: 24–30.

Karol, R. (2002B Mar). Better new business development at DuPont – II. *Res. Technol. Manag.* 45: 47–56.

Larsen E., People: the key to successful project management, *Chem. Eng. Prog.* 2004 100 Sep, 55.

Madden J. and Stewart R. 1996 Retrieved from the web at https://www.projectsmart.co.uk/recommended-reads/one-hundred-rules-for-nasa-project-managers.pdf

Merrow E., A review of cost estimation in new technology implications for energy process plants, Rand, 1979, R2481-DOE.

Merrow E., Understanding cost growth and performance shortfalls in pioneer process plants, Rand, 1981, R2569-DOE.

Merrow E., Cost growth in new process facilities, Rand, 1983, P6869-DOE.

Merrow E., Understanding process plant schedule slippage and startup costs, Rand, 1986, R3215-DOE.

Pavone, A. (2006 June). How to prepare a process design basis. *Chem. Eng.* 113: 48–53.

Peachey, B. (2007). Project management for chemical engineers. *Educ. Chem. Eng.* 2 (I. 1): 14–19.

Portny, S. (2022). *Project Management for Dummies*, 480.

Read, C. (2015 May). Lay the foundation for a successful project. *Chem. Eng. Prog.* 111 (5): 45–48.

Rosentrator G., Hit the ground running on your next project, *Chem. Eng. Prog.* 2013 109 May, 45–49.

Schatz N., Evaluate project success, *Chem. Eng. Prog.* 2018 114 Jun, 43–48.

11

Storage and Bulk Transport

In walking through a plant or in considering the sequence of a process, the first items encountered are the storage facilities for incoming raw materials. While passive, these elements require the same careful consideration as any other piece of equipment. They come in various forms and dimensions, depending on the phase (or phases) of the materials, the required on-site inventory, the pressure and pressure range to be handled, and the hazard of the material. These same considerations apply to intermediate storage facilities through the process, as well as to storage of finished product.

The bulk transportation of raw materials to the plant also depends on the phase, rate, pressure, and hazard rating. The chapter reviews storage and transport in turn.

11.1 Choose the Phase of the Material to be Stored

The phase (liquid, gas, and solid) of the material to be stored is usually dictated by the form it takes at ambient pressures and temperatures. However, in some cases, it may make economic sense to convert the material to another state for storage. For instance, in the case where large quantities of gases are to be stored (greater than about 1500 std cu M), the liquefaction of the gas may become economical. However, liquefaction of gases (in cryogenic storage) often involves allowing a portion of the liquid to boil off to maintain temperature and pressure in the tank. The losses, or preventing them, from this type of storage system can make the system environmentally or economically less desirable.

Other phases such as absorption onto solids, dissolving into a liquid, or conversion to solids via chemical reaction may be considered.

11.2 Choose the Volume of Storage Required

Large use plants may utilize pipelines with minimal or no storage to supply the process. Railcars may be temporarily used to store chemicals, thus reducing or eliminating on-site storage. Hazardous chemicals may be enough of a safety liability that minimal storage is preferred or possibly on-site or *in situ* production of the material should be used. Extremely hazardous chemicals may require a secondary "de-inventory" storage tank in the event a problem develops in the primary storage system. However, for average materials, the following guidelines may be used to determine storage requirements.

Raw material storage is provided to ensure the plant never (or rarely) shuts down because the raw materials are unavailable. Thus, the reliability of the supply system must be examined.

Practical Process Design for Chemical Engineers, First Edition. Keith Marchildon and David Mody.

Baasel (1976) presented the following guidelines:

$$\textit{Amount of feedstock} \text{ that should be kept on-site} = (\text{date delivered} - \text{date ordered}) \times \text{feed rate}$$

$$\textit{Storage size} = \textit{Max} \text{ amount that COULD be present when delivery arrives} + \text{amount ordered}$$

Example:
A plant requires 20 000 lb of feedstock per day. The supplier will guarantee shipping the order in 15 days of order receipt. The time to ship the material is anywhere between 2 and 5 days. The mode of transport is 36 000 gal jumbo railcars. The specific gravity of the material is 0.85.

Solution:
Examine the two possibilities (no delays and maximum delay)

Maximum delay:

- Our plant takes 3 days to process our order (over a long weekend).
- The supplier ships 15 days after receipt of the order.
- The shipping takes 5 days to travel to site.

Thus, the amount of feedstock that should be on-site when the order is placed should be

$$(3 \text{ days} + 15 \text{ days} + 5 \text{ days}) \times 20\,000 \text{ lb d}^{-1} = 460\,000 \text{ lb}$$

No delays:

- A jumbo railcar (36 000 gal) could arrive in 2 days.

If the railcar arrives in 2 days, when we had 460 000 lb of material on-site at the time of order, then the amount of storage we need on-site is

$$\begin{aligned} &460\,000 \text{ lb} - 2 \text{ days' transit} \times 20\,000 \text{ lb d}^{-1} + 36\,000 \text{ gal } (0.85 \times 62.4 \text{ lb ft}^{-3}/7.48 \text{ gal ft}^{-3}) \\ &\quad = 676\,000 \text{ lb of storage (34 days)} \end{aligned}$$

The logistics of when a second railcar must be ordered is left to the reader.

Reliable and efficient supply chains and storage are topics of the following references:

Cabezas et al. (2015), Fiksel (2010), Fuller (2009), Goti et al. (2011), Thomas & Tong (2009), and Vazquez-Esparragoza and Chen (2016).

11.3 Choose a Design Pressure

In the situation where a liquid is to be stored, there are two ways to approach the choice of design pressure:

- by determining the maximum expected pressure in the tank due to fluid thermal cycling and by designing above that, which prevents vapor losses but may require a custom design pressure tank, or
- by arbitrarily choosing a design pressure for a standard tank and then dealing with the losses in some other way (i.e. a vapor recovery system).

The normal day-to-night thermal cycling of a tank causes the pressure in the tank to rise and fall. If the tank has an opening to atmosphere, the increased pressure is dissipated by the gases (rich with the fluid being stored) in the tank escaping. As the tank cools, the pressure drops and air is drawn in to maintain the atmospheric pressure. The disadvantages of this practice are the emissions to atmosphere as well as the introduction of possibly undesired oxygen, humidity, or other gasses into the tank.

In such a freely vented tank, the factors that affect how much gas is expelled through the vent are as follows:

(1) the change in the temperature in the vapor space,
(2) the change in the vapor pressure of the liquid (thus the change in the stored fluid temperature, which is likely different than the vapor space temperature), and
(3) the change in the liquid level due to the change in the stored fluid temperature/density.

The above disadvantages can be mitigated by providing a *conservation vent*, consisting of a path that opens to allow gas to leave at a higher-than-atmospheric pressure, and another path that opens to allow air to enter at subatmospheric pressure. See Figure 11.1 for this device.

The design's low-pressure threshold which determines the disk weight on the conservation vent's vacuum side should be slightly below atmospheric (i.e. ½ oz/sq in). This is because standard tanks have limited resistance to crushing pressure. From this low-pressure limit, we calculate the highest anticipated pressure based on the factors mentioned. If the conservation vent's high-pressure limit exceeds this value (and assuming the tank can handle this pressure),

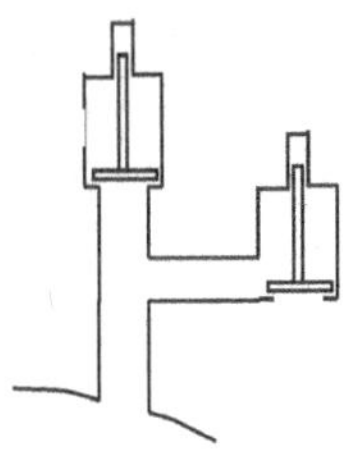

(a)

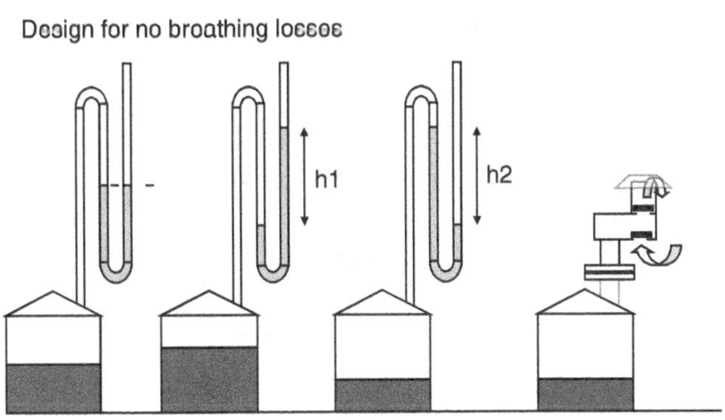

(b)

Figure 11.1 (a) Conservation vent, (b) Reduction of breathing losses via tank design pressure.

Table 11.1 Typical manufacturer's list of API tank sizes.

A Selection of Typical API Field Constructed Tanks									
Tank diameter		**Approx. capacity**		**Tank height**			**Total volume**		
Ft	**m**	**gal ft^{-1}**	**m^3/m**	**ft**	**m**	***L/D***	**US Gal**	**US Barrels**	**cu M**
15	4.6	1320	16.4	18	5.5	1.2	23 800	567	90
20	6.1	2350	29.2	18	5.5	0.9	42 300	1007	160
25	7.6	3670	45.6	18	5.5	0.7	66 100	1574	250
25	7.6	3670	45.6	24	7.3	1.0	88 100	2098	333
30	9.1	5290	65.7	24	7.3	0.8	127 000	3024	481
35	10.7	7190	89.3	30	9.1	0.9	216 000	5143	818
45	13.7	11 900	147.8	36	11.0	0.8	428 000	10 190	1620
70	21.3	28 800	357.6	54	16.5	0.8	1 550 000	36 905	5867
100	30.5	58 700	728.8	36	11.0	0.4	2 110 000	50 238	7987
120	36.6	84 500	1049.2	30	9.1	0.3	2 540 000	60 476	9615
200	61.0	190 000	2359.1	18	5.5	0.1	4 230 000	100 714	16 012

there will be no vapor losses (Figure 11.1b). Regardless, conservation vents help minimize these losses.

If the material is valuable or if it is hazardous, then the outflow should be directed to a holding tank, possibly with a condenser so that the material can be returned to the tank. The EPA states tanks can economically justify vapor recovery systems to avoid breathing losses (Anonymous 2006). If oxygen is undesired in the tank, then the inflow should be connected to an "inerting" gas, probably nitrogen.

11.4 Selecting a Tank Type

The choice of tank depends partly on the required design pressure and partly on the amount of material to be stored.

The reader is cautioned that there is sometimes confusion in stating whether the stated volume of a container is determined by its dimensions or by the volume of fluid (i.e. gas at high pressure) that can be stored in the container.

For instance, a container 2 ft (0.61 m) in diameter and 20 ft (6.1 m) long has a dimensional volume of about 1.8 cu M, but can store 300 std. cu M of gas when filled to a pressure of 2450 psig. In the Table 11.1, the dimensional volume is used to state the capacity of the containers. The table is published by the American Petroleum Institute.

11.5 Storage of Gases

Gases, due to their low density, tend to be stored under pressure to minimize the cost of the container and plant real estate. The type and dimensions of a tank for a gas depend on the compressed volume and the pressure.

11.5.1 Small Quantities

Range 0–1000 m^3.

For low pressure, thin-walled tanks or drums can be used.

At higher pressure, the familiar upright thick-walled bottles are used, as are horizontal "bullet tanks" which are cylindrical with semispherical ends. The choice of container size depends on the consumption rate.

All pressure vessels must be built by shops that follow the American Society of Mechanical Engineers (ASME) VIII code and are certified as so doing. Design pressures are allowed up to 3000 psig and, for vessels, above 10 000 psig.

11.5.2 Midsize Quantities

This range is typically with a maximum of 35 000 m^3. Tanks in the shape of spheres or spheroids are commonly used. Spheroids are typically used for pressures of 30 psig or less. Spheres take pressures up to 200 psig. Sphere diameters typically range from 32 to 120 ft. Once again, the vessel must be built to satisfy the ASME VIII code.

11.5.3 Mid to Large Quantities

Even though the gas has been compressed, the required storage volume becomes difficult to handle. At this point, consideration should be given to storage as a liquid.

11.5.4 Very Large Quantities

It may be possible to store gases in underground caverns, where natural geography allows. Guidance is provided in the Engineering Data Book of the Gas Processors Suppliers Association.

Another possible approach is to convert the gas into another, denser, substance.

11.5.5 High Consumption Rates; Hazardous Gases

Instead of the usual bulk transport – and – storage model, consideration should be given to

(1) bringing in the gas by pipeline, or
(2) producing the gas on-site. This choice may be the most economical. For hazardous gases, this choice minimizes the possibility of human contact and emissions.

11.6 Storage of Liquids

Here, we look at the storage of liquids external to processes. The term of storage is usually more than one day, and the tanks may be part of a tank farm, with facilities for inter-tank transfer and for blending.

The combination of tank operating pressures (as dictated by the vapor pressure of the fluid) and required storage volume drives the selection process (Figure 11.2).

A major distinction is the vapor pressure of the liquid, with different vessels being required for pressures:

- less than 1.5 psia,
- less than 11.1 psia

all classified as low-pressure storage. Vapor pressures near or exceeding atmospheric pressure require storage in pressure vessels.

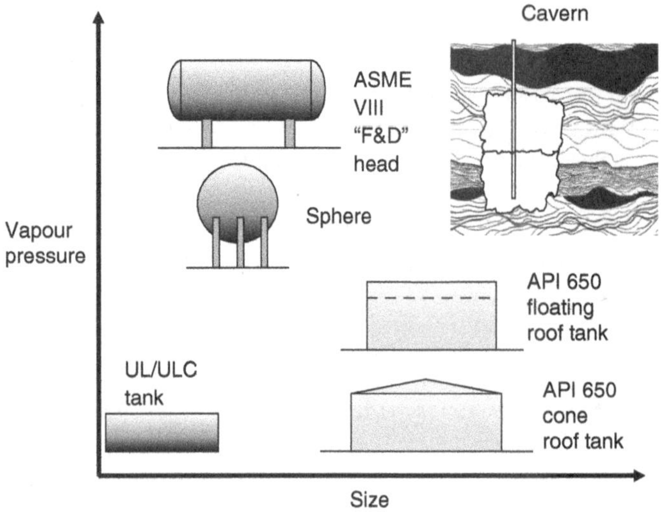

Figure 11.2 Storage tank variation.

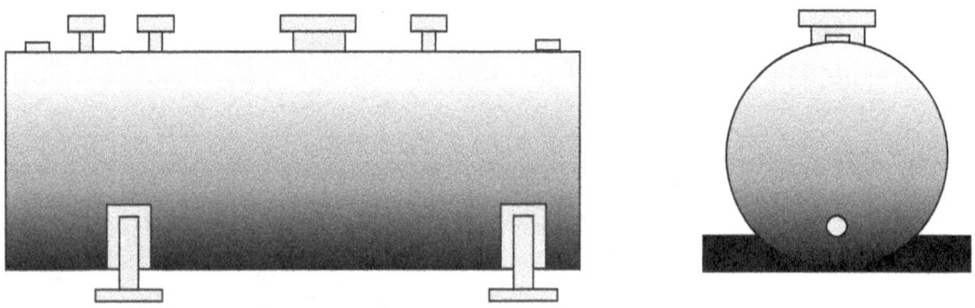

Figure 11.3 UL tank.

11.6.1 Low-Pressure Storage

Many industrial groups have published standards for liquid storage tanks. These publications are intended to produce vessels that meet government regulations. Two sets of standards which are in frequent use are those of Underwriters Laboratory and of the American Petroleum Institute.

The *UL-142 standard* is used in constructing small low-pressure tanks for flammable and combustible liquids (Figure 11.3). Domestic examples are fuel storage at service stations and oil storage tanks for home heating. They are constructed to withstand a pressure of up to 5 psig, but this is just for leak testing. The UL-2085 standard is for a double wall, thus eliminating the need for dikes. These Underwriters Laboratory standards satisfy the legally required regulation 30 of the NFPA (National Fire Protection Association).

For the construction of larger low-pressure tanks, the *American Petroleum Institute* provides designs and standards (i.e. API 650 and 620). The simplest is the vertical vessel with a conical roof (Figure 11.4), used for liquids of low vapor pressure, generally less than 1.5 psia. The head space is a mixture of air, or inert gas, and vapor of the substance. These tanks are typically up to 300 ft in diameter, but no actual limit is stated. This class of storage can also be used for liquids with higher vapor pressure, up to about 11.1 psia. The cutoff for this class of vessel is a total pressure of 14.9 psig, above which the pressure-vessel regulations apply.

Figure 11.4 API tank.

Figure 11.5 Floating roof storage tank.

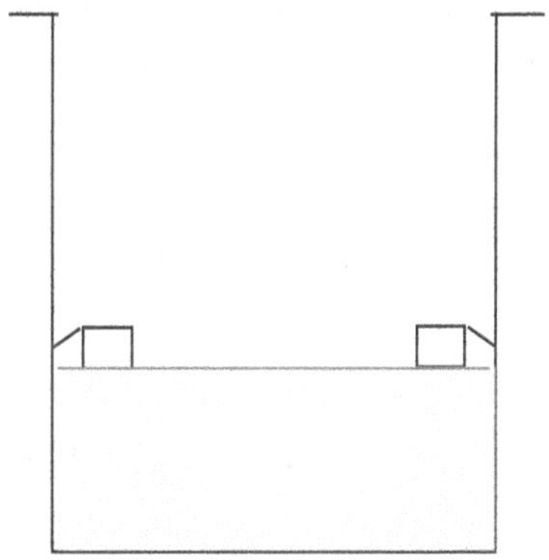

If the liquid has a vapor pressure in excess of 0.75 psia, then the Environmental Protection Agency (EPA) requires a vapor recovery system if the storage quantity exceeds 40 000 US gallons. Other conditions apply to smaller vessels. See EPA's "NSPS Subpart Kb" for details on vapor recovery system requirements for storage tanks (EPA 2023).

Also contained in the API 620, 650 standards is the *floating roof* storage tank (Figure 11.5), which does not require a vapor recovery system. The tank has no headspace as the roof of the tank floats on top of the liquid and rises and falls as the liquid level changes; thus, vapor is not vented when liquid levels rise. The lack of headspace ensures there are no "breathing" losses from the tank. A seal (of which there are various types) ensures negligible evaporation of the liquid even with fluids that have low atmospheric boiling points. This type of tank design is required by the EPA in the United States for fluids with vapor pressures greater than 1.5 psia and less than 11.1 psia.

A variation is the internal floating roof tank (Figure 11.6), a conical roof atop the floating roof vessel, to protect from weather. Consideration to vent this secondary headspace may be required to eliminate flammability issues.

In Table 11.1 are listed an API list of some typical tank sizes.

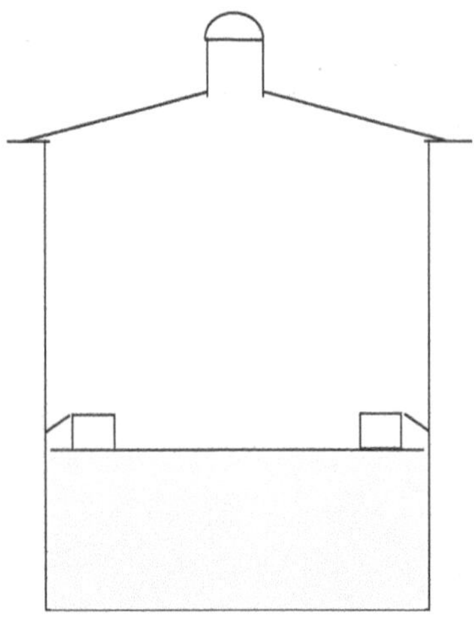

Figure 11.6 Internal floating roof tank (hybrid cone/floating roof tank).

The tanks so far described are all welded, and the welds have to be checked and certified. A different approach is to build the tank in the field by bolting together prefabricated sections. Suitable for fluids with low vapor pressures, these tanks can be erected by hand and easily dismantled and transported elsewhere later.

Various companies publish tables of API tank sizes and capacities that companies use as standard sizes, but generally the tanks are custom made for the application.

11.6.2 High-Pressure Storage

For operating pressures from 15 psig upward to about 250 psig, spheres and spheroids may be used. Designs are generally to ASME Section VIII Div 1 code (Figure 11.7).

Drums or "bullet tanks" which are cylindrical with flanged and dished (torispherical) heads are suitable for high-pressure applications. Standards for pressures up to 3000 psig are covered by ASME Section VIII Div 1. Designs for pressures greater than 10 000 psig are covered by ASME VIII Div 3.

Underground storage is particularly useful for high-vapor-pressure fluids and large volumes of liquid (equivalent of 100s of storage tanks) or gas. Pressures of 3000 psig or higher for compressed natural gas are not unheard of. The storage requires particular geological formations such as salt caverns or previous oil and gas production locations. No formal standards exist; thus, this storage option will require extensive geological design and evaluation. Thus, the solution is not available to every plant. An overview is "Deep-Well Storage in Salt Caverns - Lambton County" (Tom Hamilton 2006), a Profile of Underground Natural Gas Storage Facilities and Market Hubs (Anonymous, Foster Assoc Inc. 1995). See also the Compressed Natural Gas storage and consult the Gas Processors Suppliers Association handbook for further information.

11.6.3 Storage Suggestions

It may be possible to reduce vapor pressure and facility cost by cooling the liquid.

The above descriptions have been for storage periods of greater than a day. Within an operation, smaller tanks closer to the process are used as day storage. A small version of some of the above tank styles may be appropriate, likely "bullet tanks", in horizontal or vertical orientation.

For after-process waste disposal of liquids, a lined outdoor pond may be suitable, where liquid can evaporate. This may be an intermediate step on the way to a treatment facility.

Figure 11.7 Spherical high-pressure storage tank.

11.6.4 Tank Management

Storage is such an important aspect of processes of all types that process operators have written extensively about it. Ambrouche et al. (2002) presented some general rules. Gorgi & Jari (2006), Jenkins (2009), Kenkre (2017), Montgomery IV (2002), Mukherjee (2006), and Pullarcot (2007) provided education on the subject of tankage.

Although storage is normally the passive part of an industrial operation, there is still maintenance to be done and safety to be respected. Cerulli & Franks (2002) and Pagcatipunan (2003) presented ideas on tank cleaning. Ritchie (2009) spoke about the prevention of storage tank fires. Gollin (2010) and Kinsley Jr (2001) discussed purging and inerting. Mikkola & Lieb (2007) addressed corrosion checking.

An ongoing part of tank management is the measurement and analysis of tank contents. An anonymous article from 2011 discussed tank farm logistics. Agrawal (2007) discussed efficient sampling. Blundell (2018) talked of overfill protection. Doane (2007), Hambrice & Hotard (2004), Mallon (2019), Sivaraman et al. (2010), Skaug (2008), and Syrnyk & Seiler (2007) discussed the measurement of liquid amount in vessels.

One of the concerns of liquid storage, referenced in the above descriptions, is the handling of vapor generated by vapor pressure. This aspect is treated by Bahadori (2009), Drennen (2019), Foiles (2004), Norouzi & Mofrad (2008), and Peress (2001). The effect on stored liquid is discussed by Anjan & Khan (2014) and Peress (2003).

11.7 Solid Storage

Solid storage is typically done either by piling on the ground (possibly inside a building, i.e. in the case of hydroscopic materials), or in bins or silos.

Baasel (1976) suggested the following:

1. It is cheapest to build bins with a cylindrical cross section.
2. Provide one large bin wherever possible rather than multiple small bins to save on supports, materials, fabrication costs, and conveyors.
3. Bins larger in diameter than 11 ft 6 in are difficult to transport by road and should thus be avoided if possible. A practical length is about 30 ft.
4. Coarse, uniform-particle size materials flow easily (i.e. plastic pellets). Fine, relatively uniform materials are almost fluid (i.e. kitchen starch). The greater the distribution of particle sizes in a mixture, the greater the tendency to compact and to resist flow.
5. To ensure materials freely flow out the bottom of a bin (to avoid bridging), make the bottom an eccentric cone with one straight vertical side.

Experience has shown that the cone angle should always be the greater of the “angle of slide” or the “angle of repose”.

Angle of slide – The angle of slide is measured by a simple test, whereby the material is placed on a flat plate made from the materials and same finish as the bin is to be constructed from. The flat plate is tipped up, and the angle of the plate at which the material begins to slide is noted as the angle of slide.

Angle of repose – The angle of the pile when the material is poured onto a flat surface.

For materials that may be hydroscopic, sticky, or fuses together (i.e. ice), seek advice from experts such as Jennike and Johanson (http://www.jenike.com) or Jerry Johanson (http://www.jrjohanson.com).

More detailed information may be obtained from the following references:

Amrouche et al. (2002), Baasel (1976), Gas Processors Suppliers Association 13th edition (2020), and Steve (2000).

11.8 Bulk Shipping

The ideal transportation method for materials and chemicals is dependent upon:

- the volume of material to be used on a weekly or monthly basis,
- the pressure required (for gases),
- the state of the material to be used (i.e. if liquid nitrogen is required for freezing, then vapor delivery is of little use), and
- proximity to existing pipelines, proximity to rail, water, or roads, and suppliers of the material.

The common bulk shipping methods for gases, liquids, and solids are discussed further below.

11.8.1 Cylinder

- Usually transported by trucks, cylinders provide a convenient method of moving small volumes of gases (up to 10 m^3 per cylinder).
- Where slightly larger volumes of gases are required, liquefied gas transported in dewars (insulated vessels) is utilized (nitrogen liquid has 4× the density of nitrogen gas at 2450 psig, hydrogen a factor of 5×).

11.8.2 Ship "Container"

- Although not a "mode" of shipment, containers may be shipped by road, rail, or water.
- There are a series of standardized sizes for "containers", but all containers are 8 ft wide. The most widely used containers are the general purpose dry van (DV) containers having a nominal length and height of 20′ × 8.5′, 40′ × 8.5′, and 40′ × 9.5′.
- The capacity of a 20′ dry cargo container is 24 000 kg (52 900 lbs.), and that of a 40′ dry cargo container is 30 480 kg (67 200 lbs.). The containers themselves weigh 2400 kg and 3900 kg, respectively.
- Containers are available for carrying bulk gases, liquids, bulk solids, and refrigerated products.

11.8.3 Truck

- Generally, a transit distance within 1000 km using road freight is competitive compared to rail and air freight.
- Maximum weight allowable on Canadian roads is a complex calculation based upon tire widths, axel distances, the number of tires, and the time of year. However, the weight is generally in the 18 000–34 000 kg range.
- Bulk gases delivered by a tank truck (usually hydrogen or helium) are utilized when consumption rates are 25 000–150 000 std ft^3 per month.
- Liquefied gases may be transported where higher volumes of gas must be handled (usage rates 30 000 to several million std ft^3 per month).

11.8.4 Rail

- Railcars are typically 40–89 ft long, and each car is limited to a weight of 120 metric tons (typical range 60–120 metric tons). When handling containers, a typical 50-car train can haul 3 million kg.
- Hopper cars have typical volume capacities of 4750–5150 cu ft.
- General information about rail transportation can be found at the CN website at http://www.cn.ca/en_index.shtml.
- The guidelines for transportation of dangerous goods can be found at the transport Canada website at http://www.tc.gc.ca/tdg/menu.htm.
- Railcars can be insulated (for liquefied gases or liquids that must kept warm), and they will have pressure ratings for pressurized gases.

11.8.5 Ship

- Suitable where easy access to water is available.
- Suitable for large volumes and especially heavy cargo.

- Economical for large distances.
- Ships commonly utilize containers (approximately 180 million twenty foot equivalent unit (TEUs) are handled by the world's ports every year, pre-2021).
  - A TEU is a 20-foot equivalent unit based on the size of a standard 20-foot long shipping container.
  - Containers are available for carrying bulk gases, liquids, and bulk solids.
- Ships are generally limited to 900 ft in length and 105 ft in width (to fit the Panama Canal) and can carry 2000–4500 TEUs. Non-Panama Canal ships can handle over 24 000 TEUs.
- Liquid-carrying ships are measured in their cargo-carrying capacity expressed in deadweight tons (DWTs). Capacity of ships typically vary from 10 000 to 320 000 DWT. The TI Oceania has a capacity of 441 893 DWT.

11.8.6 Pipeline

- Commonly used method of delivering fluids and gases (i.e. tap water or natural gas to houses and industry).
- Provides the lowest cost per lb transportation charge for large capacities.
- Where an existing pipeline infrastructure is nearby, economic and inherent safety (minimal site inventory) advantages exist.
- Pipelines exist for water, natural gas, oil, oxygen, nitrogen, and hydrogen (the later three in the Gulf Coast area).
- For instance, natural gas pipelines transport on average 442 million cu M d^{-1} (2019, 15.7 billion cu ft) in Canada.
- Steam distribution from central heating centers is less common today due to the use of natural gas instead, but can be economical in certain situations.
- Liquid pipelines are designed with velocities up to 10 ft s^{-1} and maximum pressures to 1000 psig. Gas pipelines have higher velocities.

11.8.7 Conveyor Belt

- Conveyor belts are typically used in mining applications where large masses of material must be transported over reasonable distances.
- Conveying distances of 8–98 km have been commercially proven.
- Example: 750 metric tons h^{-1} over 6 km distances, energy use = 0.4 kW per metric ton h^{-1} (0.68 BTU per lb), at an electrical cost (0.07 \$ kW^{-1} h) (USD) of about 3.1E-5 \$ kg^{-1} of material.
- See http://www.conveyor-dynamics.com/

11.8.8 Air

- Air freighters like the Boeing 747-8F can carry loads weighing up to 140 metric tons or about 30 000 cu ft of cargo.
- Generally, air freight is perceived as being expensive as compared to other forms of transportation.

11.8.9 On-site Generation

Although not really a mode of transport, this is commonly grouped with transportation methods for comparison purposes.

Table 11.2 Comparison of transportation methods by capacity.

	Gases		Liquids/solids	
	Unit capacity	**Yearly**	**Unit capacity**	**Yearly**
Truck	1500 cu M per truck[a)]		30 000 kg per truck	110 million kg yr^{-1} [b)]
Containers			30 000 kg per container	
Rail			120 000 kg per car	624 million kg yr^{-1} [c)]
Ship	Usually liquefied		98 000 000 kg per ship	5100 million kg yr^{-1} [d)]
Pipeline	3.50E + 07 kg d^{-1} [e)]	12 775 million kg yr^{-1}	2 600 000 kg h^{-1} [f)]	22 800 million kg yr^{-1}
Conveyor	n/a		750 000 kg h^{-1}	6600 million kg yr^{-1}
On-site generation				

Notes
a) At 2450 psig.
b) Assume 10 trucks per day.
c) Assume 2 trains, composed of 50 cars per week.
d) Assume 1 ship per week.
e) Not necessarily typical of all pipelines.
f) liq, 4 ft s^{-1}, SG = 0.9, 36 in dia.

- On-site generation of standard gases (nitrogen, oxygen) can provide for significantly larger consumption rates: oxygen plants of 100–135 000 std ft^3 h^{-1} (72 000–100 million std ft^3 per month), and typical nitrogen plant sizes are 5000–160 000 std ft^3 h^{-1} (Table 11.2)

References

Agrawal, S.S. (2007 Jun). Advances in tank quality measurements can help cut operational costs. *Hyd. Proc.* 86 (6): 67–70.

Amrouche, Y., Davè, C., Gursahani, K. et al. (2002 Dec). General rules for aboveground storage tank design and operation. *Chem. Eng. Prog.* 98 (12): 54–58.

Anjan, S. and Khan, K. (2014 Feb). Update on ambient heat transfer for storage tanks. *Hyd. Proc.* 93 (2): H77–H81.

Anonymous (2011 Feb). Automation module improves tank farm logistics. *Hyd. Proc.* 90 (2): 2023.

Anonymous, Foster Assoc Inc., A Profile of Underground Natural Gas Storage Facilities and Market Hubs, 1995, retrieved 2023.

Anonymous, Installing Vapor Recovery Units on Storage Tanks, 2006, EPA website retrieved 2023.

Baasel, W.D. (1976). *Preliminary Chemical Engineering Plant Design*. Elsevier Nort Holland.

Bahadori, A. (2009 Jun). Minimize vaporization and displacement losses from storage containers. *Hyd. Proc* 88 (6): 83–84.

Blundell, B. (2018 Jul). Better practices for tank overfill prevention. *Hyd. Proc.* 97 (7): 47–48.

Cabezas, H., Heckl, I., Bertok, B., and Friedler, F. (2015 Jan). Use the P-graph framework to design supply chains for sustainability. *Chem. Eng. Prog.* 111 (1): 41–47.

Cerulli, G.F. and Franks, J.W. (2002 Feb). Making the case for clean in place. *Chem. Eng.* 109 (2): 78–82.

Dhodapkar, S. and Jacob, K. (2002 Mar). Smart ways to troubleshoot pneumatic conveyors. *Chem. Eng.* 109 (3): 95–98.

Doane, R.C. (2007 Dec). Accurate wetted areas for partially filled vessels. *Chem. Eng.* 114 (13): 56–57.

Drennen, T.W. (2019 Dec). Protect tanks from overpressure and vacuum. *Chem. Eng. Prog.* 115 (12): 24–30.

Environmental Protection Agency, 2023, https://www.federalregister.gov/d/2021-00678/p-110 Retrieved 19 Jan 2021.

Fiksel, J. (2010 May). Evaluating supply chain sustainability. *Chem. Eng. Prog.* 106 (5): 28–36.

Foiles, R. (2004 Oct). To vent or not to vent. *Chem. Eng.* 111 (11): 58–61.

Fuller, B.A. (2009 Oct). Managing transportation safety and security risks. *Chem. Eng. Prog.* 105 (10): 25–29.

Gas Processors Suppliers Association (2020). *Engineering Data Book*, 13the. Apr 30.

Gollin, M. (2010 May). Purging and inerting large-volume tankage and equipment – jet mixing concept. *Hyd. Proc.* 89 (5): 35–41.

Goti, A., Zabaleta, N., Garcia, A. et al. (2011 May). Optimize reordering of critical raw materials and parts. *Hyd. Proc.* 90 (5): 79–81.

Hambrice, K. and Hotard, R. (2004 Jun). Policing liquid levels. *Chem. Eng.* 111 (6): 32–36.

Hamilton, T., Deep-Well Storage in Salt Caverns - Lambton County, 2006, retrieved from https://www.bgc.bg/upload_files/file/salt1.pdf, 2023.

Gorgi, A.H. and Jari, H.K. (2006 Oct). Improve selection and sizing of storage tanks. *Hyd. Proc.* 85 (10): 95–101.

Jenkins, S. (2009 Nov). Aboveground and underground storage tanks. *Chem. Eng.* 116 (12): 32.

Kenkre, P. (2017 Mar). Designing atmospheric storage tanks. *Chem. Eng.* 124 (3): 77–82.

Kinsley, G.R. Jr., (2001 Feb). Properly purge and inert storage vessels. *Chem. Eng. Prog.* 97 (2): 57–61.

Mallon, D. (2019 Sep). Automatic gauging addresses tank farm challenges. *Hyd. Proc.* 98 (9): 55–58.

Mikkola, C. and Lieb, J. (2007 Jul). Similar service assessment of aboveground storage tanks. *Hyd. Proc.* 86 (7): 115–117.

Montgomery, G.J. IV, (2002 Jul). Test your tank smarts. *Chem. Eng. Prog.* 98 (7): 96.

Mukherjee S., Understanding atmospheric storage tanks, *Chem. Eng.* 2006 Apr, 113*(4)*, 74–81, 84.

Nomady, T. (2018 Aug). Selection of silos for bulk storage. *Chem. Eng.* 125 (8): 60–62.

Norouzi, S. and Mofrad, S.R. (2008 Nov). What you should know about liquid thermal expansion. *Hyd. Proc.* 87 (11): 67–69.

Pagcatipunan, C. (2003 Aug). Options for automated tank cleaning. *Chem. Eng.* 110 (8): 27–32.

Peress, J. (2001 Aug). Estimate storage tank emissions. *Chem. Eng. Prog.* 97 (8): 44–45.

Peress, J. (2003 Apr). Estimate emissions from batch heating. *Chem. Eng. Prog.* 99 (4): 28–31.

Pullarcot, S.K. (2007 Feb). Optimum specification and quality requirements for pressure vessels. *Hyd. Proc.* 86 (2): 123–127.

Renfro, J., Stephenson, G., Marquez-Riquelme, E., and Vandu, C. (2014 May). Use dynamic models when designing high-pressure vessels. *Hyd. Proc.* 93 (5): 71–76.

Ritchie, R. (2009 Nov). Prevent storage tank fires. *Hyd. Proc.* 88 (11): 35–39.

Sandstrom, C.E. (2003 Apr). The economics of storing process solutions. *Chem. Eng.* 110 (4): 36. 39–40, 42, 44, 47.

Sivaraman, S., Bertotto, A., and Comstock, D. (2010 Aug). Estimating tank calibration uncertainty. *Hyd. Proc.* 89 (8): 55–61.

Skaug, C. (2008 Jun). Increased plant safety puts focus on tank gauging. *Hyd. Proc.* 87 (6): 135–138.

Steve, E. (2000 Jul). Sizing up the storage bin. *Chem. Eng.* 107 (7): 84–88.

Syrnyk, P. and Seiler, D. (2007 Aug). Use liquid ultrasonic meters for custody transfer. *Hyd. Proc.* 86 (8): 91–96.

Thomas, C. and Tong, D. (2009 Oct). Agile supply chain planning. *Hyd. Proc.* 88 (10): 35–39.

Torzewski, K. (2008 May). Acid storage. *Chem. Eng.* 115 (5): 57.

Vazquez-Esparragoza, J. and Chen, J. (2016 Jan). Use discrete event simulation as decision support for storage and shipping. *Hyd. Proc.* 95 (1): 41–48. 2016 Feb. *95(2)*, 55–58.

Von Essen, J.A. and Ricks, B. (1999 Nov). Design agitated slurry storage tanks to minimize costs. *Chem. Eng. Prog.* 95 (11): 51–55.

12

In-Plant Transfer of Liquids and Liquid Mixtures with Gases and Solids

The transfer of materials – whether they be liquids, gases, or solids – around a plant is a technology unto itself. For liquids, the complications are as follows:

(1) Internal molecular structure manifests in a variety of viscosities and general rheological stress–strain properties.
(2) The liquid, while being the main component, may be carrying a significant amount of gas or solid.

These issues affect the choice of liquid movers and the choice of auxiliary hardware.

Sections 12.1 and 12.2 of this chapter discuss the flow of liquids and their admixtures, principally their flow patterns and pressure drop. This is a prelude to the consideration of liquid movers, which need to "know" what flows and pressures they must provide.

Section 12.3 discusses movers that are *dynamic*, which transform high-speed internal flow into pressure to drive a liquid through pipes and other equipment. Centrifugal pumps are by far the major representative of this class.

Section 12.4 discusses *positive displacement* movers, which impel a liquid by simply pushing on it. Some other lesser-used pumps are also included.

Section 12.5 discusses the ancillary items that are part of liquid flow systems, e.g., piping, fittings, valves, and seals.

A very helpful reference for the topics of this chapter is the Crane *Flow of Fluids Handbook*.

12.1 Flow of Liquids in Single Phase: Newtonian and Non-Newtonian

The distinction between Newtonian and non-Newtonian liquids is based on the form of their stress–strain behavior. It is easily visualized by considering a flat plate moving parallel to a stationary plate with liquid in between. The shear rate is the plate velocity divided by the distance between plates, and the shear stress is the force required to move the plate divided by its area parallel to flow. Within a flowing fluid, every point has an internal stress and strain – related to each other according to the rheological nature of the material.

12.1.1 Single-Phase Newtonian Flow

The definition of a Newtonian liquid is that the shear rate and shear strain are related linearly, starting at (0,0). There are two key properties – the density and the viscosity, with viscosity

Practical Process Design for Chemical Engineers, First Edition. Keith Marchildon and David Mody.

being the constant ratio between the shear stress and shear rate, units equal to $(\text{N m}^{-2})/[(\text{m s}^{-1})\ \text{m}^{-1})] = \text{N m}^{-2}\,\text{s} = \text{Pa s}$.

Fluid flow at constant or near-constant temperature is described by Bernoulli's energy balance,

$$P_1/\rho + (g/g_c)\,Z_1 + {u_1}^2/(2\,g_c) + \eta\,W_f = P_2/\rho + (g/g_c)\,Z_2 + {u_2}^2/(2\,g_c) + h_f \tag{12.1}$$

where

P is the pressure,
ρ is the local density,
Z is the elevation,
g is the acceleration due to terrestrial gravity,
g_c is Newton's second law conversion factor, kg m s^{-2} N^{-1} (= 1),
u is the fluid local velocity,
W_f is the mechanical energy input being received by the fluid, the work term, that increases pressure and/or velocity,
η is the efficiency factor, and
h_f is the rate of energy loss due to friction.

In a specific application, any of the terms may be important and should be estimated before being deleted. If we leave out the elevation, velocity, and work terms, then the equation reduces to

$$P_2 - P_1 = -\rho \times h_f \tag{12.2}$$

or the more familiar

$$\Delta P = 4f\,(L/D)\,\rho\,u^2/(2\,g_c) \tag{12.3}$$

for pressure drop in a circular pipe.

The *Fanning friction factor*, f, depends on the Reynolds number, Re, and, for turbulent flow, the internal surface roughness of the pipe. This dependence is shown in Figure 12.1. The bulk of the graph is for turbulent flow. Laminar flow is shown by the left-hand line, which is simply $f = 16\,\text{Re}^{-1}$. For accuracy, it is advised to use the original Moody graph, which is reproduced in references such as Perry's Chemical Engineers' Handbook.

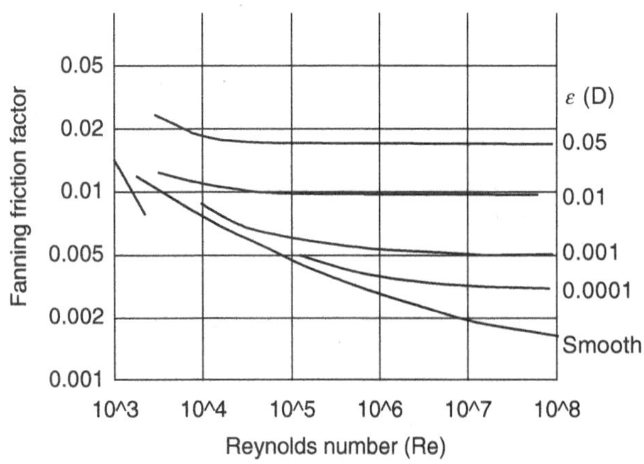

Figure 12.1 Moody chart for the Fanning friction factor.

Table 12.1 Pipe wall protuberances.

Pipe	ε (inches)
Drawn tubing	0
Commercial steel pipe	0.000 15
Cast-iron pipe	0.000 85
Concrete pipe	0.01
Corroded or encrusted pipe	Much higher

The *Darcy friction factor* is identical to the Fanning friction factor but includes a 4× factor in its value. Some references (i.e. Crane Technical paper 10) will use the Darcy friction factor and the Darcy equation, which excludes the four multiplier in Eq. (12.3).

The value of ε is the average height of protuberances in the pipe's inner wall. Table 12.1 suggests values.

To enable the use of the chart for calculation purposes, Colebrook (1939) developed the following formula for turbulent flow:

$$1/{f_{\mathrm{D}}}^{0.5} = -2 \times \log 10 \left\{(\varepsilon/D)/3.7 + 2.51/\left(\mathrm{Re} \times {f_{\mathrm{D}}}^{0.5}\right)\right\} \tag{12.4}$$

in which f_{D} is the *Darcy friction factor*, which is numerically exactly four times the Fanning friction factor. After dividing the value from Eq. (12.4) by 4, it is seen that there is a satisfactory match to Moody's plot. The implicit form of Eq. (12.4) is a small, easy-to-get-around impediment in its application.

The following authors have further studied this important prediction: Churchill (1977), Gilmont (2006), Cordero (2008), and Goudar (2008). The equation of Churchill is explicit.

For cross sections other than circular, there are relations that calculate a pseudo-diameter to use as the denominator D in Eq. (12.3).

The first part of Eq. (12.3) is sometimes written as

$$N = 4 f \,(L/D)$$

so that

$$\Delta P = N\,[\rho\, u^2/(2\, g_{\mathrm{c}})] \tag{12.5}$$

For other pipe elements, Eq. (12.5) is also used but with different values of N. The grouping $u^2/(2\ g_{\mathrm{c}})$, units $\mathrm{N\,m\,kg^{-1}}$, is often referred to as a *velocity head*, so N is the number of velocity heads given up by the passage of liquid through an element. The upstream velocity is used for u in calculating the velocity head (Table 12.2).

12.1.2 Non-Newtonian Liquid Flowing as a Single Phase

For Newtonian liquids, the viscosity was defined as the ratio between the shear stress, γ, and shear strain, that is,

$$\text{Dynamic viscosity, } \mu = \gamma/(\mathrm{d}u/\mathrm{d}y) \tag{12.6}$$

Table 12.2 Frictional coefficients (velocity heads) for turbulent flow.

Element	*N*
Sudden enlargement of cross section	(1 – upstream x-section/downstream x-section)^2
Sudden contraction of cross section	<0.1 for laminar flow; 0.4°× (1 – downstream cross section/upstream x-section)
Globe valve, wide open	10
Gate valve, wide open	0.2
Gate valve, half open	5.6
Tee	1.8
Elbow, 90°	0.9
Elbow, 45°	0.4

Non-Newtonians do not have this simple relation. For some liquids, especially molten polymers, the following simple variation is a good approximation:

$$\text{Effective viscosity} = A_0/(1 + A_1 \times |\,du/dy\,|) \tag{12.7}$$

Another common expression is

$$\gamma = K \times (du/dy) \times |du/dy|^{n-1} \tag{12.8}$$

where K and n are constants. For Newtonian liquids, n is 1 and K is the viscosity.

Many rheologists have studied these flows, but the early work of Dodge and Metzner (1959) is still highly considered. They applied Eq. (12.8) to turbulent flow and derived the following expression for use as the Fanning friction factor in Eq. (12.3) (Figure 12.2):

$$f_F^{-0.5} = 4.0 \times n^{-0.75} \times \log 10 \left[\text{Re}' f_F^{(1-n)/2}\right\} - 0.40 \times n^{-1.2} \tag{12.9}$$

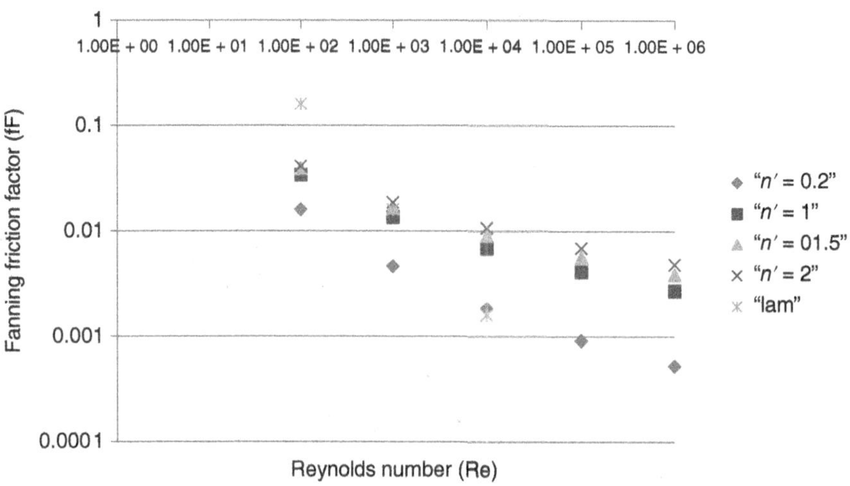

Figure 12.2 Power law Fanning friction factors.

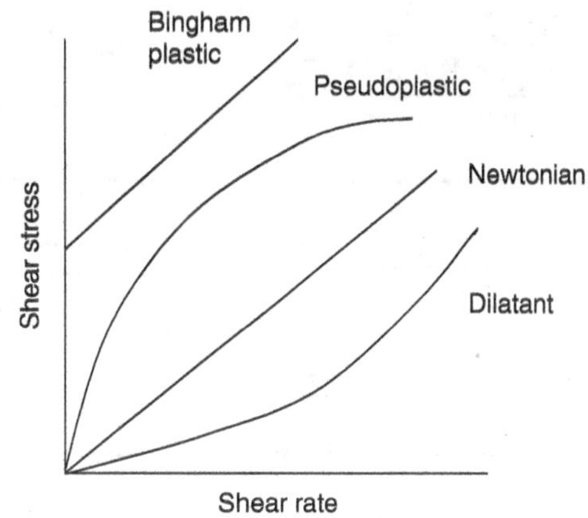

Figure 12.3 Variation in the stress–strain behavior of liquids.

The Reynolds number, Re′, still dimensionless, is defined as

$$\mathrm{Re}' = D^n V^{2-n} \rho/(g_c K 8^{n-1}) \tag{12.10}$$

where K and n come from Eq. (12.8). The expression reduces to the standard definition of Re for Newtonian fluids, where n is 1 and K is the viscosity.

For Reynolds number less than about 3000, the formula for laminar flow should be used,

$$F_F = 16/\mathrm{Re}' \tag{12.11}$$

Applying this procedure requires knowledge of the two constants, K and n, which could be measured by viscometry in the lab. Those tests would also tell if, in fact, the behavior actually follows Eq. (12.8). It may be necessary to conduct tests at full scale if the liquid displays highly and non-Newtonian characteristics that have no existing correlations.

Figure 12.3 shows some of the types of fluid stress–strain.

12.2 Two-Phase Flows

Two-phase flows are common in the chemical and process industries, but not commonly treated in the introductory fluid-flow instruction. There are three combinations:

- liquid and gas,
- gas and solid, and
- solid and liquid.

The gas–solid combination is treated in Chapter 14, Transportation and modification of solids. The other two are discussed in this section. Also discussed is the flow of liquid under gravity.

12.2.1 Two-Phase Liquid–Gas Systems

In the case of a gas and a liquid flowing together, there are several different patterns of flow ranging from *fully stratified* where the two phases move in separate top and bottom layers with minimal influence on each other to *spray/dispersed* where the liquid is completely dispersed in the gas. Some of these flow patterns are shown in Figure 12.4.

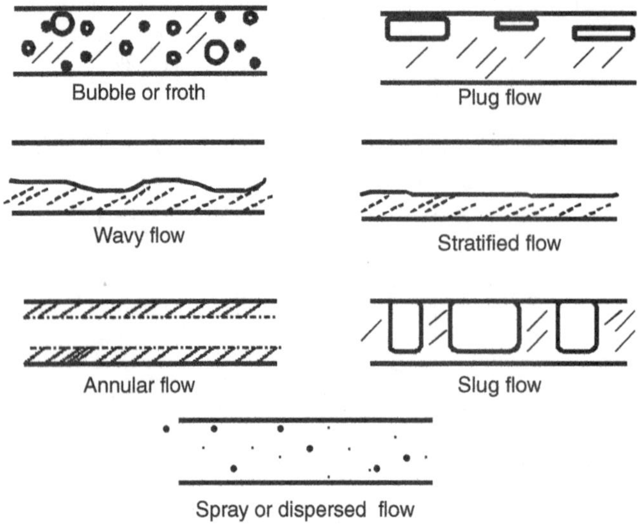

Figure 12.4 Flow regimes in the gas–liquid flow.

It is obviously desirable to be able to predict the nature of the flow in particular cases, and the standard way of doing so is to locate the flow on the map (Figure 12.5) proposed by Baker (1954). By experiment with water and air, Baker found the flow type to be a function of two parameters:

- the weight–flow ratio of liquid to gas $G_{\mathrm{liquid}}/G_{\mathrm{gas}}$ and
- the mass flow rate of gas G_{gas}. The plot, with these coordinates, shows the boundaries between the various flow regimes.

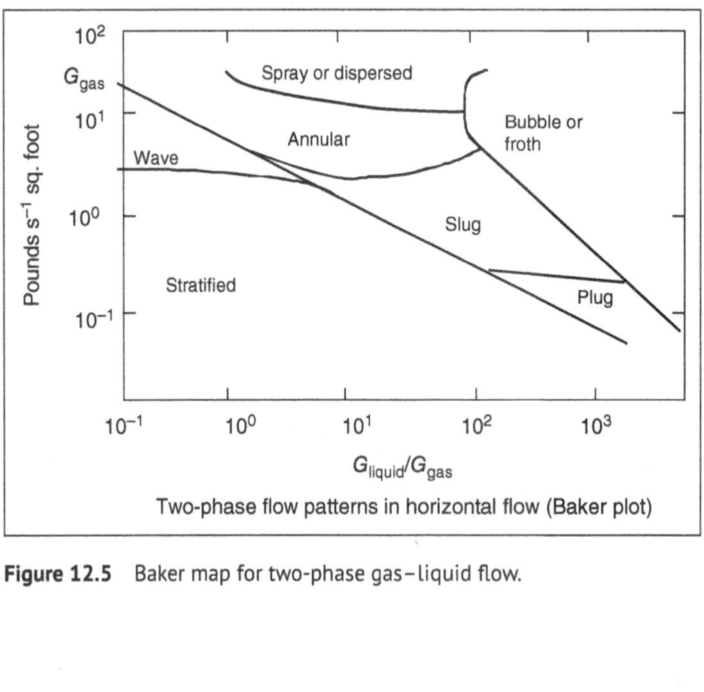

Figure 12.5 Baker map for two-phase gas–liquid flow.

In the event where the fluids are not air and water, the coordinates become

$$(G_{\text{liquid}}/G_{\text{gas}}) \times [(\rho_{\text{gas}}/0.075) \times (\rho_{\text{liquid}}/62.3)]^{1/2} \times (73/\sigma) \times \left[(\mu_{\text{liquid}}/1) \times (62.3/\rho_{\text{liquid}})^2\right]^{1/3} \tag{12.12a}$$

and

$$G_{\text{gas}} \times [(0.075/\rho_{\text{gas}}) \times (62.3/\rho_{\text{liquid}})]^{1/2} \tag{12.12b}$$

In most cases, the *slug* region is the one to avoid. This type of flow consists of successive zones of liquid and gas following each other, with the liquid "slugs" being propelled at large velocity by the gas. Serious vibration and pipe stress can result. *Steam hammer* in steam pipes is a common example.

For design purposes, it is necessary to predict the pressure drop under two-phase conditions. Various methods exist, some of them specific to one or the other flow regime, but one method, that of Lockhart and Martinelli (1949), provides a reasonable general prediction. The method makes use of the curves in Figure 12.6.

The procedure is as follows:

1. For each of the two components, calculate the pressure gradient that would occur if that component flowed by itself through the pipe, $(dP/dL)_{\text{liquid}}$ and $(dP/dL)_{\text{gas}}$.
2. Form the parameter X equal to square root of $(dP/dL)_{\text{liquid}}/(dP/dL)_{\text{gas}}$.
3. From Figure 12.6, read off the appropriate multipliers, Φg and ΦL.

$$\text{Two phase pressure gradient} = \Phi\text{L} \times (dP/dL)_{\text{liquid}} \text{ or } \Phi\text{g} \times (dP/dL)_{\text{gas}} \tag{12.13}$$

The choice of lines depends on whether the fictitious single-phase flows would have been both viscous (i.e. laminar) or both turbulent. If one is viscous and the other is turbulent, set the value of ϕ halfway between the upper and lower prediction.

Lockhart and Martinelli also provided a graph for the *in situ* volumetric fractions of liquid and gas. It must be realized that the ratio of volumetric flow rates and the ratio of local (*in situ*) volumes are generally not the same because the speeds of the two phases are generally different.

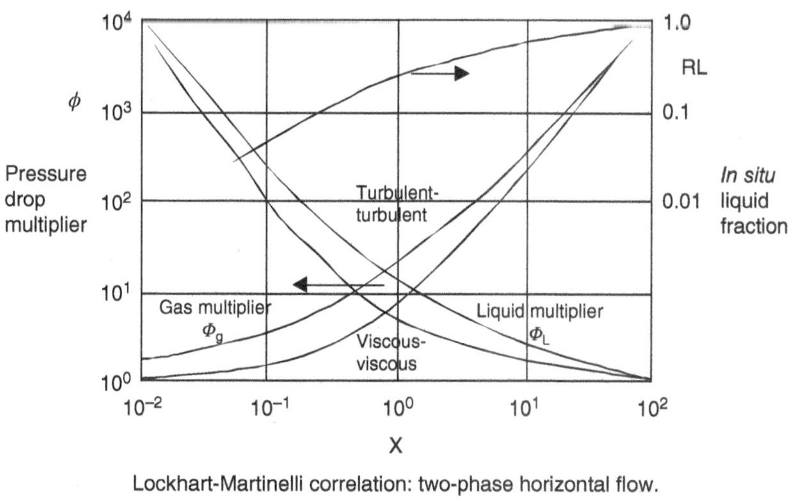

Figure 12.6 Lockhart & Martinelli multipliers for gas–liquid flow.

Figure 12.6 provides a prediction of **RL**, the *in situ* liquid volumetric fraction, also as a function of the parameter X. The concept of a *homogeneous flow* where gas and liquid move as a single fluid with a single velocity is generally not correct.

For the use of this plot on the computer, the following empirical equations have been fitted (2nd Author's Note: There appears to be some equations below that do not predict the graph correctly. Please check the output before using):

For the local liquid volumetric fraction, RL,

$z = \log10(1.32/X)$;

if X is less than 1.32, $y = 1.0\,z + 0.556\,\exp(-1.36\,z)$;

if X is greater than 1.32, $y = 0.556\,\exp[0.4394\,(z - (-z)\text{^}2.117)]$,

$\mathrm{RL} = 10\text{^}(-y)$.

For the liquid pressure drop multiplier, ΦL,

$xx = \log10(X)$;

if X is less than 1.0,

$z = A\,\{[B - C\,xx]/[B - xx]\}$,

$y = -2\,xx + D\,\exp(z)$,

where

$A = 1.098\,1$, $B = 0.025\,4$, $C = 0.82$, and $D = 1.271$ when both phases are turbulent;

$A = 1.390\,6$, $B = 0.066$, $C = 0.75$, and $D = 1.075$ when one phase is turbulent and the other is laminar;

$A = 1.929\,1$, $B = 0.124\,3$, $C = 0.85$, and $D = 0.845$ when both phases (in single flow) are laminar.

If X is greater than 1.0,

$y = A\,\exp[(-B)\,(xx + xx\text{^}C)]$,

where

$A = 1.271$, $B = 0.475\,5$, and $C = 1.835$ when both phases (in the single-phase flow) are turbulent;

$A = 1.075$, $B = 0.469\,9$, and $C = 1.713$ for one phase turbulent and one laminar;

$A = 0.845$, $B = 0.437\,8$, and $C = 1.639$ for both phases laminar.

$$\Phi\mathrm{L} = 10\text{^}y \quad (12.14)$$

For **vertical flow,** Lockhart and Martinelli's equations provide a reasonable prediction of frictional pressure loss for vertical as well as horizontal flow, but, of course, in the vertical case the effect of changing static head pressure must also be accounted for. Their equations also describe the liquid *in situ* fraction except in the case of slug flow (where the local fractions would be expected to be the same as the volumetric flow fraction). Vertical flow has regimes somewhat similar to those of horizontal (Baker plot), but a different plot was developed by Charles and Oshinowa (1974) (Figure 12.7). The coordinates and the chart are as follows:

$$R_v = \text{volumetric flow rate of gas/volumetric flow rate of liquid}$$

$$F_{rtp} = \text{two phase Froude number} = (\text{two phase velocity})^2/g/\text{pipe diameter}$$

$$\Lambda = \mu_l/\mu_g \times \left[(\rho_{water}/\rho_l)^{1/4} \times (\sigma_{water}/\sigma_l)^{3/4}\right]$$

where g is the acceleration due to gravity and σ is the surface tension.

Once again, the slug flow region is preferably avoided.

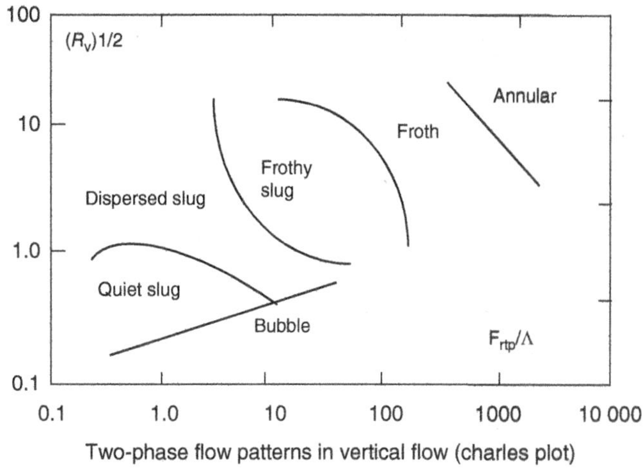

Figure 12.7 Vertical flow of liquid–gas mixtures, Charles & Oshinowa.

12.2.2 Two-Phase Liquid and Solid Systems – Slurries

The closeness of densities distinguishes the liquid–solid combination from the other two combinations, gas–liquid and gas–solid. It has its own distinct technology. We examine the flow in horizontal pipelines, and we assume that the density of the solids is greater than the density of the liquid.

In most cases, the material of interest is the solid and the water is its carrier. However, this assumption does not affect the following treatment.

If liquid flow were zero or very low, the solids would settle into a nonmoving zone at the bottom of the pipe. As the liquid speeds up, the bed begins moving. At higher speeds, particles are swept up by currents and ride along, at liquid velocity, until they subside into the bed. Thus, everything is moving but at two separate speeds. At still higher liquid velocity, all particles are swept from the bottom of the pipe and the flow becomes homogeneous, acting as a single phase.

The effect of liquid velocity on these regimes is shown qualitatively in Figure 12.8. The term "saltation" is used to describe the action in the heterogeneous range. In practical applications, it is not necessary to achieve fully homogeneous flow, which occasions high pressure drop and high energy.

To be conveyed, the solid particles do need to be lifted off their bed. Durand (1953) determined that a minimum value of the pipe-flow Froude number is required. The effect of the particle size and particle loading (%CV) on this minimum number is shown in Figure 12.9.

$$\mathrm{Fr_L} = \text{Froude number} = \sqrt{\{u_L{}^2/[2\,g\,D\,(s-1)]\}} \tag{12.15}$$

where

u_L is the velocity of the liquid,
g is the acceleration due to terrestrial gravity, 9.8 m s^{-2} or 32.2 ft s^{-2},
D is inside diameter of the pipe, and
s is the ratio of densities of the solid to the liquid.

From the chart, $(\mathrm{Fr_L})_{\text{minimum}}$ is obtained so that the minimum liquid velocity to achieve significant saltation is

$$(u_L)_{\text{minimum}} = (\mathrm{Fr_L})_{\text{minimum}}\,[2\,g\,D\,(s-1)]^{0.5} \tag{12.16}$$

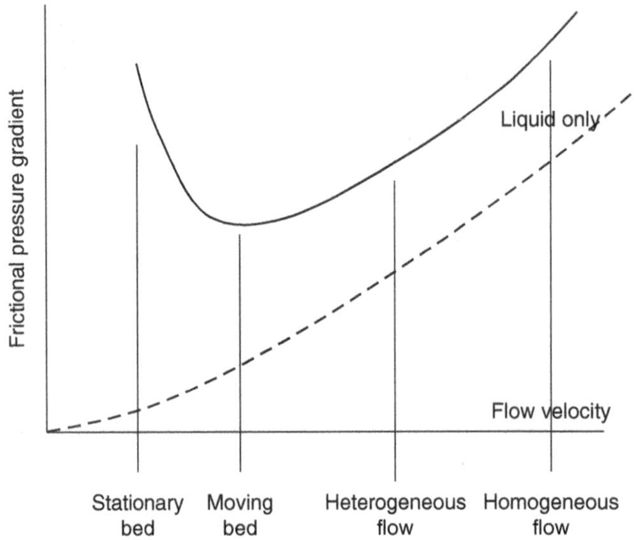

Figure 12.8 Two-phase liquid–solid flow regimes.

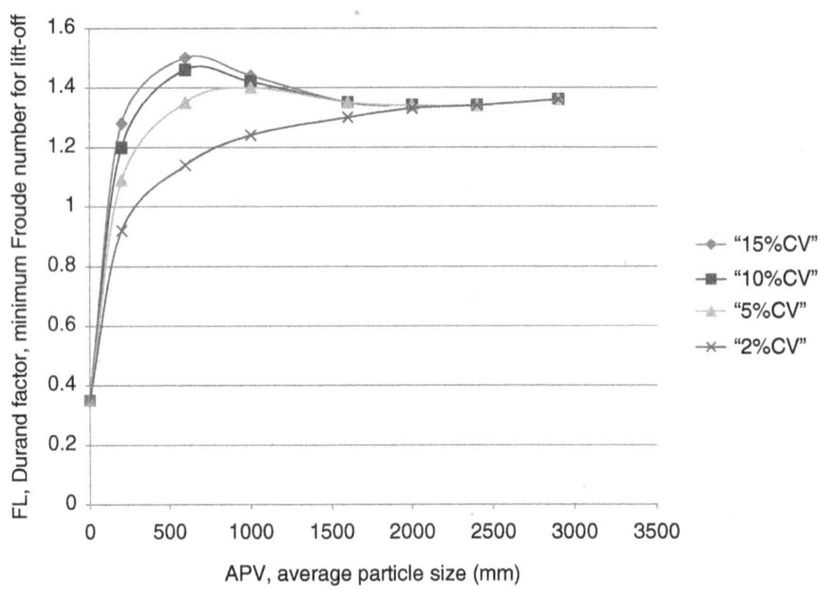

Figure 12.9 Required Froude number to begin saltation.

Many studies have been done to measure and mathematically describe the pressure gradients of combined liquid and solid flow. Turian & Yuan (1977) presented individual equations for the various regimes based on the examination of the data of numerous workers. Lahiri & Ghanta (2009B) have attested to the value of this correlation. By comparing these two works, it should be noted that the solid concentration in Turian is expressed as a percentage and in Lahiri as a fraction.

Turian and Yuan's equation is for the increase in the Fanning friction factor above the factor of water alone, which can be quite significant. The general form for all regimes is

$$f_F - (f_F)_{water} = A \times \text{Conc}^B \times (f_F)_{water}{}^E \times C_D{}^G \times \{u^2/[D g (s-1)]\}^H \tag{12.17}$$

Table 12.3 Coefficients for the Turian and Yuan gas–liquid frictional pressure drop (Eq. 12.17).

Regime	A	B	F	G	H
Stationary bed	0.4036	0.7889	0.7717	−0.4054	−1.096
Saltation	0.9857	1.018	1.046	−0.4213	−1.354
Heterogeneous	0.5513	0.8887	1.200	−0.1677	−0.6938
Homogeneous	0.8444	0.5024	1.428	+0.1516	−0.3531

where

Conc is the solid concentration in volume percent,
C_D is the particle drag coefficient,
A, B, E, G, and H are constants for each regime and are listed below in Table 12.3, and
C_D is the dimensionless ratio of force on a particle by fluid flow around it,

$$C_D = \text{Force}/[\,0.5\,\rho\,A_P\,u^2/g_c] \quad (12.18)$$

where A_P is the area of the particle projected in the direction of motion. Values for this coefficient are obtained from handbooks such as Perry's.

Alderman & Heywood (2004A, 2004B), Duffy (2015), and Lahiri & Ghanta (2008, 2009A, 2009B) discussed suspension rheology and flow.

12.2.3 Liquid Flow Under the Influence of Gravity

Bernoulli's equation (12.1), includes gravity terms $(g/g_c)Z$ which account for the effect of fluid height or head on the mechanical energy balance. These terms are insignificant for low-density fluids, e.g., gases, but may be important for liquids. In the present discussion, the assistance of gravity in moving a liquid is examined. Gravity may reduce the requirement for pumping; it may be sufficient by itself to move the liquid, and it may require throttling to control the flow.

Gravity flow can be obtained in full pipes or partially full pipes. Circumstances usually dictate which is appropriate. The one rule is to make a design decision: filled or not filled. Letting the system determine the state may lead to instability (e.g. "fill-dump") or to restriction of flow due to vapor locking.

Two questions are addressed here: outflow from vessels and flow in partially filled pipes.

12.2.3.1 Vessel Outflow

Hills (1983) distinguished the two cases: filled-pipe outflow (sketches "A1" and "A2" in Figure 12.10) and open-pipe outflow (sketches "B1" and "B2" in Figure 12.10). Either may be acceptable, but a choice must be made and must be designed for. Hill defined a quantity as

$$J_L = [4\,Q/(\pi\,d^2)]/(g\,d)^{1/2} \quad (12.19)$$

where Q is the flow rate in, say, cubic meters per second,
d is the outflow pipe diameter in meters, and
g is the acceleration due to terrestrial gravity, 9.8 m s^{-1}.

Examination of the right-hand side shows J_L to be a Froude number,

$$J_L = U/(g\,d)^{1/2} \quad (12.20)$$

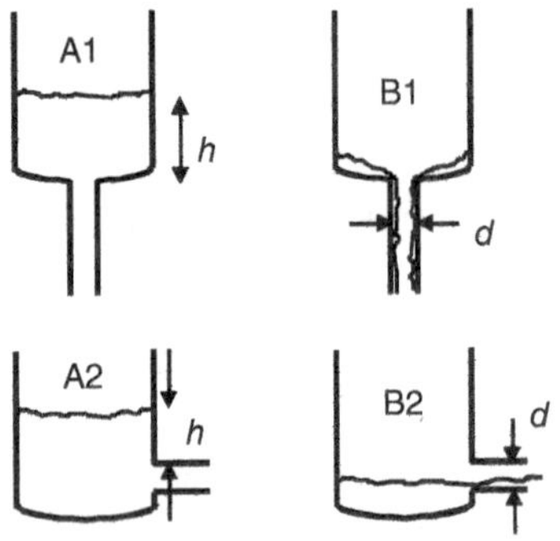

Figure 12.10 Outflow from vessels.

The objective in cases "A1" and "A2" is to avoid the entrainment of gas into the liquid outflow, i.e., to *avoid* a two-phase flow. For situation "A1", bottom-flow, Hill specified the criterion,

$$J_L < 1.6\,(h/d)^2 \tag{12.21}$$

where h is the height of liquid above the bottom of the vessel. Here, the decision is where to set the liquid height: Inverting Eq. (12.21) leads to

$$h > 0.892\,[Q^2/(g\,d)]^{0.25} \tag{12.22}$$

Yu (1997) had a slightly different proposal for (12.21),

$$J_L < 3.25\,(h/d)^{2.5} \tag{12.23}$$

leading to

$$h > 0.688\,(Q/g^{1/2})^{0.4} \tag{12.24}$$

Hills' equation for situation "A2" is similar to Eq. (12.21) but appears to have an error in the index of ($2\,h/d$). Replacing Hills' ½ by 2 gives

$$J_L < (2\,h/d)^2 \tag{12.25}$$

where h is the height above the top of the outflow pipe.

See also Rochelle & Briscoe (2010) and Ukil & Mathew (2011) for the elimination of air entrainment during tank drainage. The tank drainage problem in general is treated by Kossik (2000), Steve (2012), and Yu (2003).

12.2.3.2 Flow in Partially Filled Pipes

Two typical questions are as follows:

(1) for a given liquid flow, will the pipe be filled or partly filled;
(2) what pipe size is needed to convey a required flow at a desired level of fill.

Durand and Marquez-Lucero (1998) stated that, for a fully horizontal pipe (Figure 12.11), the criterion for partial filling is

$$Q < 10.2 \times d^{2.5} \tag{12.26}$$

where Q is the volumetric flow in gallons per minute and
d is the pipe internal diameter in inches.

Durand and Marquez-Lucero presented the following formula relating volumetric flow to the geometry of the space occupied by the liquid in a partially filled pipe (Figure 12.12):

$$Q\,(\text{cubic inches/minute}) = (A^3\,g/T)^{1/2} \tag{12.27}$$

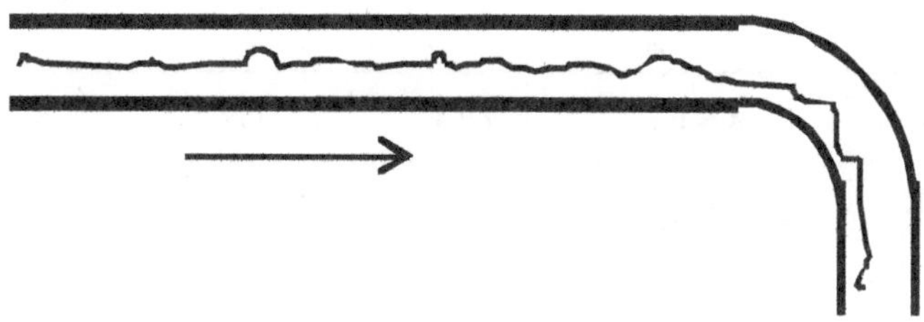

Figure 12.11 Partially filled horizontal pipe.

Figure 12.12 Dimensions of partial filling.

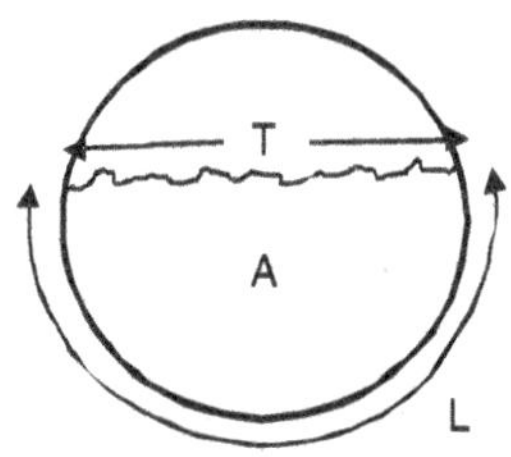

where A is the cross-sectional area of the liquid flow in square inches,

g is the gravitational constant, 1.39E6 in/min^2, and
T is the chord length.

Useful formulae for the geometric calculations, based on the fraction of filled height K, are as follows:

$$\text{Chord length} = T = 2\,d\,[K\,(1 - K)]^{0.5} \tag{12.28}$$

$$\text{Wetted perimeter} = l_w = \alpha\, d/2 \tag{12.29}$$

where

$$\alpha\ \text{(radians)} = \text{angle subtended by filled cross section} = 2\ \arccos\,(1 - 2\,K).$$

Davis (1942) presented the formula

$$Q = 7.3\, d^{2.56}\, K^{1.84} \tag{12.30}$$

also for an un-sloped pipe.

Steve (2000) presented a formula for a sloped pipe,

$$Q = [S\, L_p/(n^2\, L_e)]^{1/2}\, [2673.9678/(l_w/A^{2.5})]^{2/3} \tag{12.31}$$

where Q is the flow, gal min^{-i},
S is the downward slope of the pipe in inches per foot,
L_p is the physical length of the pipe,
n is the Manning coefficient, typically 0.011–0.015,
L_e is the equivalent length of the pipe, taking into account the pressure drop due to fittings,
L_w is the wetted part of the pipe interior perimeter in feet, and
A is the cross-sectional area of the filled section in square feet.

A comparison of predictions is made in Table 12.4. For the prediction of Steve, a slope of 1-in-40, i.e., 0.3 inches per foot, is assumed, and also it is assumed that L_e equals L_p.

Table 12.4 Flow in partially filled horizontal pipes.

Pipe		$K = 0.2$	$K = 0.3$	$K = 0.4$	$K = 0.5$	$K = 0.6$
Diameter		0.2	0.3	0.4	0.5	0.6
inches						
2	Dur&Mar	1.21	2.66	4.64	7.11	10.08
	Steve	1.73	3.87	6.66	9.88	13.28
	Davis	2.23	4.70	7.98	12.02	16.82
3	Dur&Mar	3.33	7.34	12.79	19.61	27.78
	Steve	5.10	11.41	19.64	29.13	39.15
	Davis	6.29	13.26	22.52	33.95	47.48
4	Dur&Mar	6.84	15.07	26.25	40.25	57.03
	Steve	10.99	24.57	42.29	62.74	84.31
	Davis	13.14	27.70	47.03	70.91	99.17
5	Dur&Mar	11.94	26.33	45.87	70.31	99.62
	Steve	19.92	44.56	76.67	113.76	152.86
	Davis	23.26	49.04	83.27	125.54	175.58
6	Dur&Mar	18.84	41.53	72.35	110.91	157.14
	Steve	32.40	72.45	124.68	184.99	248.56
	Davis	37.09	78.22	132.79	200.21	280.02

Predicted volumetric flow of liquid in partially filled pipe.

12.3 Liquid Movers – Dynamic

As defined, the dynamic movers generate kinetic energy which is then converted to exit pressure in accordance with Bernoulli's mechanical energy balance. They are limited to low-viscosity liquids, say up to 45 centipoise, and usually to non-flashing liquids. The pressure they generate is low compared with positive-displacement devices. Significant solids can be handled but with appropriate design changes. In spite of these limitations, dynamic pumps – and in particular centrifugal pumps – are ubiquitous in chemical processing.

12.3.1 Centrifugal Pumps

Centrifugal pumps are by far the most common liquid movers. As shown in the schematic in Figure 12.13, a high-speed rotating impeller imparts kinetic energy to the liquid, which had entered at the central "hub", and this energy converts to pressure at the inside walls of the casing and at the start of the exit piping.

A photograph of a typical impeller is shown in Figure 12.14.

For a given speed, a pump delivers a fixed energy per unit weight to the liquid. Within the pump, this energy is kinetic and is the $u^2/(2\ g_c)$ term in the Bernoulli balance. The units are

Foot × pound force/pound of liquid passing through or, Newton × meters/kilogram of liquid

This energy converts to pressure potential, the *P*/density term in the Bernoulli balance, so that

Pressure = energy × density

At the same time, in Figure 12.15, the pressure equals Head × density, so

Energy × density = Pressure = Head × density

Consequently, Head is a measure of pump energy and is fixed regardless of liquid density. It is the ordinate in the following plots of performance versus volumetric throughput.

Unfortunately, the impeller-imparted energy does not all end up in the outflow because there is energy loss in friction through the vanes. This loss increases with volumetric flow. The key indicator of pump performance is the *pump curve*, two examples of which are shown in Figure 12.16. One pump is designed for higher output pressure, the other for higher flow. The feature of these curves

Figure 12.13 Centrifugal pump.

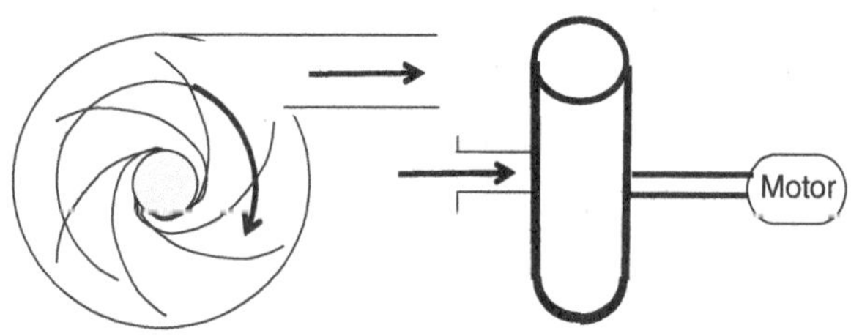

Figure 12.14 Centrifugal impeller.

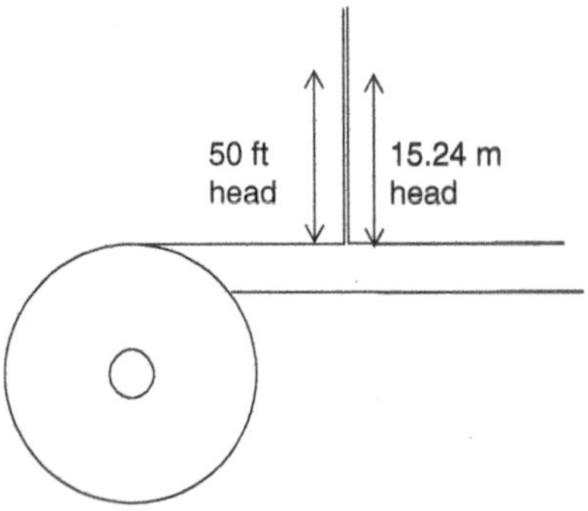

Figure 12.15 "Head" leaving pump.

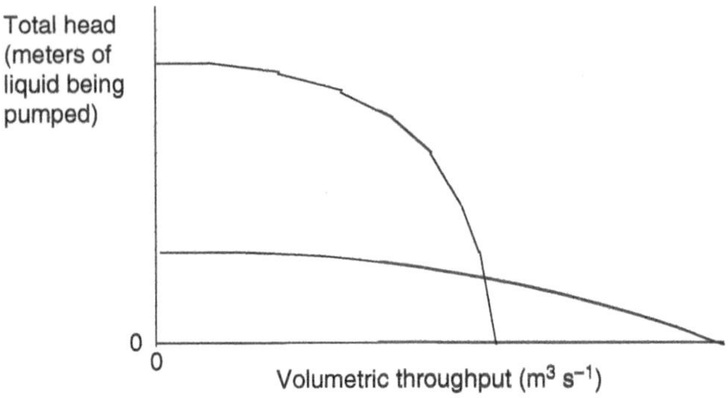

Figure 12.16 Centrifugal flow curve.

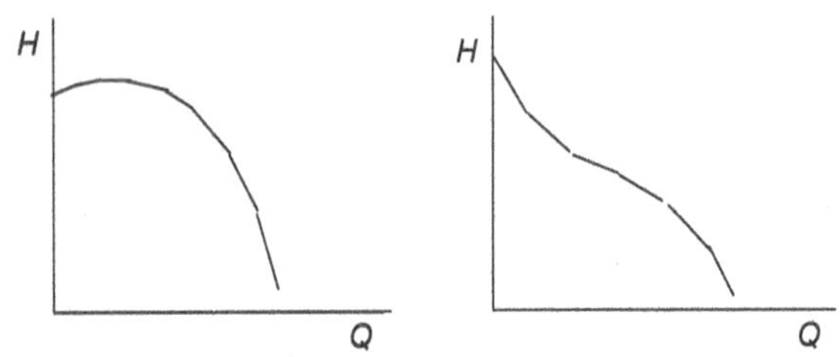

Figure 12.17 Other flow curve types.

is the drop-off of pressure (or Head) with increased flow. Some pumps have variations in the shape of the curves, as shown in (Figure 12.17), depending on the design, especially of the impellers.

A pump curve is specific to a given pump size, to a given impeller diameter, and to a given speed. The prime use of these curves is in matching the pump to the combination of required increase with the flow rate. Figure 12.18 shows the pump curves for three different speeds, along with the *system curve* which shows how the downstream frictional losses increase with flow [for instance, see Eq. (12.3)]. The intersection of the system curve with the three pump curves shows the combination of head and flow that can be delivered. If none of these intersections is satisfactory, then the pump may need to be larger or of a different design.

Along with the pumping characteristics, the pump has three other important features:

- the power required to run it,
- the efficiency with which the power is turned into actual useful pumping, and
- the required pressure at the *inlet*.

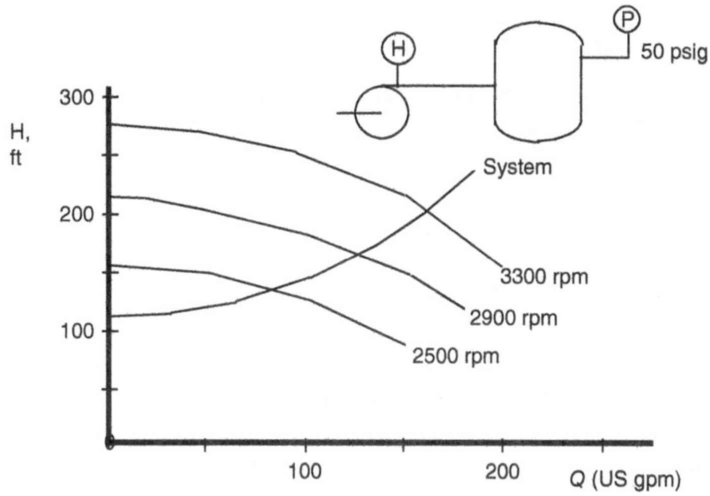

Figure 12.18 Pump and system characteristics.

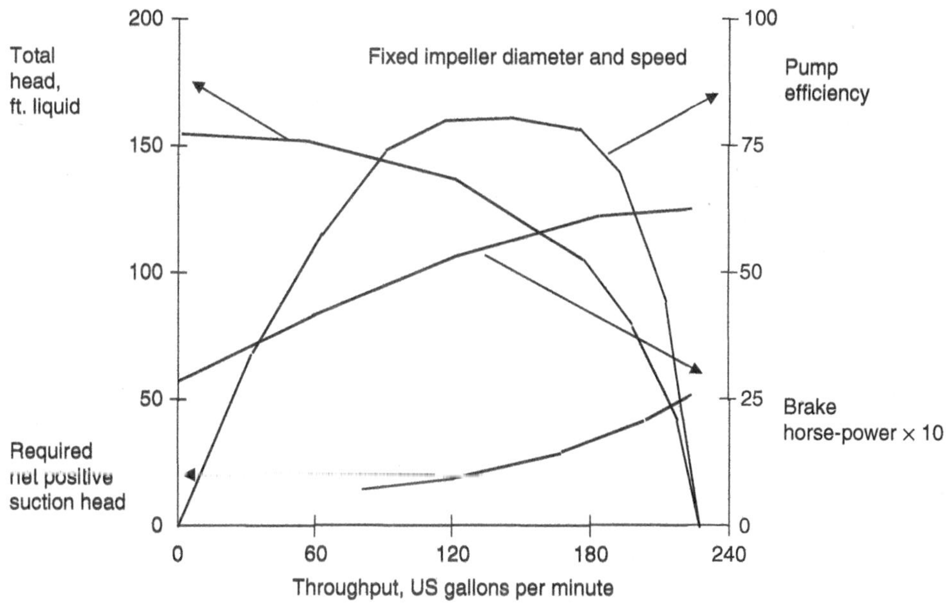

Figure 12.19 Performance characteristics for a centrifugal pump.

Figure 12.19 incorporates curves for these three variables for a given pump at a particular speed. Other pumps, and this pump at a different speed, would have similar plots but with different numbers.

The thermodynamic prediction for power to raise pressure across the pump is

$$\text{Pumping power} = \text{volumetric throughput} \times \text{pressure rise} = Q\,\Delta P \tag{12.32}$$

The brake power, on the other hand, is the measured power required to drive the pump and includes all losses due to friction with packing and other inefficiencies. In Figure 12.19.

$$\text{Pumping efficiency} = Q\,\Delta P/\text{brake power} \tag{12.33}$$

Since Figure 12.19 is expressed in "traditional" units, it will be instructive to

(1) convert the units to *Systeme Internationale* and
(2) recalculate the efficiencies at different conditions for comparison with the existing curve.

Head (m) = Head (ft) × 0.304 8,

$$\text{Pressure rise (Pa)} = \Delta P = \text{Head}\ (m) \times \text{density}\ (\text{kg mass}^{-3}) \times \text{kg force/kg mass}\ (1) \times N/\text{kg force}\ (9.8) = 9800 \times \text{Head}\ (m), \text{ assuming the fluid is water}$$

Volumetric flow = $Q\ (\mathrm{m^3\,s^{-1}}) = Q\ (\mathrm{USGPM}) \times 0.0000631$

Brake power (watts) = Brake horsepower × 745.6

See Table 12.5.

One observation is that the best efficiency point (BEP), occurs when the head has decreased by about 20% from its maximum (i.e. at zero flow). This is a fairly general conclusion. Pump experts recommend operating close to the peak efficiency. Fortunately, the efficiency curve plateaus over a wide range of conditions.

The other line in the plot is the *required net positive suction head.* It gives the minimum pressure condition ahead of the pump, below which the pump malfunctions. The malfunction consists of vapor formation at the pump entrance which has two adverse effects:

decrease in efficiency because of the space taken by vapor, and
collapse of vapor bubbles causing damaging cavitation inside the pump.

The net positive suction head (NPSH) is a differential pressure, is calculated as the pump inlet pressure in absolute pressure minus the vapor pressure of the liquid, and is converted to liquid head (ft or m). It is recommended that the actual, or "available", NPSH be the greater of 3 or 4 ft, or 20%, than the "required" NPSH from the chart. If the intended positioning of the pump results in too low a value, then the remedial action may be to

find a different pump,
lower the location of the pump,
increase the diameter of the upstream piping,
add a suction inducer (which creates other problems),
cool the fluid, and
increase the suction pressure.

Meeting NPSH requirements can be difficult if the feed is from a boiling reactor or if there are dissolved gases in the liquid. Wood et al. (1998) discussed the latter situation.

Figure 12.20 illustrates the calculation of the net positive suction head.

Table 12.5 Centrifugal pump calculations.

Q USGPM	*Q* cu m s^{-1}	Brake power HP	Brake power watts	Head ft	Head *m*	Pressure rise Pa ΔP	Calculated efficiency	Chart efficiency
60	0.003 79	3.8	2833	150	45.72	448 056	0.599	0.6
120	0.007 57	5.5	4101	135	41.15	403 270	0.744	0.76
150	0.009 47	5.7	4250	120	36.58	358 484	0.799	0.8
180	0.011 4	6.2	4623	100	30.48	298 704	0.737	0.752
210	0.013 3	6.3	4697	57	17.37	170 226	0.482	0.45

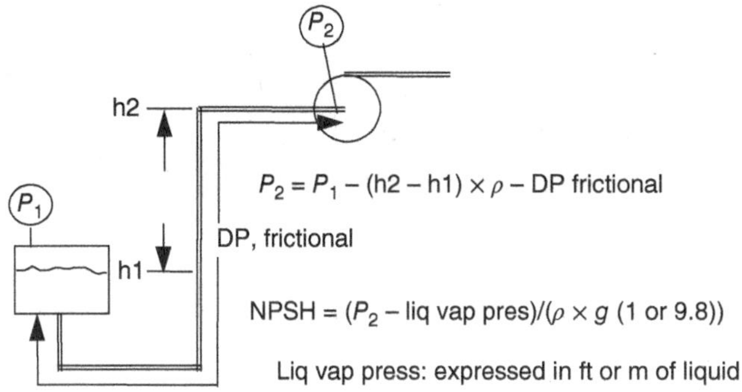

Figure 12.20 Calculation of the net positive suction head.

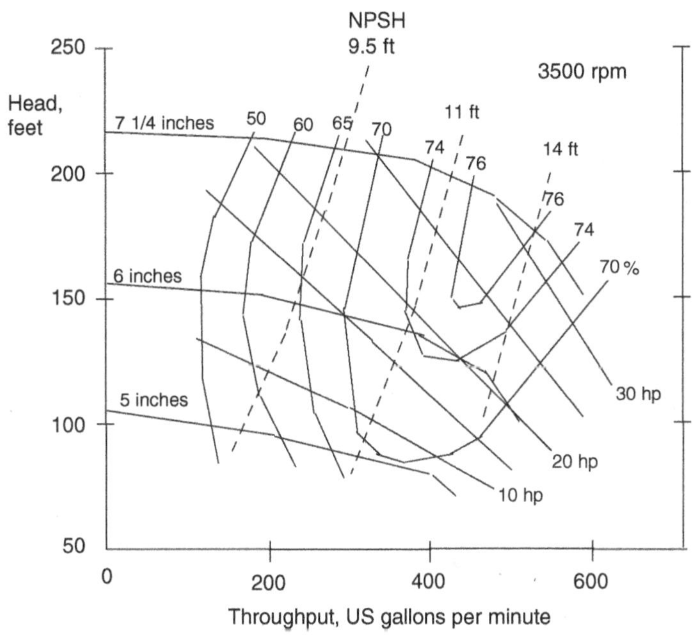

Figure 12.21 Effect of the impeller diameter on pump characteristics.

A much rearranged version of Figure 12.19 is given in Figure 12.21, showing the effect of the impeller diameter within a single casing. Obviously, the small impeller delivers a much reduced result of both the head rise and quantity of flow. However, if those conditions are all that is required, then changing the impeller is much cheaper than buying a whole new pump. As noted, the data are for just a single pump speed.

The effects of the pump speed and pump diameter have been studied, and *affinity laws* have been identified,

$$Q_{max} \text{ is proportional to } D^3 \text{ and to speed}^1 \tag{12.34}$$

$$H_{max} \text{ is proportional to } D^2 \text{ and to speed}^2 \tag{12.35}$$

where D is the diameter of the impeller, which is assumed to contain a full-sized impeller.

If we eliminate D from the two relations, we determine that

$$Q_{max}{}^{1/2} \times \text{speed}/H_{max}{}^{3/4} \tag{12.36}$$

is a constant for the proportionalities of Eqs. (12.34) and (12.35).

This ratio is constant for a given style of pump (i.e. of its impeller) and is independent of size.

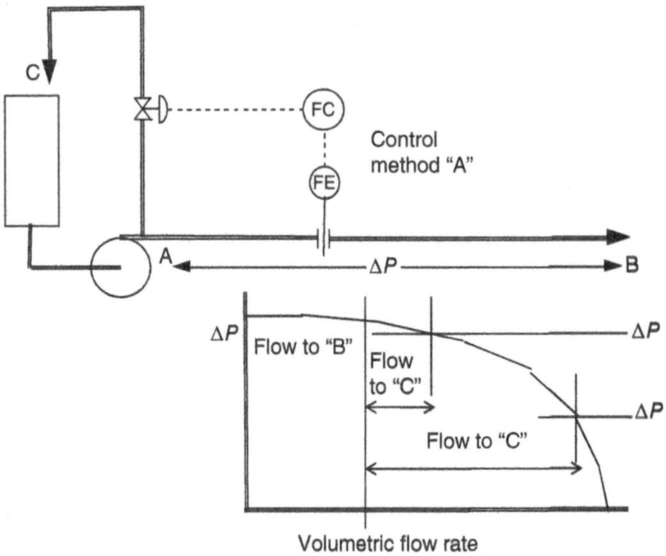

Figure 12.22 Control of net flow via a recycle valve.

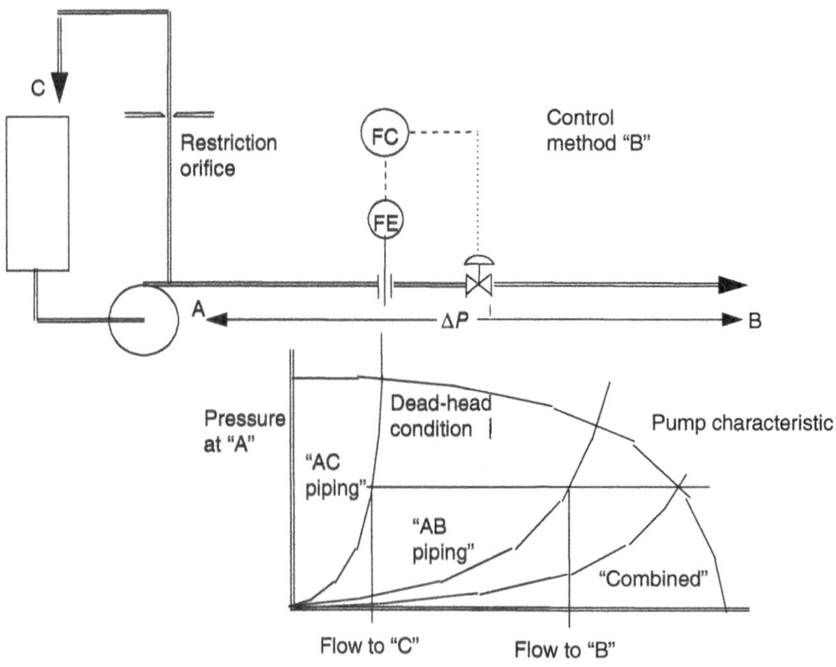

Figure 12.23 Control of flow via a mainline valve, with recycle relief.

This ratio is known as the *specific speed*. In units of US gallons per minute for Q and feet of liquid for H, the specific speed of commercially available pumps ranges from about 500–15 000. The designer of a system should calculate the specific speed based on the task planned for the pump. This value will help in selecting the style of pump. If the value lies outside the range of availability, then perhaps a multiple-stage pump or a non-centrifugal pump is a better choice.

Although centrifugal pumps are not positive displacement in nature, they may still be used as the final elements in the *control of flow or level*. The preferred methods, however, do not involve varying the speed of a centrifugal pump: Not only would this introduce complexity into the drive but it is also harmful to the pump if the flow drops too low. Most vendors recommend flows no lower than 25% of the maximum-efficiency flow. Rather, the most common practice is to provide a recycle loop to handle part of the pump output. The sketches show two possible configurations. Method "B" is preferred because it maintains a base flow even if the main line "AB" gets "dead-headed", i.e., completely shut off. The objective in both cases is for the pump to operate at a constant point on its pump curve (Figures 12.22 and 12.23).

12.4 Liquid Movers – Positive Displacement and Other Pumps

Positive displacement means pockets of fluid are trapped and forced downstream by the direct mechanical action. Pumps of this type are used in three situations:

where the liquid is viscous,
where the precise pumping rate is desired, and
where very high pressure rise is required.

The principle of the piston-driven *reciprocating pump* is easily visualized. Figure 12.24 is a sketch of a mechanically actuated diaphragm-driven pump. Once again the inlet and outlet valving acts to allow filling from a low-pressure source and emptying into a higher-pressure destination. The ability of this valving may limit the viscosity that can be handled.

The pumping rate of reciprocating pumps can be varied in two ways:

by varying the speed of reciprocation or
by varying the stroke length.

Sometimes a set of pumps are "ganged" together on a common drive, in situations where a group of liquids are being supplied to a single destination at a variable overall rate but always in the same

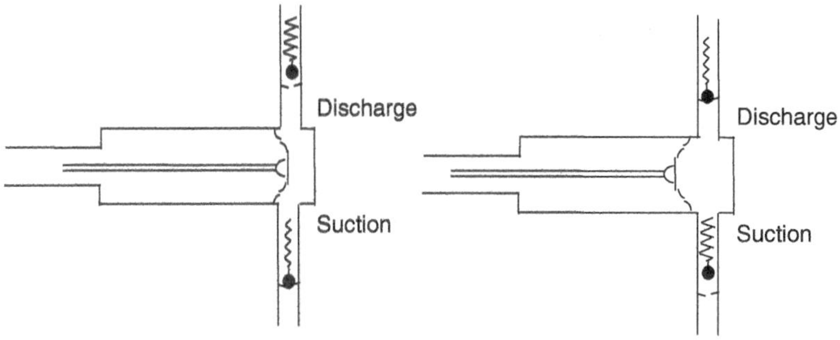

Figure 12.24 Reciprocating-diaphragm positive-displacement pump.

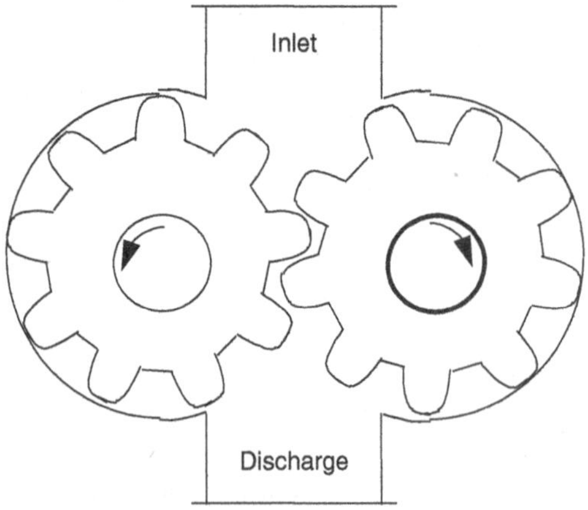

Figure 12.25 Rotary gear pump.

proportion to one another. The proportions are adjusted from time to time, as desired by adjusting the stroke lengths of the individual pump heads.

Reciprocating pumps deliver a flow that pulsates at the frequency of the reciprocation. This characteristic may or may not be acceptable. If it is not, then it can be partially redressed by installing a pulsation dampener downstream of the pump. In specifying the strength of the piping downstream of the pump, it must be remembered that the instantaneous flow of liquid can reach values up to π times the average flow.

Experience is that reciprocating pumps must work against an adverse (rising) pressure gradient in order that the valves open and close properly.

Rotary gear pumps, as seen in Figure 12.25, are commonly used for high-viscosity liquids. Two or more gears trap liquid in the space between the gear teeth and the casing wall and convey it from the inlet to outlet. Obviously, it is essential to minimize paths through which liquid could flow backward, i.e., between the intermeshing gears, over the tips of the gears, and over the top and bottom faces of the gears. This is especially important if the pump is raising the pressure significantly. The pumping rate can be described by an equation of form

$$Q = \alpha \times \text{RPM} - \beta \times \Delta P/\mu$$

where Q is the volumetric throughput, μ is the liquid viscosity, ΔP is the pressure rise generated by the pump, and α and β are constants that depend on the style, size, and clearances of the pump.

If clearances are too high, the pump loses much or all of its pumping efficiency. Gear pumps are widely used in the polymer industries, where viscosities of thousands of poise (and even of pascal-seconds) are encountered and where pressures of thousands of psi (tens of megapascal) are required to force these liquids through pipes and vessels.

The *screw pump* in Figure 12.26 is related to the gear pump in that it acts by pushing liquid along the inner surface of the casing, in this case the screw barrel. The most common embodiment is

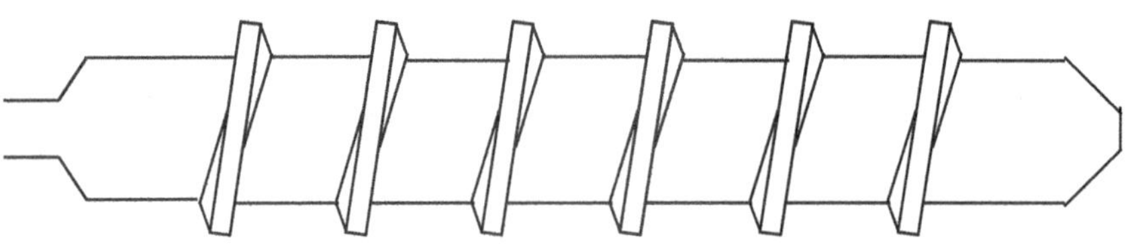

Figure 12.26 Auger for a single-screw pump.

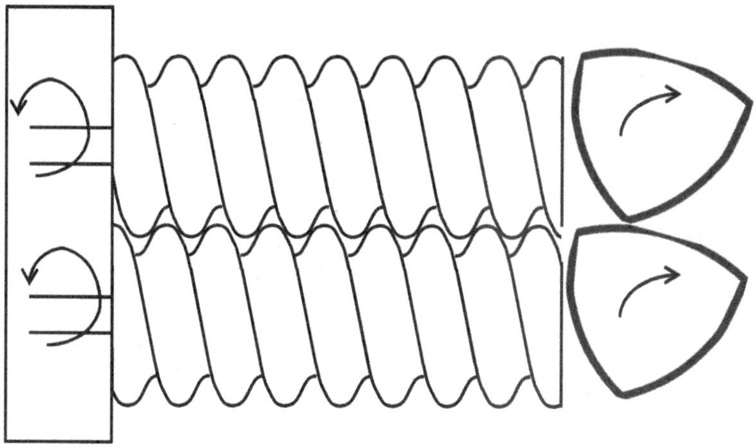

Figure 12.27 Twin screws, corotating.

a single screw in a single barrel, but other models make use of two screws in parallel intersecting barrels, where the screws may corotate (Figure 12.27) or counter-rotate. Screw pumps are not considered as positive displacement because the liquid is able to flow back along the screw channels. The screw equation is similar to that of the gear pump,

$$Q = \alpha \times \text{RPM} - \beta \times \Delta P/(\mu \times L)$$

where L is the filled length of the screw. The constant β depends on both the cross-sectional area of the screw channels and the clearances between the screw and barrel. Screw pumps can still pump against significant pressures in spite of the backflow term. As the viscosity increases, the backflow term decreases, which is why screw pumps are widely used for viscous liquids.

Liquids containing gas or vapor can pose a problem, but centrifugal pumps are able to handle some fraction of gas in a liquid stream and, in fact, specially designed pumps are said to be able to tolerate up to 50% by the volume of gas. Naturally, there is a drop in the efficiency and liquid pumping capacity, but this may be acceptable. However, a cavitating liquid is something different because the vapor collapses within the pump and produces points of very high and damaging stress. This was explained earlier. Cavitation is avoided by providing an adequate suction head.

Prang (1997) recommended a counter-rotating *rotary-screw pump* for mixtures of gas and liquid, where the volumetric fraction of gas goes as high as 90%. Even at such a high loading, there is enough liquid to seal the clearances between the screw and screw and between the screw and wall.

Corrosive liquids: If an exotic metal is required to handle the liquid, it may be less expensive to use a pump of plastic construction. Another choice is to use a reciprocating pump with an inert diaphragm. A *peristaltic* pump, shown in Figure 12.28, may be the answer because the liquid never leaves the tubing, which is made of inert plastic. Diaphragm and peristaltic pumps are members of the class of *seal-less pumps*. There are other styles of seal-less pumps, and they too may be suitable candidates if leakage is a major concern.

Slurries: Centrifugal pumps have some tolerance for slurries and, in fact, some manufacturers have models specifically for that application. Akhtar (1996) presented data on the loss of head generation and of efficiency that may be expected when a liquid contains particulates of various types and loadings. Ondrey (2005) presented slurry-pump standards.

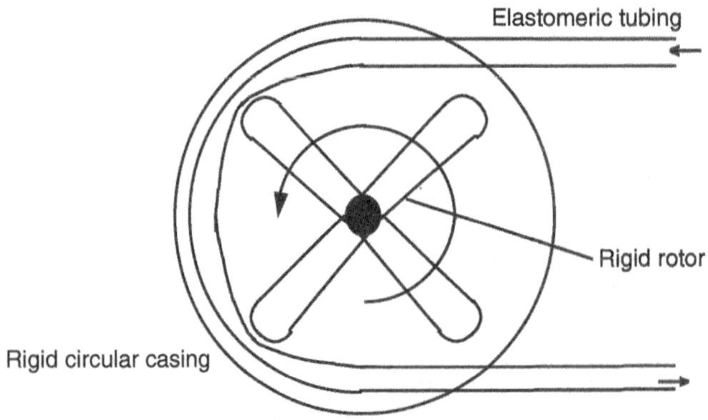

Figure 12.28 Peristaltic pump.

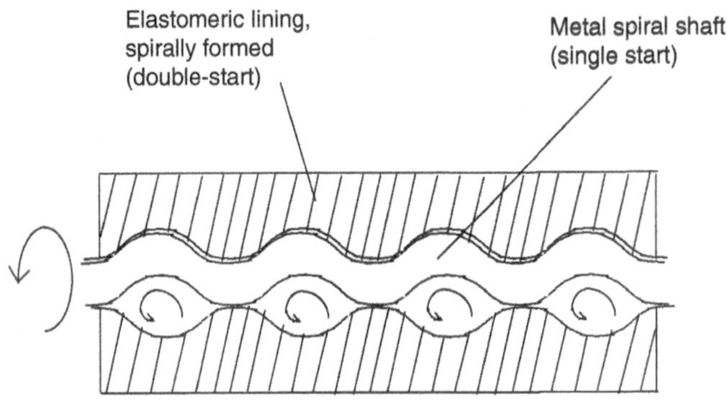

Figure 12.29 Moyno progressive cavity pump.

Three specialized types of pumps are recommended for handling solid-containing liquid streams: *progressive cavity, peristaltic, and double diaphragm*.

The principle of the *progressive cavity* was formulated in 1929 in France by Rene Moineau, hence the well-known Moyno pump. As shown in Figure 12.29, the device consists of

a barrel, an elastomeric lining molded with the shape of a two-start helix, bonded into the barrel, and

a metal one-start helical rotor turning within the barrel and lining.

The rotating screw forms pockets with the lining, and these pockets move progressively toward one or other end of the barrel depending on which way the screw is turning. The pump is able to handle high loadings of even abrasive solids and can pump liquids with viscosity up to 1000 pascal-seconds, 10 000 poise. Displacement is positive although there is back-leakage (calculable) if pumping against a pressure. The Moyno and other brands of progressive cavity pumps have other attractive features, although they often find themselves pumping mixtures of waste that are less than attractive. Whitmore (1998) is a reference for this type of pump.

Sealless pumps address a weakness of externally driven pumps: namely the possibility of leakage along the driveshaft. Most pumps use either a packing or a mechanical seal to minimize or eliminate leakage, see Section 12.5. A different approach is to eliminate the leakage path altogether and to turn the driveshaft either magnetically (see Figure 12.30) or electrically (see Figure 12.31).

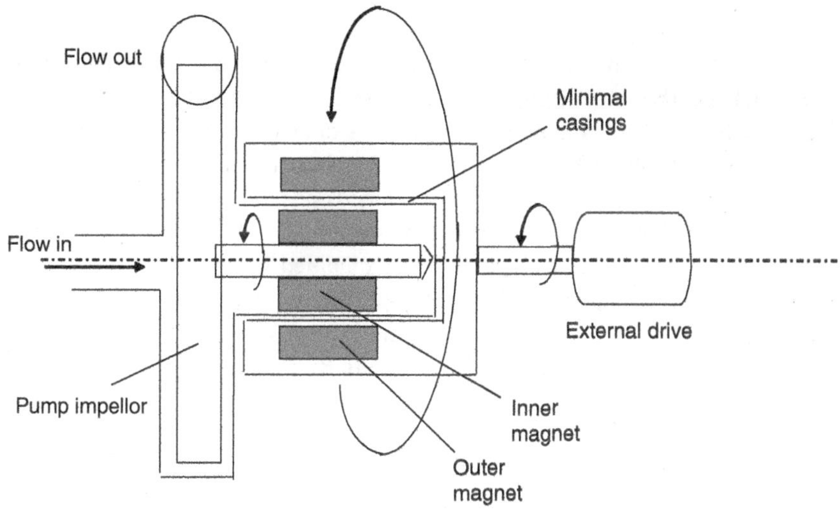

Figure 12.30 Magnetically driven sealed pump.

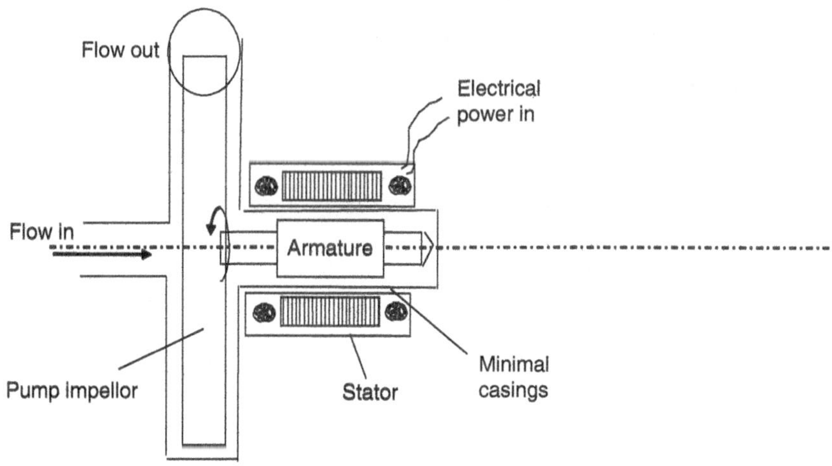

Figure 12.31 Electrically driven "canned-motor" pump.

Although these pumps are a small fraction of centrifugal pumps, they are an effective solution when the fluid is too toxic or too flammable to be allowed into the atmosphere.

Information on these two types of sealless pumps is given by Brodersen (2001), Carr (1995), Fegan (1996), Meyer (2001), Nasr (1996), Ondrey (2002), Shelley & Ondrey (1998), Vetter (1996), and Vanetti (1997).

The "canned" pump, with internal electrical circuits, is safer in atmospheres that may be flammable or explosive.

12.5 Ancillary Equipment

The transport of fluids requires piping, valves, and other fittings. Devices also exist to *prevent* the undesirable transport of fluids, i.e., gaskets at pipe connectors and seals at the driveshafts of pumps.

12.5.1 Conduit

The path to convey fluids comes in many materials and many sizes. It also comes in two general forms: *pipe* and *tube*. By far, the carbon steel pipe is the workhorse in industry and its many dimensions are listed in tables such as in Perry's handbook. Diameters range from 1/8 inch (nominal) to 30 inches, wall thickness from 0.049 to 0.625 inch. It is rigid and comes in standard lengths of 20–40 ft. Sections are attached by screwed ends, welding, or flanged ends. Elbows and tees are used for the change in direction, and reducers for the change in diameter.

Tubing is generally thinner walled, is bendable, and can be obtained in straight sections or in coils of hundreds of feet. Its inner walls are smoother than pipe. Pieces are joined by soldering or compression fittings. It is specified by its outer diameter, generally from 1/32 to 9 inches. The wall thickness is usually given by BWG (Birmingham Wire Gauge), with BWG being converted to inches in Figure 12.32.

To obtain a rough indication of the required inside dimension, Table 12.6 may be consulted.

These velocity ranges are a safe compromise between the excessive pressure drop, erosion of conduit walls, and stagnation. One consideration is to oversize pipe or tube with regard to future plant expansion. Pressure and pressure drop are calculated using one of the formulae in Section 12.1.

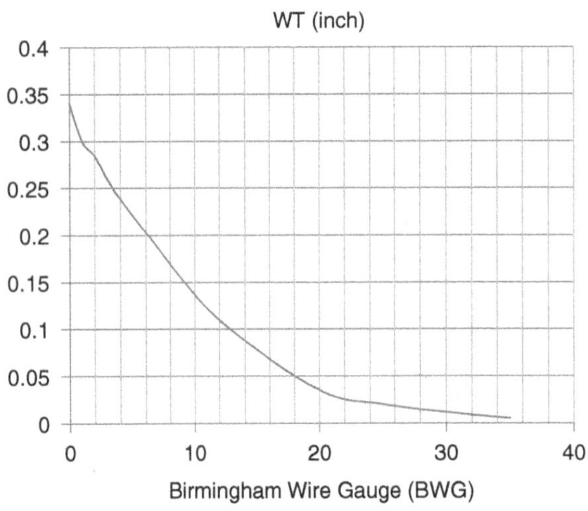

Figure 12.32 BWG-inch conversion.

Table 12.6 Typical velocities in conduits.

Fluid	Flow type	Recommended velocities ($m\,s^{-1}$)
Thin liquid	Gravity flow	0.15–0.30
	Process line	1.0–2.4
Viscous liquid	Pump inlet	0.06–0.15
	Pump discharge	0.15–0.6
Steam		9–15
Air or gas		9–30

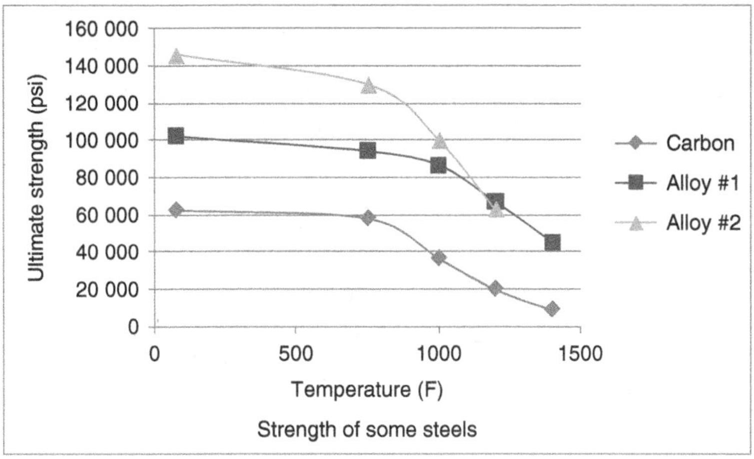

Figure 12.33 Tensile strength of materials.

The outside dimension is probably best left to a piping designer, who can calculate the wall thickness required to contain the inner pressure. One (among several) formula for this relation is

$$Pb = 2 \times Wt \times St/[OD - (0.8 \times Wt)] \tag{12.37}$$

where

Pb is the bursting pressure, psi,
Wt is the wall thickness,
St is the minimum tensile strength, psi, and
OD is the outside diameter.

For a 2-inch schedule 40 pipe, this works out to

$$2 \times 0.154 \times 60\,000/[2.375 - (0.8 \times 0.154)] = 8207\text{ psi}$$

The intended maximum operating pressure in the line should be one-third or one-fourth of that pressure, i.e., 8207/4 = 2051 psi.

The tensile strength depends on the piping material as well as on the temperature. As shown in Figure 12.33, there begins a marked drop after about 700 °F (370 °C).

There are many references on the subject of piping systems. A selection is Maiti (2019), Huitt (2008A, 2008B, 2007A, 2007B, 2007C, 2007D), Shelley (2005), and Bandel & Lawson (2002). Recommendations for pumping of hazardous fluids are presented by Huitt (2010) and Grossel (2008).

12.5.2 Valves

The flow of a fluid often needs to be controlled or else to be manipulated in order to control some other variable. Valves and fluid movers are the two methods of exerting control, valves being the passive means and fluid movers being the active means. Valves are considered in this section. Gordon (2009) provided a primer for valve selection. Sahoo (2004) and Herrera (2015) are also helpful.

Valve bodies are often referred to according to the pressure for which they are rated, but it is better to refer to the "Class", which describes the whole spectrum of temperature-wise pressure capability (Figure 12.34).

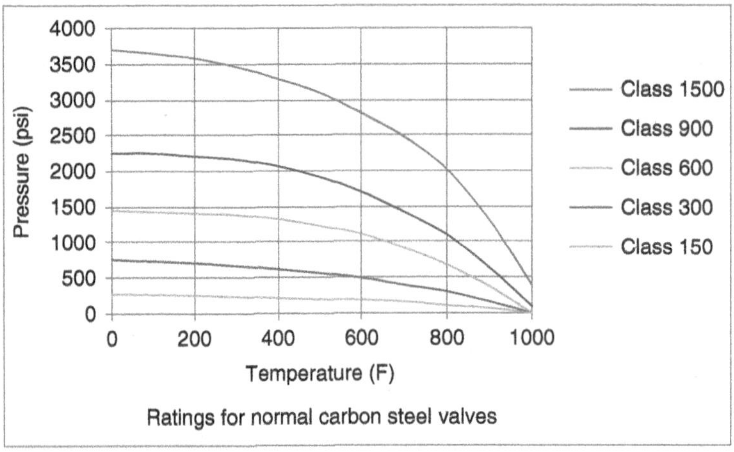

Figure 12.34 Ratings for normal carbon steel valves.

Various types of valves are shown in Figures 12.35–12.37. Coulson & Richardson (1993) and Torzewski (2008) explained the different functions of these valves, and a summary is given here.

Gate Valve

- Used in 70–80% cases.
- Diameter of the opening = Diameter of pipe.
- Does not change the flow direction.

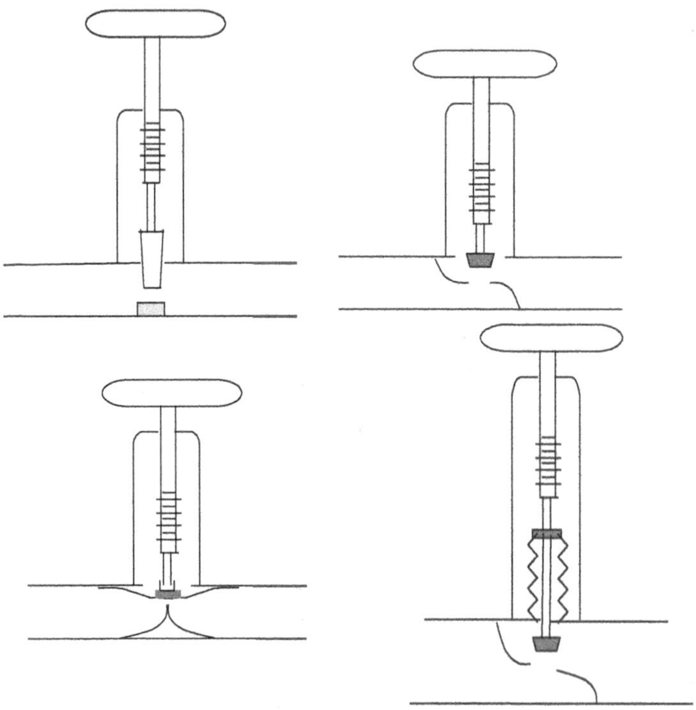

Figure 12.35 Valves: gate, globe, diaphragm, and bellows-seal.

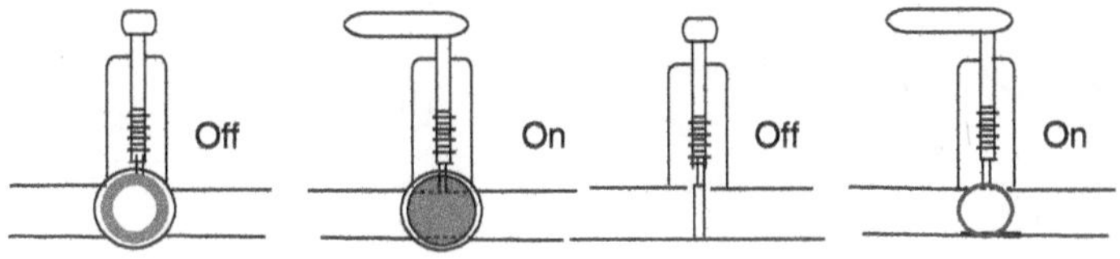

Figure 12.36 Valves: ball and butterfly.

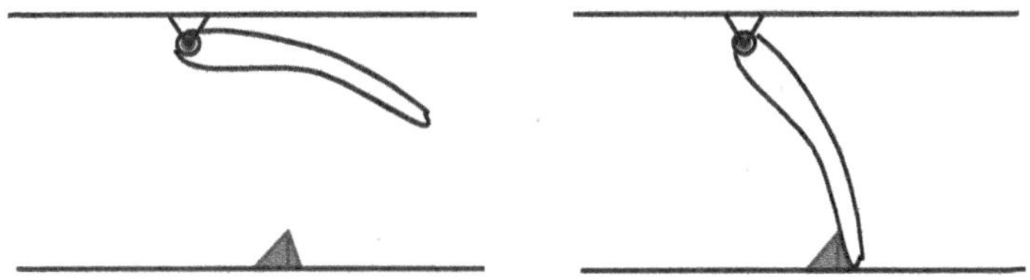

Figure 12.37 Swing check valve.

- Small pressure drop (Δp).
- Works best when fully open/fully closed, i.e., used as a shut-off valve.

Globe Valve

- Better than the gate valve in controlling the fluid flow.
- Large pressure drop (Δp).

Diaphragm Valve

- Good for particulate-bearing fluids.

Bellows Seal Valve

- Eliminates leakage up the stem.

Ball Valve

- Quick opening/closing.
- Used for flow regulation, or as a shut-off valve.
- Temperature limitation as the elastomeric seat contacts the flow.

Plug Valve

- Similar to ball valves.
- Can be built with multi-ports.
- Galling of metal surfaces may occur.

Butterfly Valve

- Quick action.
- Minimal resistance to flow.
- Can be made for large bores.

Check Valve

- Provides flow in one direction, limits flow in opposite direction. Considered to not prevent pressure from moving backwards though.
- Various types, same result.

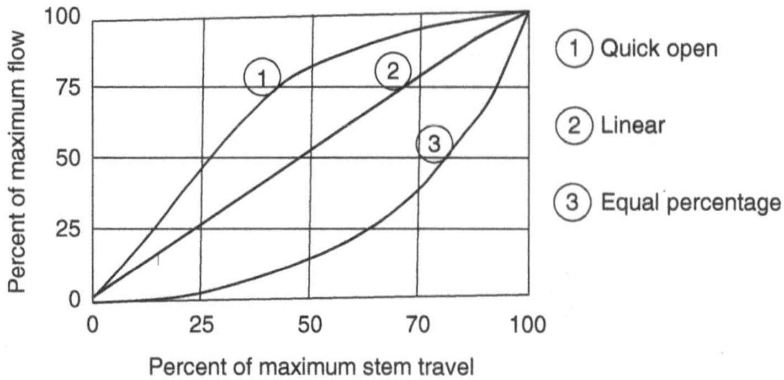

Figure 12.38 Typical valve characteristics.

Every stem-operated valve has a characteristic relation between the flow rate and amount of stem travel. Three forms of relation are shown in Figure 12.38.

These variations are due to different designs of the valve *trim*, the interior movable element that regulates the flow. Assuming the valve is used as the final element in a process control scheme, the different characteristics have different superiorities depending on what is being controlled: liquid level, flow, or pressure.

Control systems using valves are discussed by Brucken (ball valves) (2002), Driskell (1983), Niesen (2008), Noel & Lyons(2001), Purcell (2001), Roth & Stares (2001), Stepanek (2002), and Yu (2000, 2007, 2008). Sizing control valves is discussed by Connell (1987).

The liquid flow capacity of valves is most simply described by Eq. (12.38),

$$Q = C_V\,[\Delta P/\mathrm{SG}]^{0.5} \tag{12.38}$$

where

Q is the volumetric flow of liquid through a fully open valve, US gallons per minute,
C_V is the valve sizing coefficient, a number supplied by the vendor,
ΔP is the pressure difference between the upstream and downstream of the valve, psi, and
SG is the ratio of liquid density to density of water.

This equation is subject to various amendments. Darby & Molavi (1997) provided a correction for viscosities that are significantly above that of water. The stream flow through a valve is captured in Figure 12.39.

Complications arise when there is a possibility of partial volatilization of the liquid. While the liquid may be safely above its vapor pressure P_V at points "1" and "2", it may fall below that pressure at the *vena contracta*. There are three possible consequences:

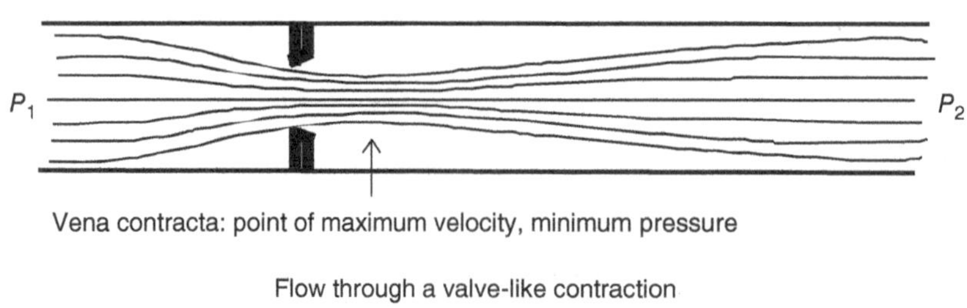

Figure 12.39 Critical point in flow through a valve.

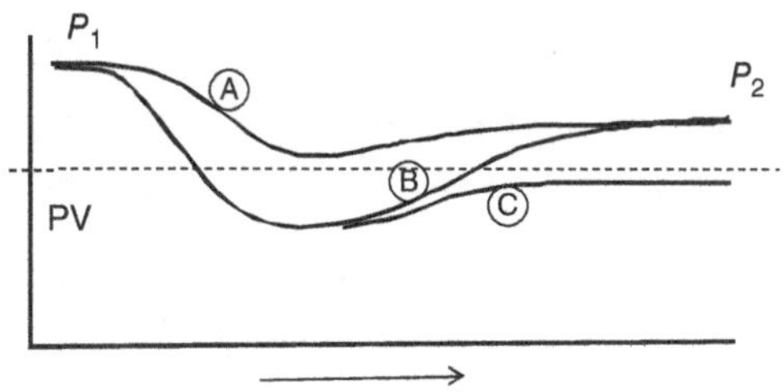

Figure 12.40 Variety of valve pressure profiles.

1. The formation of vapor may accelerate the velocity into a range where sonic effects limit the flow. This situation is called *choking*, and the overall rate of flow Q becomes independent of the overall pressure differential, P_1–P_2.
2. As the pressure recovers, the vapor bubbles collapse and cause erosion-producing *cavitation*.
3. Alternatively, the pressure does not recover to a value greater than the vapor pressure and the flow persists in a two-phase state, i.e., *flashing* has occurred.

These situations are illustrated in Figure 12.40. Profile "A" is free of the above effects and simply follows Eq. (2.8). Profiles "B" and "C" show vapor formation at the vena contracta. Profile "B" illustrates cavitation, and profile "C" illustrates flashing.

Manuals such as the Fisher *Control Valve Handbook* provide the methods and constants for predicting the behavior of specific valves.

Pressure relief valves: The function of some valves is to provide relief from excessive pressure in a system. The valve is normally closed, but it opens to allow the material to escape when a set internal pressure is exceeded. Special treatment is required if the material is in two-phase or is viscous. There is a wealth of articles on these important valves. Listed here is a selection:

Brosius & Dial II (1997), Darby & Molavi (1997), Darby (2000), Darby et al. (2001), Darby et al. (2002), Darby (2003), Diener & Schmidt (2004), Fauske (1999), Mofrad (2008), Ouderkirk (2002), Simpson (1991), and Van Ness (1997).

More general references on valves are the following.

Casado Flores (2006) on cavitation in butterfly valves, Chidrawar (2000) on bellows-seal valves, Coulson and Richardson on valve types, Fehr (1998) and Hayes (1995) on pump and valve combinations, Frenck (2001) and Gibson (1999), Kiesbauer et al. (2006) on cavitation, Morgenroth (1980) compared valve types, Perusek & Orlando (2005) on diaphragm valves, Sahoo (2004) on valve choices, and Torzewski (2008) illustrated valve types.

A valve of a different sort is the ***steam trap***, which is able to distinguish between the two phases, allowing condensate to pass and barring steam vapor. There are various designs. These devices often malfunction, and some references deal with the problem: Sahoo (2005), Page (2006), and Marshall (2007).

References

Akhtar, S.Z. (1996 Nov). Sizing pumps for slurries. *Hyd. Proc.* 75 (11): 161–170.

Alderman, N.J. and Heywood, N.I. (2004A Apr). Improving slurry viscosity and flow curve measurements. *Chem. Eng. Prog.* 100 (4): 27–32.

Alderman, N.J. and Heywood, N.I. (2004B May). Making accurate slurry flow curve measurements. *Chem. Eng. Prog.* 100 (5): 35–41.

Anonymous (1995 Feb). Pumping options for difficult fluids. *Chem. Eng.* 102 (2): 95–96.

Baker, O. (1954). Simultaneous flow of oil and gas. *Oil Gas J.* 53 (12): 185. 1957, *55(45)*, 154.

Bandel, F. and Lawson, J. (2002 Sep). Working the kinks out of piping design: minimize change orders by accounting for a multitude of details at the design stage.(Cover Story). *Chem. Eng.* 109 (10): 56–64.

Brodersen, S. (2001 Mar). Magnetic-drive pumps: technology and new applications. *Hyd. Proc.* 80 (3): 45–48.

Brosius, M.G. and Dial, J.L. II, (1997 Jul). Properly model relief systems. *Chem. Eng. Prog.* 93 (7): 78–82.

Brucken, F.W. (2002 Aug). Using ball valves in control applications. *Hyd. Proc.* 81 (8): 55–60.

Carr, D.M. (1995 Aug). Where sealless pumps make sense. *Chem. Eng.* 102 (8): 78–82.

Casado, F.E. (2006 Aug). Avoid cavitation in butterfly valves. *Hyd. Proc.* 85 (8): 63–66.

Charles, M.E. and Oshinowa, T. (1974 Feb). Vertical two-phase flow, part 1 flow pattern correlations. *Can. J. Chem. Eng.* 52 (1): 25–35.

Chidrawar, S. (2000 Jul). Choosing a bellows seal valve for zero leakage. *Hyd. Proc.* 79 (7): 63–64.

Churchill, S.W. (1977 Nov). Friction factor equation spans all fluid flow regimes. *Chem. Eng.* 84 (24): 91–92.

Colebrook, C.E. (1939). Turbulent flow in pipes with particular reference to the transition region between the smooth and the rough pipe laws. *J. Inst. Civil Eng.* 11 (4): 133–156.

Connell, J.R. (1987 Sep). Realistic control-valve pressure drops. *Chem. Eng.* 94 (9): 123–127.

Cordero, G.O. (2008 Jul). An improved experimental correlation for Darcy friction factor. *Hyd. Proc.* 87 (7): 97–99.

Coulson, J.M. and Richardson, J.F. (1993). Chemical engineering. In: *Chemical Engineering Design (Chemical Engineering Technical Series)*, 2e (ed. R.K. Sinnott), 954. Oxford, UK: Pergamon Press.

Crane, *Flow of Fluids Handbook*, Technical Paper 410, 2018.

Darby, R. (1997 Jun). Control valves: match the trim to the selection. *Chem. Eng.* 104 (6): 147–152.

Darby, R. (2000 Spring). Evaluation of two-phase flow models for flashing flow in nozzles. *Process. Saf. Prog.* 19 (1): 32–39.

Darby, R. (2003 Sep). Size safety relief valves for any conditions. *Chem. Eng.* 110 (9): 42–50.

Darby, R. and Molavi, K. (1997 Summer). Viscosity correction factor for safety relief valves. *Process. Saf. Prog.* 16 (2): 80–82.

Darby, R., Meiller, P.R., and Stockton, J.R. (2001 May). Select the best model for two-phase relief sizing. *Chem. Eng. Prog.* 97 (5): 56–64.

Darby, R., Self, F.E., and Edwards, V.H. (2002 Jun). Properly size pressure-relief valves for two-phase flow. *Chem. Eng.* 109 (6): 68–74.

Davis, D.S. (1942). Nomograph for flow from partially filled pipes. *Ind. Eng. Chem.* 34 (1): 52.

Diener, R. and Schmidt, J. (2004 Dec). Sizing of throttling device for gas/liquid two-phase flow, part 1: safety valves. *Process. Saf. Prog.* 23 (4): 335–344.

Diener, R. and Schmidt, J. (2005 Mar). Sizing of throttling device for gas/liquid two-phase flow, part 2: control valves, orifices, and nozzles'. *Process. Saf. Prog.* 23 (1): 29–37.

Dodge, D.W. and Metzner, A.B. (1959). Turbulent flow of non-Newtonian systems. *AICHE J.* 15 (2): 189–204.

Driskell, L. (1983 Sep). Predicting through control valves. *Chem. Eng.* 90 (9): 94–100.

Duffy, J. (2015 Jan). Controlling suspension rheology. *Chem. Eng.* 122 (1): 34–39.

Durand, R. (1953). Basic relationships of the transportation of solids in pipes--experimental research. *Proc. 5th Minneapolis Int. Hydr. Conv., Int. Assoc. Hydr. Res.* 89–103.

Durand, A.A. and Marquez-Lucero, M. (1998 Mar). Determining sealing rates in horizontal pipe runs. *Chem. Eng.* 105 (3): 129–134.

Durand, A.A., Guerrero, C.A.A., and Ronces, E.A. (2002 Mar). Optimize pipeline design for non-Newtonian fluids. *Chem. Eng.* 109 (3): 62–69.

Fauske, H.K. (1999 Feb). Determine two-phase flows during releases. *Chem. Eng. Prog.* 95 (2): 55–58.

Fegan, D. (1996 Oct). Magnetic-drive versus canned-motor pumps. *Chem. Process.* 53–60.

Fehr, M. (1998 Jul). Improve valve and pump sizing. *Hyd. Proc.* 77 (7): 53–57.

Fisher Controls International Inc (1999). *Control Valve Handbook*, 3rde. Marshalltown.

Frenck, J.P. (2001 May). Making the most of valves. *Chem. Eng.* 108 (5): 66–73.

Gibson, W.B. (1999 Feb). Valves widen the field of operation. *Chem. Eng.* 106 (2): 41–45.

Gilmont, R. (2006 Jun). Pipeline pressure drop: a new design correlation. *Chem. Eng. Prog.* 102 (6): 34–41.

Gordon, B. (2009 Mar). Valves 101: types, materials, selection. *Chem. Eng. Prog.* 105 (3): 42–45.

Goudar, C.T. (2008 Aug). Comparison of the iterative approximations of the Colebrook-White equation. *Hyd. Proc.* 87 (8): 79–83.

Grossel, S. (2008 Feb). Pump hazardous liquids safely. *Chem. Eng.* 115 (2): 36–42.

Gulyani, B.B. (1999 Aug). Simple equations for pipe flow analysis. *Hyd. Proc.* 78 (8): 67–70.

Hardee, R.T. and Sines, J.L. (2012). *Piping System Fundamentals: The Complete Guide to Gaining a Clear Picture of Your Piping System*, 2nde (ed. R.T. Hardee and J.L. Sines), 95–270. Lacey, WA: Engineered Software Inc. Press.

Hayes, W.R. (1995 May). Match valves with pumps to improve system performance. *Hyd. Proc.* 74 (5): 45–49.

Herrera, R. (2015 Jul). Valve selection best practices. *Chem. Eng.* 122 (7): 34–42.

Heywood, N.L. and Alderman, N.J. (2003 Apr). Developments in slurry pipeline technologies. *Chem. Eng. Prog.* 99 (4): 36–43.

Hills, P.D. (1983 Sep). Designing piping for gravity flow. *Chem. Eng.* 90 (9): 111–114.

Huitt, W.M. (2007A Feb). Piping for process plants, part 1: the basics. *Chem. Eng.* 114 (2): 42–47.

Huitt, W.M. (2007B Mar). Piping for process plants, part 2: flanges. *Chem. Eng.* 114 (3), 114(3): 56–61.

Huitt, W.M. (2007C Jul). Piping for process plants, part 3: design elements. *Chem. Eng.* 114 (7): 50–57.

Huitt, W.M. (2007D Oct). Piping for process plants, part 4: codes and fabrication. *Chem. Eng.* 114 (10): 68–76.

Huitt, W.M. (2008A Apr). Piping for process plants, part 5: installation and cleaning. *Chem. Eng.* 115 (4): 48–58.

Huitt, W.M. (2008B Jun). Piping for process plants, Part 6: testing and verification. *Chem. Eng.* 115 (6): 48–54.

Huitt, W.M. (2010 Jun). Piping design for hazardous fluid service. *Chem. Eng.* 117 (6): 36–42.

Kiesbauer, J., Vnucec, D., Roth, M., and Stoffel, B. (2006 Mar). Predicting cavitation damage in control valves. *Hyd. Proc.* 85 (3): 55–61.

Kossik, J. (2000 Jun). Draining time for unpumped tanks. *Chem. Eng.* 107 (6): 115–119.

Lahiri, S.K. and Ghanta, K.C. (2008 Dec). Minimize power consumption in slurry transport. *Hyd. Proc.* 87 (12): 112–118.

Lahiri, S.K. and Ghanta, K.C. (2009A Apr). Computational fluid dynamics simulation of solid-liquid slurry flow. *Hyd. Proc.* 88 (4): 99–104.

Lahiri, S.K. and Ghanta, K.C. (2009B Aug). A support vector classification method for regime identification of slurry transport in pipelines. *Hyd. Proc.* 88 (8): 71–84.

Lockhart, R.W. and Martinelli, R.C. (1949 Jan). Proposed correlation of data for isothermal two-phase, two-component flow in pipes. *Chem. Eng. Prog.* 45 (1): 39–48.

Maiti, S. (2019 Mar). Set correct design specifications for optimized piping and pipe support system. *Hyd. Proc.* 98 (3): 61–63.

Marshall, R. (2007 Feb). Steam trap installation. *Chem. Eng.* 114 (2): 31.

Meyer, E.A. (2001 Apr). Nonmetallic sealless pumps: how to pick a winner. *Chem. Eng.* 108 (4): 96–102.

Mofrad, S.R. (2008 Jan). Relief rate calculation for control valve failure. *Hyd. Proc.* 87 (1): 105–109.
Morgenroth, J. (1980 Jul). Quarter-turn plug, ball, and butterfly valves, plant. *Engineering* 24: 64–67.
Nasr, A.M. (1996 Mar). Try magnetic gear pumps instead of sealless centrifugals. *Chem. Eng. Prog.* 92 (3): 62–64.
Niesen, M. (2008 Oct). Using installed gain to improve valve selection. *Chem. Eng.* 115 (10): 34–37.
Noel, J.W. and Lyons, B. (2001 Sep). Control valves selection – rising stem or quarter turn, who's in control? *Chem. Eng. Prog.* 97 (9): 38–41.
Ondrey, G. (2002 Oct). Sealless pumps take to the field. *Chem. Eng.* 109 (10): 35–38.
Ondrey, G. (2005 Nov). Slurry-pump standards. *Chem. Eng.* 112 (12): 20.
Ouderkirk, R. (2002 Aug). Rigorously size relief valves for supercritical fluids. *Chem. Eng. Prog.* 98 (8): 34–43.
Page, G. (2006 Jan). Steam traps. *Chem. Eng. Prog.* 102 (1): 16–17.
Perry, R.H., Green, D.W., and Maloney, J.O. (1997). *Perry's Chemical Engineers' Handbook*, 7the. The McGraw-Hill Companies, Inc.
Perusek, R. and Orlando, M.D. (2005 Jan). Pristine processing – making the case for diaphragm valves. *Chem. Eng.* 112 (1): 25–30.
Polley, G.T. and Polley, H.L. (2000 Feb). Design better water networks. *Chem. Eng. Prog.* 96 (2): 47–52.
Prang, A.J. (1997 Feb). Selecting multiphase pumps. *Chem. Eng.* 104 (2): 74–79.
Purcell, M.K. (2001 Mar). Easily select and size control valves. *Chem. Eng. Prog.* 97 (3): 45–50.
Rochelle, S.G. and Briscoe, M.T. Jr., (2010 Nov). Predict and prevent air entrainment in draining tanks. *Chem. Eng.* 117 (12): 37–43.
Roth, K.W. and Stares, J.A. (2001 Aug). Avoid control valve application problems with physics-based models. *Hyd. Proc.* 80 (8): 37–48.
Sahoo, T. (2004 Aug). Pick the right valve. *Chem. Eng.* 111 (8): 34–39.
Sahoo, T. (2005 Feb). Steam trap troubles? *Chem. Eng. Prog.* 101 (2): 33–38.
Shelley, S. (2005 Mar). Pipes and fittings. *Chem. Eng.* 112 (3): 56–59.
Shelley, S. and Ondrey, G. (1998 May). Running surveillance on canned-motor pumps. *Chem. Eng.* 105 (5): 33–39.
Simpson, L.L. (1991 Aug). Estimate two-phase flow in safety devices. *Chem. Eng. Prog.* 87 (8): 98–102.
Smith, B. (2002 Apr). What makes a pump for high-purity fluids? *Chem. Eng.* 109 (4): 87–89.
Stepanek, D. (2002 Mar). Control valves for real-world service. *Chem. Eng.* 109 (3): 103–107.
Steve, E.H. (2000 May). Correctly design lines for sloped, gravity flow. *Chem. Eng. Prog.* 96 (5): 53–60.
Steve, E.H. (2012 Jun). Draining vessels. *Chem. Eng.* 119 (6): 34–40.
Torzewski, K. (2008 Aug). Facts at your fingertips: valves. *Chem. Eng.* 115 (8): 49.
Turian, R.M. and Yuan, T.-F. (1977 May). Flow of slurries in pipelines. *AICHE J.* 23 (3): 232–243.
Ukil, T. and Mathew, T. (2011 Jun). Reduce gas entrainment in liquid lines. *Chem. Eng.* 118 (6): 42–44.
Van Ness, H.C. (1997 Jul). Make the correct conversions for relief-valve sonic flows. *Chem. Eng. Prog.* 93 (7): 71–73.
Vanetti, R. (1997 Sep). Sealless pumpmakers flex their muscles. *Chem. Eng.* 104 (9): 39–43.
Vetter, G. (ed.) (1996). *Leak-Free Pumps & Compressors*. Kidlington, Oxford: Elsevier Advanced Technology.
Whitmore, K. (1998 Jul). Successfully use progressing cavity pumps. *Chem. Eng. Prog.* 94 (7): 95–96.
Wood D. W., Hart R. J., Marra E., Application guidelines for pumping liquids that have a large dissolved gas content, 1998. International Pump Users Symposium (15th: 1998). https://hdl.handle.net/1969.1/164125.

Yu, F. (1997 Nov). Line sizing for gravity flow piping. *Hyd. Proc.* 76 (11): 99–102.

Yu, F.C. (2000 Aug). Easy way to estimate realistic control valve pressure drops. *Hyd. Proc.* 79 (8): 45–48.

Yu, F. (2003 May). Generalized gravity flow line calculations. *Hyd. Proc.* 82 (5): 97–99.

Yu, F. (2005 Aug). A simple way to estimate equipment pressure drops. *Hyd. Proc.* 84 (8): 57–64.

Yu, F. (2007 Jul). Allowable liquid control valve pressure drops. *Hyd. Proc.* 86 (7): 107–113.

Yu, F. (2008 Mar). Allowable vapor control valve pressure drops. *Hyd. Proc.* 87 (3): 59–67.

13

Transfer of Gases: Compression and Vacuum

Gases and liquids share the common designation of *fluids*, and they share some of the same methods of being moved. However, there are several differences in their characteristics that make the technology different:

- Gases are compressible, unlike liquids which have close to constant density even under conditions of extreme pressure.
- In moving a gas, there may be appreciable $\int PdV$ work of compression.
- The work of compression may be accompanied by a significant temperature rise.
- Pressures can range from high vacuum, e.g., 10^{-6} atmospheres, absolute to as high as anything encountered in liquid transport, e.g., 10^4 atmospheres, giving an overall range of 10^{10}.
- If the gas is a vapor, that is, if it is below its critical temperature, then compression may bring on liquefaction.

Section 13.1 introduces the subject of compressible flow and presents equations for calculating pressure drop and maximum flow rate in gases undergoing significant pressure drop during flow.

Section 13.2 introduces fluid movers, the devices which impart some combination of momentum and pressure to impel a gas to move. The mechanical input produces heat, which causes temperature to rise. This may or may not be important. It is not usually a factor for the small differential pressure across fans and low-pressure-rise blowers but is a major factor in large compressions.

Sections 13.3 and 13.4 discuss *fans* and *blowers*, respectively, which have relatively low ratios of outgoing to incoming pressure – the *compression ratio, CR*. Many of these devices have characteristics similar to pumps for liquids.

Section 13.5 deals with *compressors*, which have more robust CRs. Their function may be to generate super-atmospheric pressure on the outgoing end or to support vacuum on the incoming end.

Section 13.6 also deals with compression but by the use of a nonmechanical device, the *ejector*.

Section 13.7 shows methods for calculating power consumption and temperature rise during compression.

13.1 Compressible Flow

The density of a gas can be highly variable, depending on temperature and (approximately linearly) pressure. This variability, when the pressure change is small, is easily gotten around in pressure-drop calculations by either assuming that the density is constant or taking an average of initial and final densities. In this case, the equations of liquid flow are carried over into gas flow. There is no exact condition where compressibility must be treated more seriously, but the general

Practical Process Design for Chemical Engineers, First Edition. Keith Marchildon and David Mody.

rule is when the velocity exceeds 0.3 of the speed of sound or when the change in pressure is above ten percent. In Section 13.1, the equations for pressure drop in compressible flow are developed.

The complexity of compressible flow is inherent in the Bernoulli mechanical energy balance for a flowing fluid:

$$P_1/\rho + (g/g_c)\,Z_1 + {u_1}^2/(2\,g_c) + \eta\,W_f = P_2/\rho + (g/g_c)\,Z_2 + {u_2}^2/(2\,g_c) + h_f \tag{13.1}$$

where P is the pressure,
ρ is the local density,
Z is the elevation
g is the acceleration due to terrestrial gravity,
g_c is the Newton's second law conversion factor, kg m s^{-2} N^{-1} (=1),
u is the fluid local velocity,
W_f is the mechanical energy input, the work term,
η is the efficiency factor, and
h_f is the rate of energy loss due to friction.

Dropping the elevation terms and the work term and taking differentials,

$$dP/\rho + d(u^2/(2\,g_c)) + 4\,(u^2/(2\,g_c))\,(f/D)dL = 0 \tag{13.2}$$

where the last term is the expression for h_f, the rate of energy loss due to friction, and f is the Fanning friction factor (see also Chapter 12 for a comparison of Darcy and Fanning friction factors).

13.1.1 Isothermal Constant-Area Pipe Flow

For incompressible flow of gases and liquids in a pipe of constant cross section, only the first and third terms are used, assuming that the velocity stays constant. The commonly used equation is as follows:

$$\text{For incompressible flow, } P_1 - P_2 = 4f\,(L/D)\,\rho\,u^2/(2\,g_c). \tag{13.3}$$

To extend to compressible flow, it is necessary to take the middle term into account because the density and velocity change as the pressure decreases. Equation (13.2) is first converted by substituting G/ρ for u, where G is the mass velocity, kg s^{-1} m^{-2}.

Then ρ is replaced by PM/(RT), assuming the gas to be ideal, where M is the molecular weight and R is the gas constant,

$$dP/(PM/(RT)) + d(G^2R^2T^2/(P^2M^2\,2\,g_c)) + 4\,(G^2R^2T^2/(P^2M^2\,2\,g_c))\,(f/D)dL = 0 \tag{13.4}$$

It can be shown that the final result is

$${P_1}^2 - {P_2}^2 = (N\,G^2RT/(g_c\,M))\,(1 + (2/N)\,\ln(P_1/P_2)) \tag{13.5}$$

where N is $4\,f\,L/D$ for a pipe of length L and diameter D. This equation is the general expression for pressure change in an isothermal flow. It reduces to Eq. (13.3) as P_2 and P_1 become closer.

Equation (13.5) requires an iterative solution if P_2 is being sought. If P_1 and P_2 are known, it is solved directly for G. To see the consequences of this relation, a specific case was calculated, showing the effect of P_2/P_1 on density, volumetric velocity, and mass velocity. Whereas in incompressible flow, the mass velocity would have kept on increasing as P_2/P_1 went to zero; in this case, G reaches a peak when the volumetric velocity reaches sonic velocity (in this case about 377 m s^{-1}) and P_2/P_1 is around 0.5. Although the results are for a specific case, the pattern can be assumed as general. The condition of peak G is called *choke flow*. The graph to the left of this point is meaningless because the mass flow is restricted at its choke point.

13.1.2 Adiabatic Constant-Area Pipe Flow

Industrial pipelines, because of their insulation, are more likely to be closer to adiabatic flow than isothermal. The same general behavior as in Figure 13.1 is observed. Lapple (1943) studied this situation and published predictions, which were later corrected by Levenspiel (1977). To take account of the change in temperature, these predictions incorporate the total energy balance (all in consistent units):

$$\text{Heat received (in this case zero because adiabatic)} = (\text{weight flow}) \times [\text{change in enthalpy} + (\text{change in volumetric velocity})^2] \tag{13.6}$$

The results are shown graphically in Figure 13.2.

The objective is to determine the mass velocity for combinations of pressure ratio, P_2/P_1, and frictional loss.

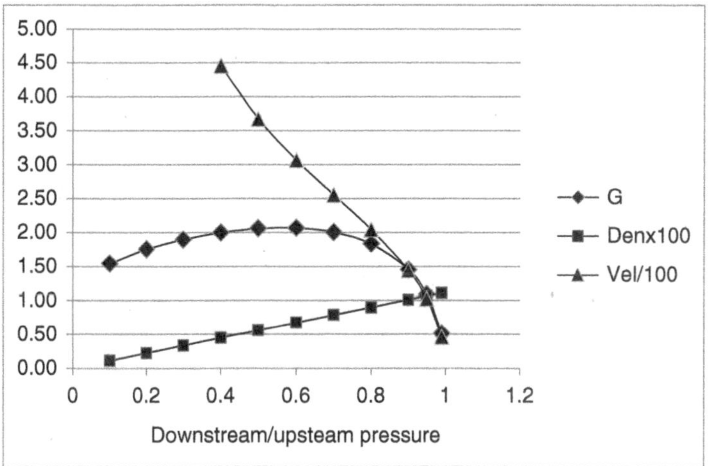

Figure 13.1 Behavior of isothermal compressible pipe flow.

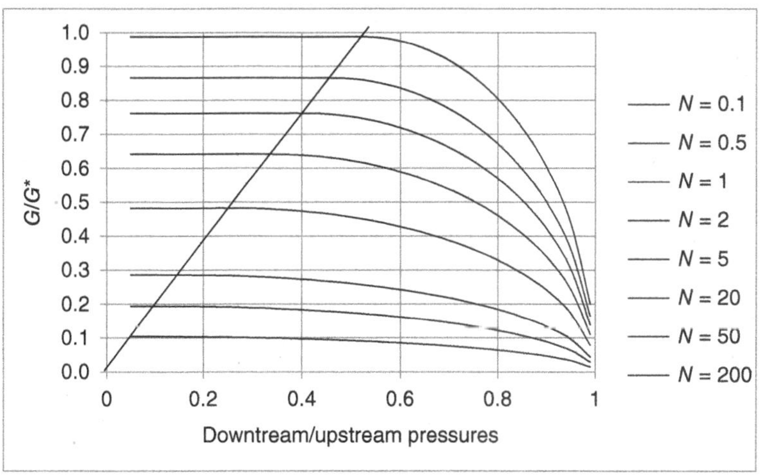

Figure 13.2 Representation of Lapple/Levenspiel charts for adiabatic pipeline flow.

The procedure for using these predictions is as follows:

(1) G^* = maximum possible mass velocity when there is no frictional resistance

$$= P_1 \times \sqrt{\{[g_c M \gamma/(R\,T_1)] \times [2/(\gamma+1)]^{(\gamma+1)/(\gamma-1)}\}} \tag{13.7}$$

where
γ is the ratio of specific heat at constant pressure to specific heat at constant volume, C_P/C_V,

(2) the maximum possible mass velocity when the frictional resistance is N velocity heads (for pipeline flow, $N = 4\,fL/D$; for orifice flow, $N = 0.63$),

$$(G_N/G^*)_{max} = 1 - \exp(-1.435/N^{0.484}) \tag{13.8}$$

(3) transition value of pressure ratio

$$\text{TRAN} = (P_2/P_1)/0.53 \text{ (see the sloped line in Figure 13.2)} \tag{13.9}$$

(4) if TRAN is less than $(G_N/G^*)_{max}$, then the flow is choked and the realizable G/G^* equals $(G_N/G^*)_{max}$, and

(5) if TRAN is greater than $(G_N/G^*)_{max}$, then

$$G/G^* = [(G_N/G^*)_{max}/(1 - X] \times \text{SQRT}\,[(1 - Y)\,(1 + Y - 2X)] \tag{13.10}$$

where X is 0.53 $(G_N/G^*)_{max}$ and Y is P_2/P_1
(the last equation has no theoretical basis and was developed simply to fit the Lapple curves).

Example, N of 5: $(G_N/G^*)_{max} = 1 - \exp(-1.435/5^{0.484}) = 0.482$.
For P_2/P_1 of 0.2, TRAN = 0.2/0.53 = 0.377, i.e., less than $(G_N/G^*)_{max}$, so G/G^* equals $(G_N/G^*)_{max}$.
For P_2/P_1 of 0.4, TRAN = 0.754, greater than $(G_N/G^*)_{max}$; $X = 0.53\,(G_N/G^*)_{max} = 0.255$,
$Y = 0.4$, $G/G^* = [0.482/(1 - 0.255)] \times \text{SQRT}\,[(1 - 0.4)\,(1 + 0.4 - 2 \times 0.255)] =$ **0.473**
which can be checked against the prediction of Figure 13.2.

13.1.3 Converging and Diverging Flow

Pipelines do occasionally narrow or widen. However, the importance of nozzles is their use in creating sonic and supersonic volumetric velocities and in the creation of vacuum. The equation for a friction-less nozzle is

$$[d(P)] \times (1 - \text{Ma}^2) = [\rho\,u^2/(A\,g_c)]\ \text{Da} \tag{13.11}$$

where
P is the pressure at a local point along the length,
Ma is the Mach number, which is local volumetric velocity u divided by the speed of sound,
ρ is the local density, and
A is the cross-sectional area of the nozzle at any point along its length.

This equation is directly derivable from Eq. (13.2), with the third (frictional) term omitted.

In a *converging* nozzle, pressure and density continuously decrease, while volumetric velocity increases. If Ma of 1 is not reached, then the nozzle acts simply as another piece of pipe.

From Eq. (13.11), it is obvious that the pressure can no longer decrease when Ma of 1 is reached. This puts a limit, a choke, on the flow.

If the exit of the nozzle is directly into a low-pressure environment (e.g. the atmosphere), then a shock wave establishes itself at the exit, where a sudden change in pressure occurs.

Figure 13.3 Converging–diverging nozzles.

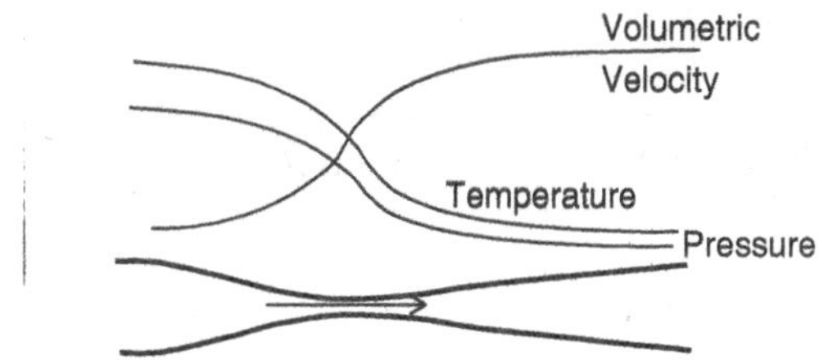

If instead the flow is guided by installing a gradually *diverging* nozzle after the choke point, the pressure keeps on decreasing but only gradually and the volumetric velocity rises – into the supersonic range. With proper geometry, it is possible to create a zone of subatmospheric pressure which can be used to draw a vacuum. See Figure 13.3.

For more information on compressible flow, consult Demneh & Mesbah (2008), Kenkre (2013), Kumar (2002), Shackelford (2015), Teng et al. (2014), and Yu (1999).

13.2 Gas Movers - General

Gas movers come in three types: fans, blowers, and compressors. A key parameter is CR, the CR, $P_{\text{Discarge}}/P_{\text{Inlet}}$, P_2/P_1, where the pressures are in absolute units. This parameter is often used to classify gas movers. In practice, each device operates over a range of CR and there is overlap rather than a sharp demarcation. As will be seen in the discussion of fans, the ISO definition is not based directly on CR, but it does restrict fans and blowers to a CR of 1.3 or less. However, above that value are regenerative blowers that can produce CRs of 1.7 and multistage centrifugal blowers up to 2.7. Beyond that is definitely the province of compressors. To some extent, the designation is in the eye of the beholder.

If the gas is ideal or if its compressibility factor Z is close to constant, then the ratio of absolute temperatures entering and leaving is given by

$$T_2/T_1 = (P_2/P_1)^{(\gamma-1)/\gamma} \tag{13.12}$$

where γ, i.e., C_P/C_V, is around 1.4 for many common gases.

The consequences of this relation can be seen in Table 13.1.

Two observations may be made:

- For large CRs, overheating may cause (i) damage to the equipment, (ii) higher gas volumes which lower the efficiency of the operation, and (iii) possibly dangerous conditions in the gas. Cooling is required, and in multistage compression, cooling must be done within or between stages.
- For small pressure ratios, it may be permissible to ignore these effects.

Fans are dynamic (as opposed to positive displacement) devices and come in two types, centrifugal and axial. There are fundamental differences between the two types, and they are considered separately.

Compressors are considered to start at CRs above 1.2 or 1.3. In general, the devices are mechanically driven and they comprise both dynamic and positive displacement types. Another gas mover/compressor is the *ejector*, a nonmechanical system in which a high-pressure motive gas entrains a gas stream of lower pressure and produces a combined stream of intermediate pressure. The ejector is similar to the liquid-flow *eductor* but is more powerful because of the ability of the motive gas to become supersonic and assume extremely low pressure.

Table 13.1 Temperature rise in gas compression.

Compression ratio (P_2/P_1)	Temperature ratio (T_2/T_1)	Temperature rise for T_1 of 273 °K
1.05	1.014	3.8
1.2	1.05	14.6
1.3	1.078	21.3
2	1.22	60
5	1.58	159

13.3 Fans

The ISO definition of a fan (including blowers) is that the mechanical energy, W_f, imparted to the gas is less than or equal to 25 000 J kg^{-1}. It is instructive to calculate this energy. From thermodynamics,

$$W_f = P_2V_2 - (P_1V_1 + \int P\,dV) = \int V\,dP = \int (1/\rho)dP \tag{13.13}$$

where V is the specific volume of the gas, m^3 kg^{-1}.

If the density is assumed constant,

$$W_f = (P_2 - P_1)/\rho \tag{13.14}$$

To apply the condition that W_f is 25 000, consider the case of air, at density of 1.2 kg m^{-3}, drawing from the atmosphere, at P1 of 101 325 Pa,

$$P_2 = P_1 + W_f \times \rho = 101\,325 + 25\,000 \times 1.2 = 131\,325$$

giving

$$\text{CR} = 131\,325/101\,325 = 1.3$$

The theoretical power can be calculated as

$$W \times W_f = W \times (1/\rho) \times (P_2 - P_1) = Q \times (P_2 - P_1) \tag{13.15}$$

where W and Q are the weight flow and volumetric flow, respectively. However, the actual required power is always greater – see Eq. (13.16).

13.3.1 Centrifugal Fans

Centrifugal fans (and blowers) are entirely analogous to centrifugal pumps. They use a vaned rotor to impart rotational velocity and centrifugal force to the gas, which is opposed by pressure exerted by the fan casing. The gas undergoes a ninety degree change in direction, entering at the center of the rotor and leaving tangentially from the outer periphery of the casing. A typical fan and impeller are shown in Figure 13.4.

The vanes can take one of five different forms, as shown in Figure 13.5. Qualitative performance curves are shown for each style of vane, where

Figure 13.4 Typical centrifugal fan. Source: New York Blower Company.

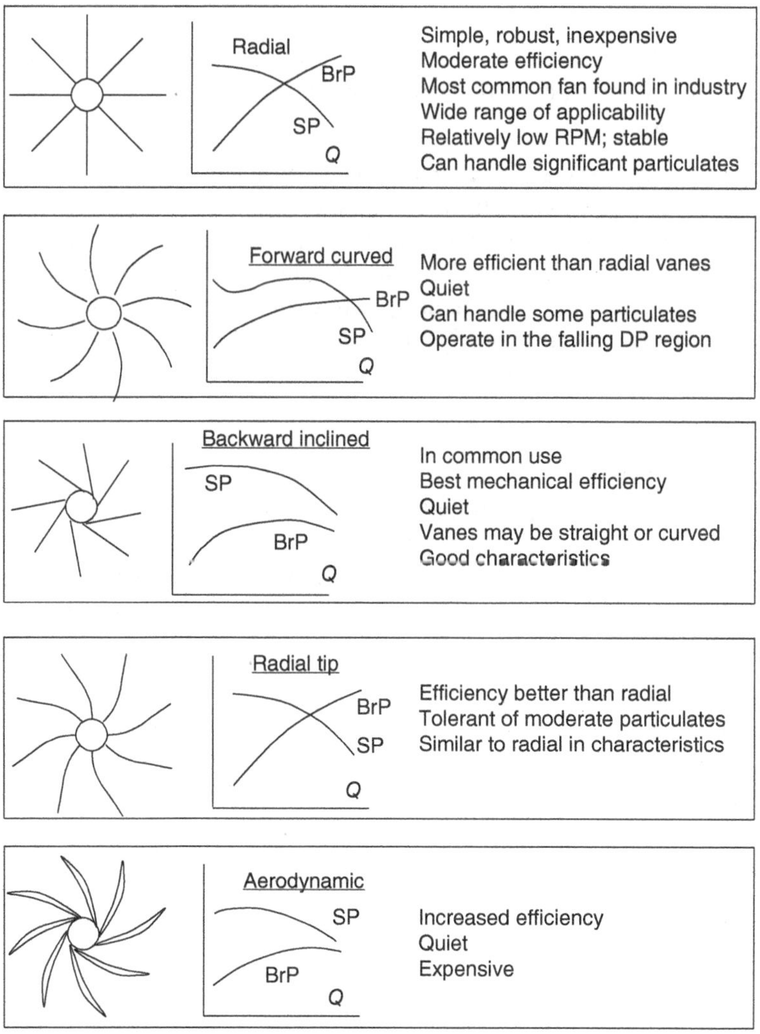

Figure 13.5 Centrifugal fan vane types.

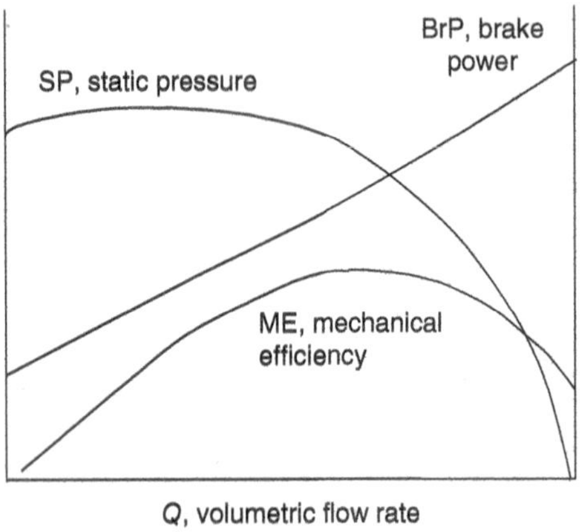

Figure 13.6 Fan characteristics.

SP is the *static pressure*, which is the term used in practice to denote the pressure rise across the fan, i.e., ΔP is the differential pressure,
BrP is the *brake power*, which is the term used in practice to denote the power delivered to the rotor impeller [as opposed to the power simply to compress the gas $(Q \times \text{SP})$], and
Q is the volumetric throughput of gas.

The behavior of a fan operating with constant impeller diameter, constant average gas pressure, and constant RPM is shown in Figure 13.6. As the throughput increases, the SP (i.e. the pressure rise across the fan) changes. For some vane types (other than radial), there is a slight increase at low Q, but then the SP gradually drops to zero, at which point the capacity of the fan is capped. Operation in the rising SP is unstable. In fact, the optimum operating point is about one-quarter of the way down the decreasing-SP curve, giving the maximum mechanical efficiency and allowing a range for fan control. The mechanical efficiency is defined as

$$\text{ME} = Q \times \text{SP}/\text{BrP} \tag{13.16}$$

units of $(\text{m}^3\,\text{s}^{-1})\,(\text{N}\,\text{m}^{-2})/W$. Other units for theoretical power $(Q \times \text{SP})$ are

$$\text{Power} = \text{volumetric flow} \times \text{static pressure} \times 0.0361 \times 144 \times 33\,000^{-1}$$

Horse-Power	cubic feet / minute	inches water	psi / inch water	sq. in. / sq.ft	horsepower / ft-lb/minute

The reason for the decrease of SP with Q is the loss of pressure due to the friction of flow between the vanes, as evidenced by the square-law dependence of SP loss versus Q.

The manufacturer of fans presents a table of fan performance from which the purchaser looks for a fan that best supplies the need. From one catalog, a sampling of data for a specific style and size of a fan is given in Table 13.2. For a desired air flow and SP, the table presents the required speed of rotation and the resulting brake power.

It is instructive to plot these data in the form of Figure 13.6. The data have been interpolated to show the effect of rotational speed (Figures 13.7 and 13.8).

Two observations stand out:

(1) the major effect of fan speed on both SP and brake power, and
(2) the presentation of data in the catalog is only for the region of likely operation, i.e., at or close to maximum efficiency, $Q \times \text{SP}/\text{BrP}$. The calculated efficiencies range from 0.54 to 0.78, with most cases near the high end.

Table 13.2 Manufacturer's fan specification.

Air flow, Q ($m^3 s^{-1}$)	Static pressure, SP pascals	Rotation speed, N (Rps)	Brake power, BrP (kW)
4.0	125	11.3	1.0
4.0	500	14.8	2.6
4.0	2000	7.3	12.0
5.2	125	14.1	1.8
5.2	500	16.8	3.7
5.2	2000	27.2	13.9
7.1	125	18.3	3.7
7.1	500	20.5	6.2
7.1	2000	28.3	17.7
8.3	125	21.3	5.5
8.3	500	23.2	8.4
8.3	2000	30.0	21.0

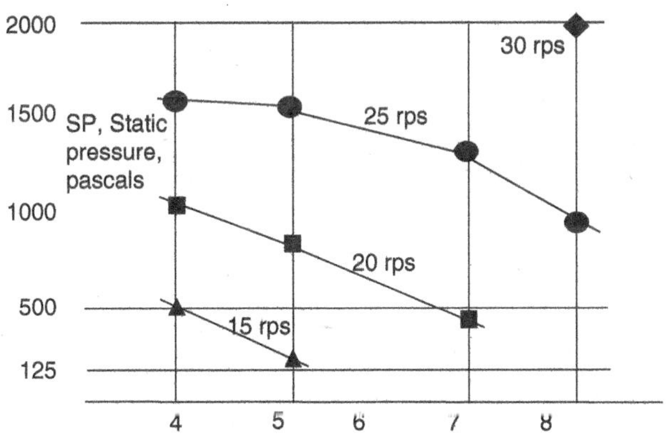

Figure 13.7 Characteristics of a commercial fan: static pressure.

There are three *laws of affinity* for centrifugal fans:

(1) Q, volumetric flow of gas is proportional to rotor speed x rotor diameter^3, $N D^3$,
(2) SP, static pressure is proportional to rotor speed^2 × gas density × diameter^2, $N^2 \rho D^2$, and
(3) BrP, brake power is proportional to rotor speed^3 × gas density × diameter^5, $N^3 \rho D^5$.

These proportionalities, for Q, SP, and BrP, are often inexact, but they give an idea of the direction and magnitude of changes. The first two are useful in devising one possible control for gas flow.

Figure 13.9 is an expansion of Figure 13.6, showing

- SP curves for three different rotational speeds and
- curves for downstream pressure drop, System Pressure Curve (SYSP), due to piping and other equipment. They are roughly proportional to the flow rate.

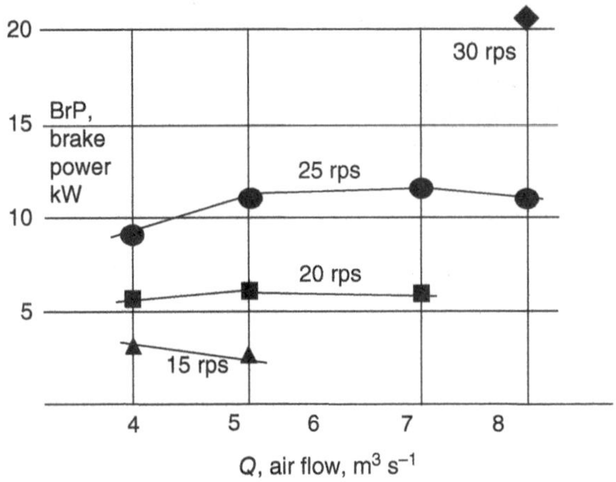

Figure 13.8 Characteristics of a commercial fan: brake power.

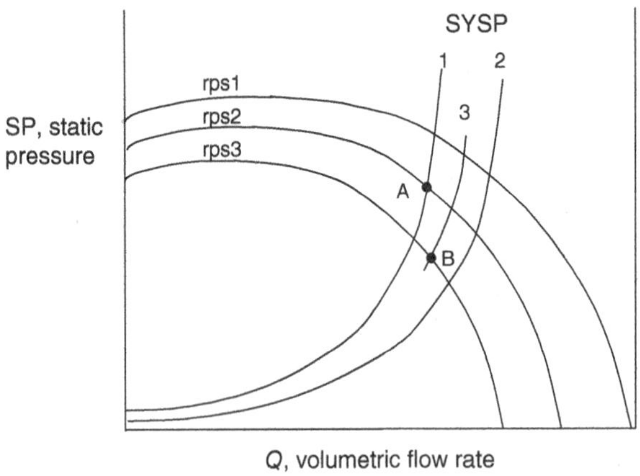

Figure 13.9 Fan and system characteristics.

If the rotational speed is rps2 and the downstream pressure follows the SYSP #1 line, then the combination of SP and Q is at point "A". If the downstream pressure shifts to SYSP #3 (perhaps by opening a valve or louver), and if it is desired to maintain the same flow, then the point of intersection must shift to point "B" and the rotational speed must shift to rps3. Mapping out these curves would help to set the constants for a control system.

This type of flow control requires a variable-speed drive. An alternative is to install an adjustable louver downstream of the fan to provide an additional source of pressure loss which can be manipulated to compensate for changes in system pressure drop, SYSP. This method costs energy. Another method is to install adjustable inlet vanes which cause swirl in the incoming flow and weaken the ability of the fan to develop pressure, which conserves horsepower.

A parameter called *specific speed* is derived from affinity laws 1 and 2 by expressing them in explicit form:

$$Q = k_Q \, N \, D^3 \text{ and } \mathrm{SP} = (k_S/g_c) \, N^2 \, \rho \, D^2$$

where the constants k_Q and k_S depend on the style of fan, mostly the shape of the vanes.

By eliminating the diameter D, the following equation results:

$$Q^{0.5} \, N \, \rho^{0.75} \, \mathrm{SP}^{-0.75} \, g_c^{\ -0.75} = k_S^{\ 0.75} \, k_g^{\ -0.5} = \text{constant} \tag{13.17}$$

The left-hand side expression, the *specific speed*, is dimensionless and provides a distinction between fans that are intended for moving the gas rather than developing pressure versus fans that are mainly required to develop pressure. It is independent of fan size but rather describes a class of fans. The purchaser should therefore estimate specific speed (using a likely value for N) and search for a sufficiently sized fan in that category.

13.3.2 Axial Fans

Axial flow fans move gas in the same direction as the axis of rotation. They tend to have higher specific speeds than centrifugal because they are used more for moving gas rather than developing pressure. They have the same *affinity laws* as centrifugal fans.

There are three general types: propeller, tube-axial, and vane-axial. Propeller types are generally completely in the open, as, for instance, in domestic fans, sometimes with a peripheral plate to reduce backflow. The SP is practically nonexistent, 250 Pa or less. They are used industrially to move large volumes of gas or vapor.

Tube-axial fans surround the rotating vanes with a tube, giving greater directionality and allowing some pressure to be developed – e.g., 3–4 inches of water, 750–1000 Pa. Figure 13.10 shows a typical tube-axial fan.

Vane-axial is a refinement of tube-axial but with an implanted set of vanes which convert much of the swirl created by the rotating vanes into useful axial motion. They also allow the *angle of incidence* (the *pitch*) of the rotating vanes to be set higher, generating more flow and pressure.

Both tube-axial and vane-axial are intended to be connected into piping systems.

Flow control of axial fans can take three approaches. The first two are the same as for centrifugal fans, namely speed manipulation and upstream or downstream louvers. A third approach, specifically for axial fans, uses the pitch of the rotating vanes to alter the SP-Q relation. Ideally, a means of changing pitch on the run would be ideal, but an expensive construction. More likely, the pitch is set ahead of time and louvers are used for fine-tuning. Pitch-control uses less energy than louvers.

Figure 13.10 Tube-axial fan. Source: New York Blower Company.

One manufacturer provides curves of SP and brake power for pitch angles of 4, 6, 10, 12, 14, showing considerable variation across this range and scope for designing and controlling the device.

Fans are discussed by Jenkins (2012, 2013) and Martin & Martin (2019).

13.4 Blowers

The distinction has already been made among the three types of gas movers. It is summarized in Table 13.3. The values for CR are specifically for supply pressure at 1 atmosphere, 101 325 Pa.

A fan at its very highest CR would be called a blower in some circles.

13.4.1 Regenerative Blowers

The *regenerative blower* and high-CR centrifugal devices are often used interchangeably, but their principle of operation is different. Centrifugal fans and blowers use their vanes to trap the gas and accelerate it toward the casing where the kinetic energy is converted into pressure. The velocity of the gas is more or less the same as the linear velocity of the vanes. The inlet and outlet are at right angles to each other, as seen in Figure 13.4. Regenerative blowers also have vanes, but they act in a different manner. As the gas enters the impeller chamber, it is picked up by a vane which accelerates it toward the wall. However, the periphery connects to a side channel which directs the gas back to the base of the vanes, at which point another vane picks it up and repeats the cycle. Each vane adds pressure to the gas, which gradually works its way around the chamber. As seen in Figure 13.11, the gas inlet and outlet are in the same plane as the rotational plane of the vanes.

The constant "regeneration" of the gas as it moves from vane to vane, and its gradual increase in pressure is what gives this blower a pressure advantage over centrifugal gas movers, as seen in Table 13.3. Centrifugal blowers tend to be used where the flow capacity is more important than pressure development, and regenerative blowers are used where the pressure development is key.

13.4.2 Rotary Lobe Blower

A photograph of the internals of a rotary lobe blower is shown in Figure 13.12. Unlike other blowers and fans, which are classed as *dynamic*, this blower is of the *displacement* type. As shown in

Table 13.3 Performance of fans and blowers.

Device	Range	Compression ratio (CR)	Static pressure, pascals	Static pressure, inches water	Static pressure (psi)
Axial fan	Top	1.01	1000	4	0.14
Centrifugal fan	Nominal top	1.07	7500	30	1.08
	Highest	1.25	25 000	100	3.6
Blower	Nominal top	1.3	30 400	122	4.4
Regenerative blower		1.74	75 000	300	11
Rotary lobe blower		2.23	125 000	500	18
Hoffman multistage blower		2.6	160 000	640	23

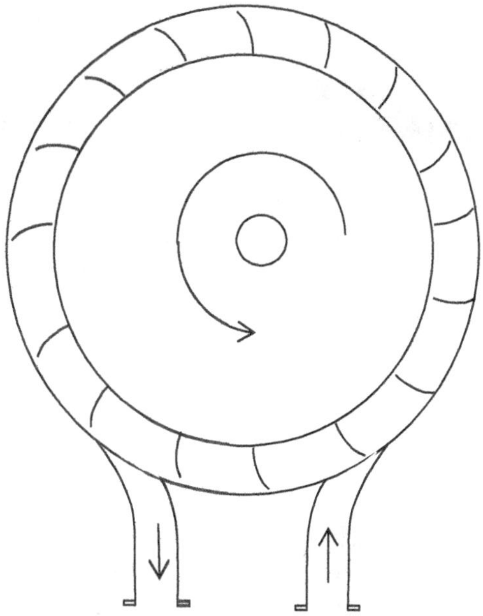

Figure 13.11 Regenerative blower.

Figure 13.12 Rotary lobe blower (Courtesy Aerzen)

Figure 13.13, the oppositely rotating twin lobes entrap pockets of incoming gas. They then reduce its volume as they turn, thus raising its pressure. The darkness of the ellipses indicates pressure. One example of this design is the well-known roots blower. The picture is of an Aerzen blower. Various other companies produce similar devices.

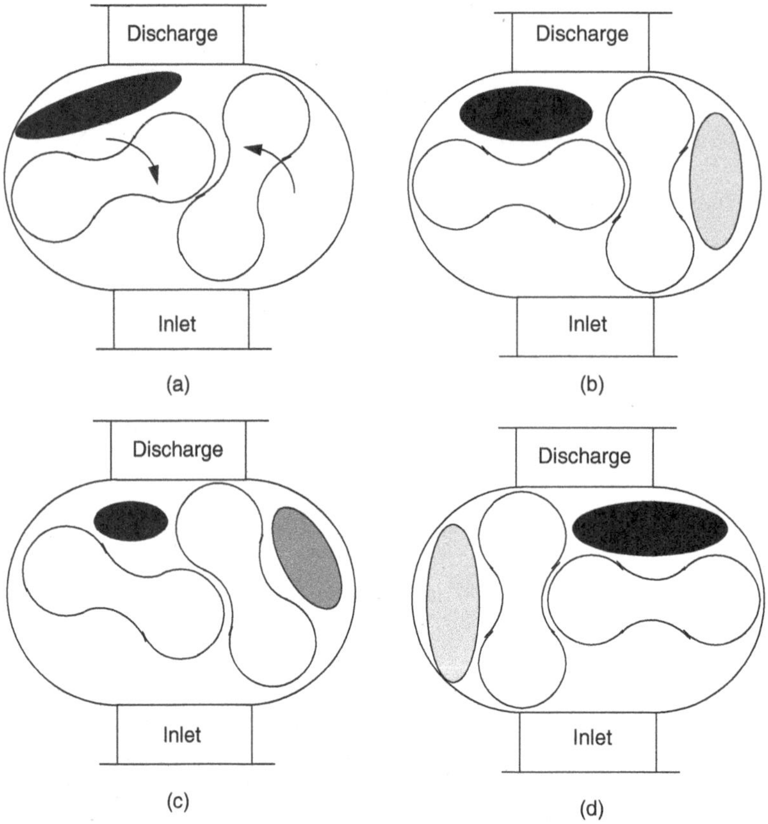

Figure 13.13 Pressure development in a rotary lobe blower.

13.4.3 Multistage Centrifugal Blowers

A blower consisting of several centrifugal units, each one feeding the next, is capable of generating high pressures relative to other blowers. Figure 13.14 shows a multistage assembly produced by the Lamson and Hoffman company.

Blowers are discussed by Bloch (2009), and lobe blowers by Blanton (2002) and Bloch (2012).

13.5 Compressors – Mechanical

The generation of flow and the generation of pressure are common to all gas movers. A distinction among them is that fans focus mainly on flow, blowers focus on both flow and pressure, and compressors focus mainly on the generation of pressure. In general, compressors handle CRs of 1.3 and higher, although regenerative, rotary lobe, and multistage centrifugal blowers reach into this territory.

A compressor may be tasked with raising gas at ambient pressure (one atmosphere for air) to a pressure above ambient. Alternatively, it may draw from a region of less than ambient and discharge to ambient, i.e., creating or maintaining a vacuum. The same equipment may be suited for both, but some are better suited for one or the other task.

Figure 13.14 Hoffman multistage centrifugal blower (Courtesy Gardner Denver).

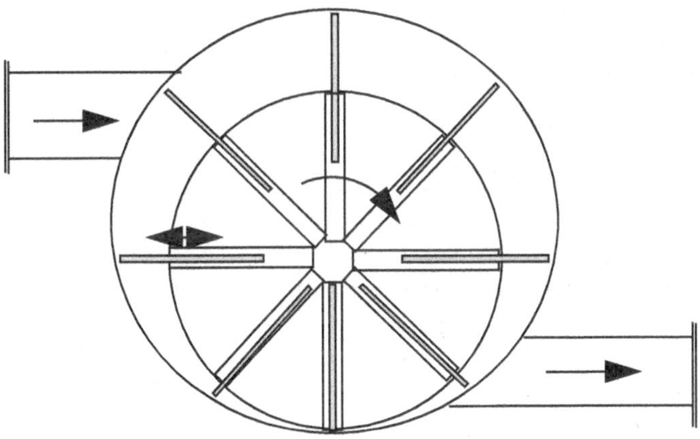

Figure 13.15 Sliding-vane compressor.

For **vacuum generation**, the three mechanical devices are the rotary lobe blower (as shown in Figures 13.12 and 13.13), the rotary sliding-vane pump, and the liquid-ring pump.

As a vacuum producer, the rotary lobe pump can sustain pressures as low as 60 mmHg abs, 8000 Pa, and 0.08 atmospheres.

A sketch of the sliding-vane pump is shown in Figure 13.15. The rotor is located off-centered from the casing, and the spring-loaded vanes move in and out. Their fullest extension is at the inlet. The trapped gas is compressed as the rotor turns, and it is expelled at the point of minimum volume and maximum pressure. A single stage can maintain a pressure of 15–20 mm. Hg abs and multiple stages can lower the upstream pressure to 0.001 mmHg abs or 0.13 Pa. Oil is usually injected to lubricate, seal, and cool.

The liquid-ring vacuum producer is shown in Figure 13.16. A rotor with vanes circulates liquid around the periphery of a noncircular chamber. The space between the liquid and the rotor varies around the chamber. Gas is introduced at two points of maximum space (points I) and is withdrawn at two points (points D) of minimum space, having shrunk and gained pressure. Vacuum as low as 100 mmHg abs is attained by a single stage. If the gas tends to liquefy, that liquid may be what

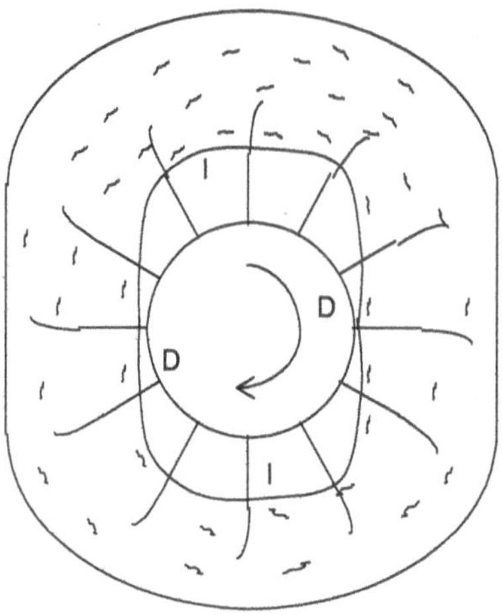

Figure 13.16 Liquid-ring compressor.

is circulated around the chamber. Sometimes a liquid-ring compressor is teamed up with another vacuum producer, providing a dumping spot for any *in situ* liquid or condensate.

The subject of vacuum generation is discussed by Collins (2012), Croll (1996), Datta (2008), Fay et al. (2000), Govoni (2017), Torzewski (2008), and Vibert (2004).

Compressors that convert ambient-pressure gas into high-pressure gas are the following: reciprocating, centrifugal, screw, scroll, and axial.

One version of a *reciprocating compressor* is a simple device where the forward motion of a piston reduces the volume of gas in a chamber and increases its pressure to equal the pressure downstream. On the backward motion, the low-pressure feed is pulled into the chamber. The piston may have a chamber on both sides, fore and aft, acting at 180° from each other. Maximum CRs are 4–8. A conceptual sketch of a single-acting pump is shown in Figure 13.17, illustrating the spring-loaded valves that control flow into and out of the compressor. Efficiencies are generally high, at 80–90%.

A weakness of piston compressors is the possibility of leakage around the piston. This limits the CR. A solution, also reciprocating, is the *reciprocating diaphragm compressor*. Instead of contacting a piston, the gas contacts a sealed flexible diaphragm, which is moved in and out by an attached

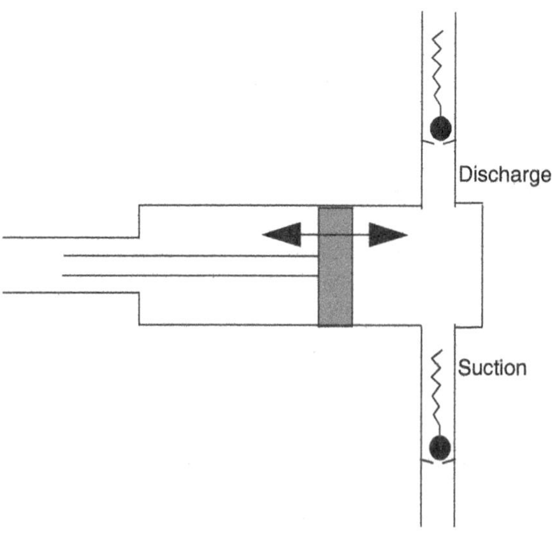

Figure 13.17 Reciprocating piston compressor.

rod or by hydraulic oil behind it. A schematic is shown in Figure 13.18. This device can deliver the highest pressure of any compressor and a higher CR than a piston unit. High-CR devices are more likely needed for low molecular weight gases which require a higher adiabatic head.

Centrifugal compressors are similar to fans and blowers of the same type. They are popular because they can be made to high capacity, have low contaminating oil injection, and are reliable. Although each stage has a low CR (less than 3), they can be staged to produce a high CR of 20 or more. A schematic is shown in Figure 13.19. There are two features that must be accounted for. One is that head of fluid, rather than pressure, is delivered by the device. Since the SP, ΔP, equals Head × gas density × g_c, centrifugal compressors are not suitable for low-density gases. The other is that the SP-versus-Q characteristic has an upward slope at low flow – see Figures 13.6 and 13.9. The throughput must be kept above this region to avoid instability – known as *surging*.

Screw compressors use two counter-rotating dissimilar elements to trap and pressurize gas and then to expel it. As seen in Figure 13.20, one screw has cavities and the other has lobes which fit into the cavities. Since the number of cavities and the number of lobes are generally not equal, the screws turn at two different rates. Timing gears are used to allow for tight fit, but without interference. With no lubrication, the CR can be up to 5, and with lubrication 20 or more.

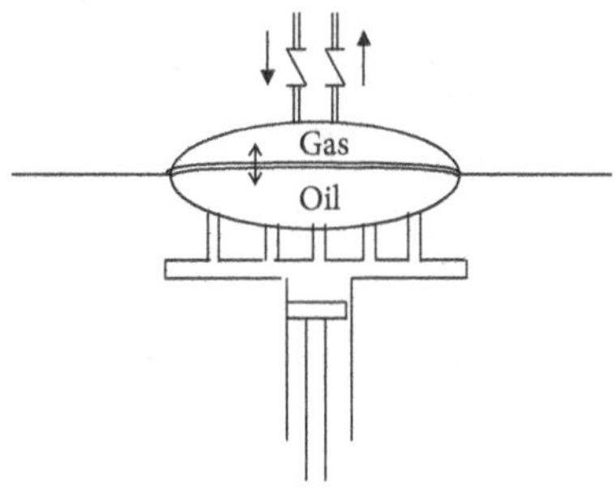

Figure 13.18 Reciprocating diaphragm compressor.

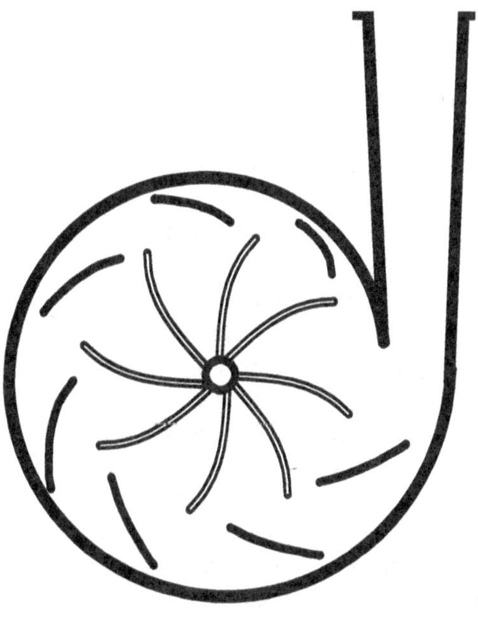

Figure 13.19 Centrifugal compressor.

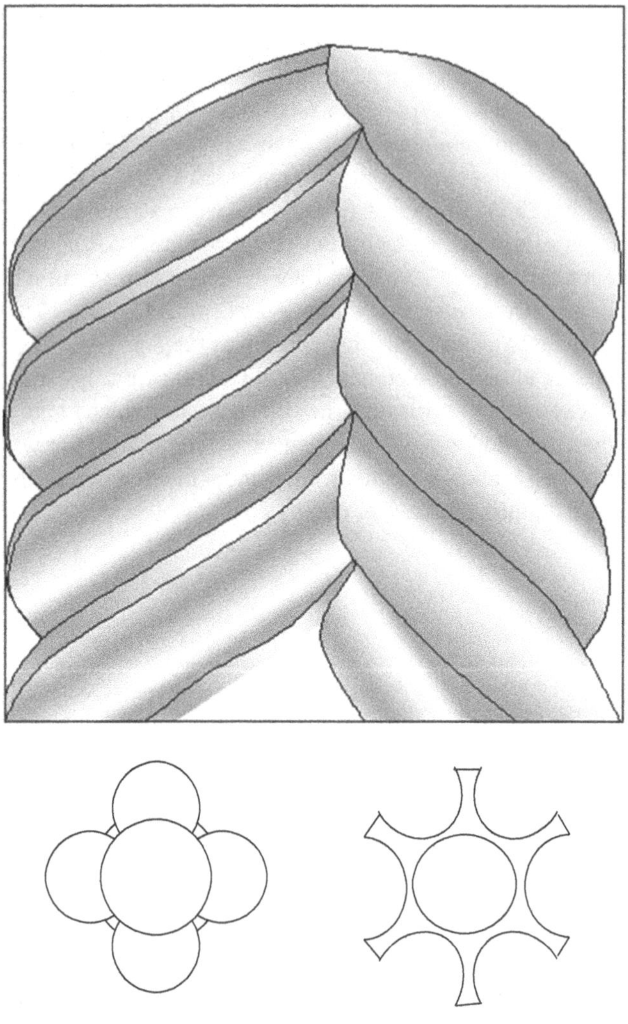

Figure 13.20 Screw compressor.

A *scroll compressor* consists of two spiral elements, one fixed and the other moving eccentrically in contact with it. Gas (or liquid) is trapped and moved from the inlet to outlet. Clearance is very tight. This is a quiet compact device, used in domestic and automotive service. CR is typically 3–4, but also 8–9 by some reports.

An *axial-flow* compressor consists of a single central rotor containing aerodynamically shaped vanes and a stator also containing aerodynamically shaped vanes. The rotating vanes, using the principle of *lift*, impart pressure and a small amount of swirl to the gas. The stationary vanes redirect the swirl into the axial direction. Axial compressors are more efficient than centrifugal, and they have the highest capacity of all compressors. See Figure 13.21 A comparison of compressors is presented in Figure 13.22.

Compressors in general are discussed by Barnett & Schramke (2000), Bloch (2012), Bloch & Geitner (2014), and Jandjel (2000). Almasi (2010) and Gong & Wan (2005) discussed reciprocating compressors. Centrifugal compressors are the subject of Almasi (2012C), Sorokes (2013), and Wilcox (2007). Screw compressors are treated by Almasi (2012A) and Sofronas (2008).

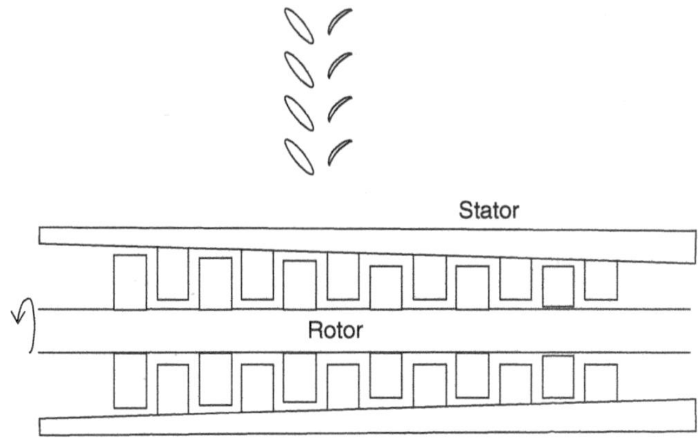

Figure 13.21 Axial-flow compressor.

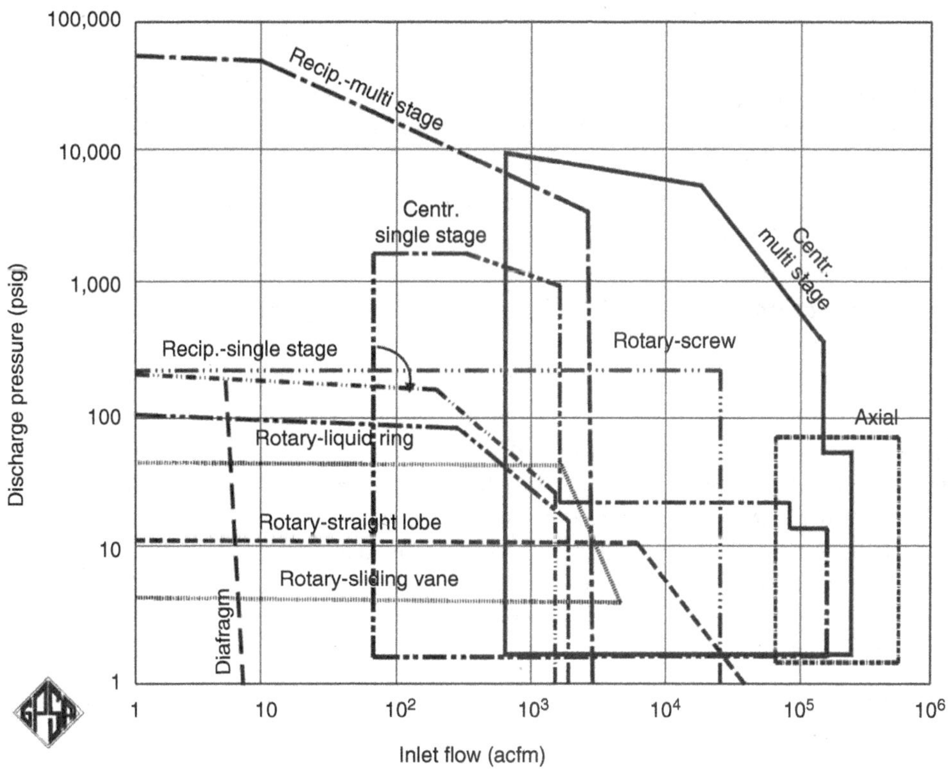

Figure 13.22 Comparison of compressors (courtesy of Gas Processors Suppliers Association).

13.6 Ejectors

An ejector uses the principles of compressible flow and allows a high-pressure stream to create a zone of low pressure (see Figure 13.3) into which a lesser-pressure stream is drawn and mixed.

The arrangement and flows are shown in Figure 13.23. Ejectors are used in two main applications

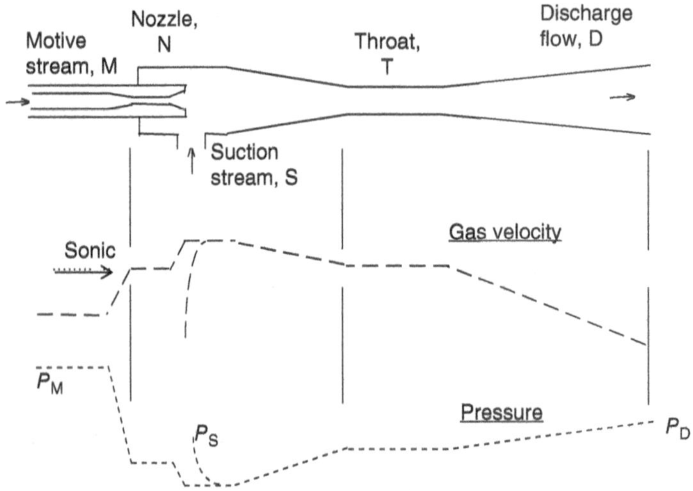

Figure 13.23 Ejector entraining a second stream.

(1) using a high-pressure gas (usually steam) to create a vacuum, and
(2) upgrading a low-pressure stream of no value by combining with a high-pressure stream to create a mid-pressure stream of value.

With steam as the motive force, there is the possibility of arranging ejectors in series to create very low absolute pressures. A two-stage system is shown in Figure 13.24. The diagram shows a number of features:

(1) intercoolers between the vessel and the first ejector and between the two ejectors, in order to liquefy most of the steam,
(2) the use of a "bleed stream" to moderate and control the vacuum, and
(3) the use of barometric legs to allow condensate to flow to an atmospheric-pressure receiver.

The capability of an ejector is obtained from the graphs in Figure 13.25. This set of curves is an extended version of the plot in Perry and Green, which is taken from the work of DeFrate & Hoerl (1959). The objective is to determine the flow of motive gas (W_M) for

a given CR, the ratio of absolute discharge pressure P_D (probably atmospheric for a single ejector) to suction pressure P_S and

a given ratio of suction pressure P_S to motive gas pressure P_M.

The procedure is as follows:

- on the right-hand side of the chart, at the intersection of the two ratios, find the required ratio of throat area to nozzle area (A_T/A_N);
- on the left-hand side, at the intersection of the A_T/A_N curve and the P_S/P_M line, read off the ratio W_S/W_M on the bottom line, suction flow/motive flow;
- knowing the required process suction flow, W_S, calculate the required flow of motive gas as $W_S/(W_S/W_M)$.

For example, if P_M, P_D, and P_S are 800, 100, and 8, respectively, in absolute units, then the chart shows the required A_T/A_N to be around 15 and W_S/W_M to be around 0.04.

Another example, for waste steam upgrading, with P_M, P_D, and P_S at 1825, 240, and 100 kPa, respectively, is the chart that indicates an area ratio of 20 and a flow ratio of suction to motive of around one. That is, one weight unit of high-pressure steam is required to boost one weight unit of low-pressure steam into a useful pressure.

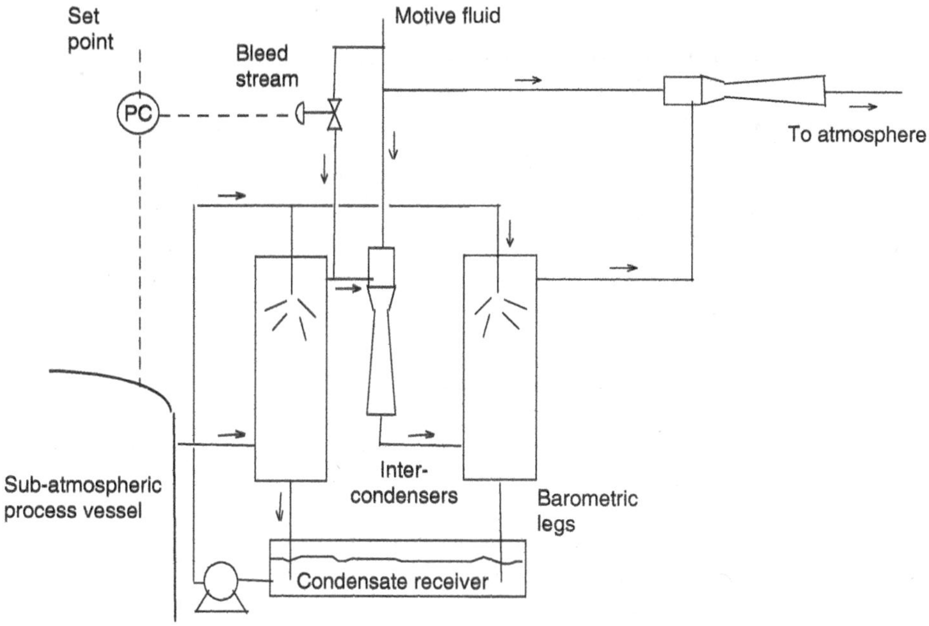

Figure 13.24 Two-stage steam ejector.

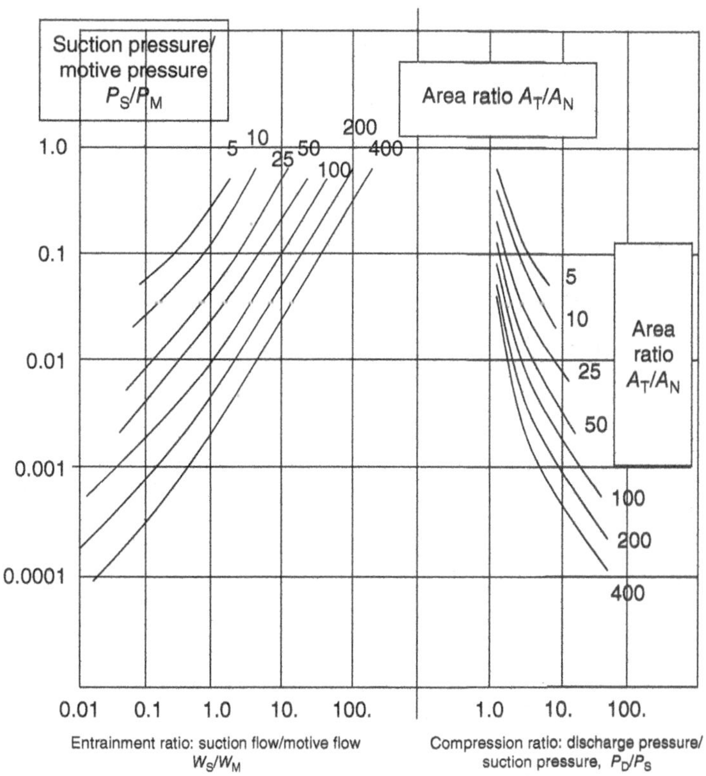

Figure 13.25 Single ejector design and performance.

The flow rate of motive gas is given by

$$W_m = 146\,000\, P_M\, A_N/T_M{}^{0.5} \text{ for air as the motive fluid and} \tag{13.18}$$

$$W_m = 113\,000\, P_M\, A_N/T_M{}^{0.5} \text{ for steam as the motive fluid} \tag{13.19}$$

The units are W_M kg h^{-1}, P_M kPa, A_N m^2, and T_M °C. In sizing an ejector, these equations get reversed to solve for A_N, the area of the nozzle.

Ejectors are the subject of papers by Lieberman & Cardoso (2016), Nayek & Venkatesh (2005), and Temur et al. (2013).

13.7 Thermodynamics of Gas Compression

The compression of a gas requires energy, and it also raises the temperature of the gas. These effects can be calculated, but they are always subject to an efficiency factor.

The energy of compression is given by the general formula

$$W_f = P_2 - P_1 - \int P\,dV = \int V\,dP = \int (1/\rho)\,dP \tag{13.20}$$

where

W_f is the energy (J) per unit (kg) of gas,
V is the volume (m^3) of unit of gas, and
ρ is the density (kg m^{-3}).

For compressible flow, it is necessary to account for the variation of density with pressure. Often the process is close to adiabatic, and the relation is

$$P/\rho^\gamma = P_1/\rho_1{}^\gamma \tag{13.21}$$

where

γ is C_P/C_V, the ratio of specific heat at constant pressure to specific heat at constant volume.

Substituting into Eq. (13.19) gives

$$\mathrm{WF}_{adiabatic} = [\gamma/(\gamma-1)]\,[P_1/\rho 1]\,\left[(P_2/P_1)^{(\gamma-1)/\gamma} - 1\right]. \tag{13.22}$$

The temperature rise is calculated from the other adiabatic relationship,

$$T_2/T_1 = (P_2/P_1)^{(\gamma-1)/\gamma} \tag{13.23}$$

which can be shown to be

$$T_{adiabatic} - T_1 = \mathrm{Wf}_{adiabatic}/C_P \tag{13.24}$$

Compression is accompanied by irreversible energy losses so that energy consumption and temperature rise are both greater than above. An efficiency factor $\eta_{adiabatic}$ may be applied such that

$$\eta_{adiabatic} = (T_{2\,adiabatic} - T_1)/(T_{2\,actual} - T_1) = \mathrm{Wf}_{adiabatic}/\mathrm{Wf}_{actual} \tag{13.25}$$

An example is shown, for P_2/P_1 of 5, γ of 1.4, T_1 of 50 °C, and η of 0.75,
from Eq. (13.12), $T_2/T_1 = 1.58$ and $T_{2\,adiabatic} = 1.58 \times (50 + 273.16) - 273.16 = 237$ °C;
from Eq. (13.14), $T_{2\,actual} = [237 + (0.75 - 1) \times 50]/0.75 = 299$ °C, i.e., 62° above $T_{2\,adiabatic}$.

If a Mollier diagram (enthalpy versus entropy) exists for the gas, then the compression can be graphed. In Figure 13.26, the adiabatic compression of steam from 30 psia (and 320 °F) to 50 psia

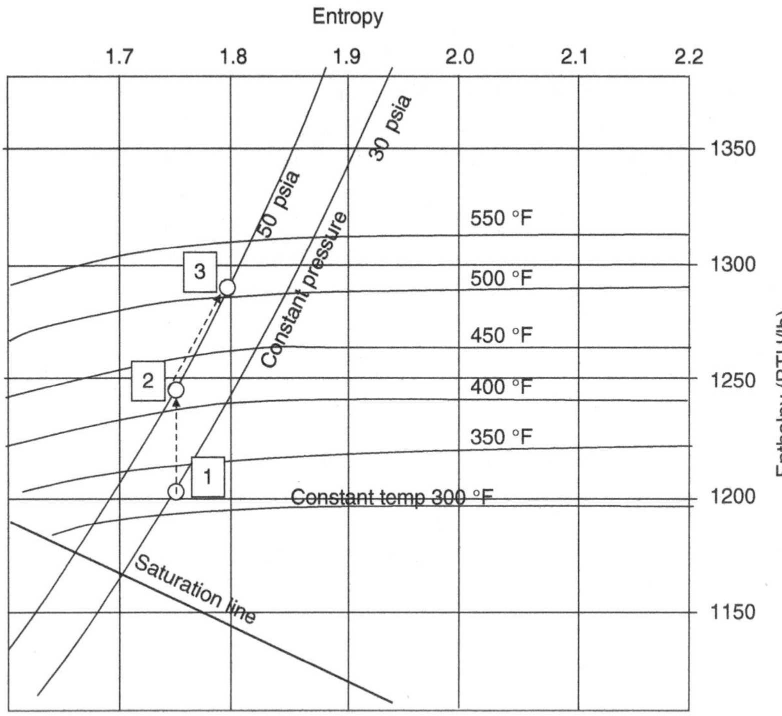

Figure 13.26 Mollier chart for gas.

along a line of constant entropy is shown. The initial enthalpy is 1200 btu lb^{-1}, and the adiabatically compressed enthalpy would be 1245 btu lb^{-1}.

If the process were ideal, the compressor would provide 45 BTU per pound and the gas temperature $T_{2\,\text{adiabatic}}$ would be 420 °F. However, if the adiabatic efficiency is only 50%, then the actual final enthalpy is 1200 + 45/0.5, i.e., 1290 btu lb^{-1}. The combination of this enthalpy with 50 psia is shown as point 3, and it indicates that the actual final temperature $T_{2\,\text{actual}}$ will be 505 °F. This is the most accurate method of calculation because it makes no general assumptions about the particular gas.

The large discrepancy between adiabatic and actual predictions has led to a different approach in which the index γ (=C_P/C_V) is replaced by an arbitrary index n in Eqs. (13.10)–(13.12), chosen to bring predictions closer to reality. Thus,

$$P/\rho^n = P_1/\rho_1{}^n \tag{13.26}$$

and

$$\text{Wf}_{\text{polytropic}} = [n/(n-1)]\,[P_1/\rho_1]\,\left[(P_2/P_1)^{(n-1)/n} - 1\right] \tag{13.27}$$

As opposed to adiabatic (an *isentropic* process), this approach is termed *polytropic*.

Since it is impossible to choose a single value of n to suit every case, a vendor will choose the mean value and then correct individual cases by a *polytropic efficiency factor*, $\eta_{\text{polytropic}}$. Then

$$\text{Wf}_{\text{actual}} = \text{Wf}_{\text{polytropic}}/\eta_{\text{polytropic}} \tag{13.28}$$

$$T_{2\,\text{actual}} = T_1 + \text{Wf}_{\text{actual}}/\text{CP} \tag{13.29}$$

References

AICE's Equipment testing procedures committee (2013 Jul). Evaluating centrifugal compressor performance. *Chem. Eng. Prog.* 109 (7): 35–39.

Almasi, A. (2010 Dec). Optimizing reciprocating compressors for CPI plants. *Chem. Eng.* 117 (13): 39–42.

Almasi, A. (2012A Feb). Choosing oil-injected screw compressors. *Chem. Eng.* 119 (2): 35–38.

Almasi, A. (2012B Mar). Condition monitoring for rotating machinery. *Chem. Eng.* 119 (3): 55–60.

Almasi, A. (2012C May). Centrifugal compressors for CPI plants. *Chem. Eng.* 119 (5): 42–45.

Almasi, A. (2012D Nov). Anti-surge valves for dynamic compressors. *Chem. Eng.* 119 (12): 43–47.

Almasi, A. (2014 Jun). Compressors for specialized applications. *Chem. Eng.* 121 (6): 47–52.

Barnett, J.M. and Schramke, T.M. (2000 Sep). Cost-effective compressor selection and specification. *Chem. Eng.* 107 (9): 70–76.

Blanton, R.E. (2002 Jul). Get the most out of your rotary lobe blower. *Chem. Eng.* 109 (7): 77–80.

Bloch, H.P. (2009 Feb). Modern blowers can be reliable. *Hyd. Proc.* 88 (2): 9.

Bloch, H.P. (2012 Jan). Consider lobe blowers combined with compressors. *Hyd. Proc.* 91 (1): 69–70.

Bloch, H.P. (2013 Jan). Update on wet and gas compressor seals. *Hyd. Proc.* 92 (1): 65–67.

Bloch, H.P. (2012 Jul). *Compressors: How to Achieve High Reliability & Availability*, 3. McGraw-Hill Education.

Bloch, H.P. and Geitner, F.K. (2014 Dec). How to select compressors for various services. *Hyd. Proc.* 93 (12): 81–82.

Bloch, H.P. and Carmody, C. (2015 Dec). Advances in dry gas seal technology for compressors. *Hyd. Proc.* 94 (12): 69–73.

Collins, D. (2012 Aug). Choosing process vacuum pumps. *Chem. Eng. Prog.* 108 (8): 65–72.

Croll, S.W. (1996 Jan). Properly specify vacuum systems. *Chem. Eng. Prog.* 92 (1): 48–49.

Datta, A.K. (2008 Oct). Better manage vacuum on gas systems. *Hyd. Proc.* 87 (10): 119–123.

Defrate, L.A. and Hoerl, A.E. (1959). Optimum design of ejector using digital computers. *Chem. Eng. Prog. Symp. Ser.* 55: 43–51.

Demneh, F.A. and Mesbah, A. (2008 May). The effect of kinetic energy change on flow in gas pipelines. *Hyd. Proc.* 87 (5): 81–84.

Fay, T., Kraus, J.N., and Levy, M.J. (2000 Sep). Improving vacuum systems. *Chem. Eng.* 107 (9): 86–91.

Ghanbariannaeeni, A. and Ghazanfarihashemi, G. (2014 Apr). Protecting against compressor pulsations. *Chem. Eng.* 121 (4): 50–55.

Gong, Y. and Wan, C.C. (2005 Apr). Get the most out of reciprocating compressors:[1]. *Chem. Eng.* 112 (4): 69–74.

Govoni, P. (2017 Sep). An overview of vacuum system design. *Chem. Eng.* 124 (9): 52–60.

Jandjel, D.G. (2000 Jul). Select the right compressor. *Chem. Eng. Prog.* 96 (7): 15–29.

Jenkins, S. (2012 Oct). Fans and blowers. *Chem. Eng.* 119 (10): 25.

Jenkins, S. (2013 Oct). Maximizing fan efficiency. *Chem. Eng.* 120 (10): 36.

Kenkre, P. (2013 Jan). Design and specification of a compressed air system. *Chem. Eng.* 120 (1): 40–48.

Kapada, M., Tellez-Schmill, R., and Ajdari, I. (2010 Aug). How the inertia number points to compressor system design challenges. *Hyd. Proc.* 89 (8): 45–49.

Knight, C. (2014 Mar). Reciprocating compressors running together – is that a problem? *Hyd. Proc.* 93 (3): 65–66.

Kumar, S. (2002 Oct). Spreadsheet calculates critical flow. *Chem. Eng.* 109 (11): 62–67.

Lapple, C.E. (1943). Isothermal and adiabatic flow of compressible fluids. *Trans. Am. Inst. Chem. Eng.* 39: 385.

Levenspiel, O. (1977). The discharge of gases from a reservoir through a pipe. *AIChE J.* 23: 402.

Lieberman, N. and Cardoso, R. (2016 Feb). Troubleshoot operation of a steam ejector vacuum system. *Hyd. Proc.* 95 (2): 59–64.

Lines, J.R. (2001 Sep). Understand freeze-condensation vacuum systems. *Chem. Eng. Prog.* 97 (9): 46–51.

Martin, V. and Martin, D. (2019 Jan). Optimize fan performance. *Chem. Eng. Prog.* 115 (1): 28–35.

Nayek, S.K. and Venkatesh, R. (2005 Dec). Use ejectors to reduce operating costs. *Hyd. Proc.* 84 (12): 69–70.

Nayek, S.K. and Chaudhari, L. (2013 Jun). Compressors: N_2 expands the applicability of dry gas seals. *Chem. Eng.* 120 (6): 50–51.

Shackelford, A. (2015 Jan). Temperature effects for high-velocity gas flow. *Chem. Eng.* 122 (1): 47–51.

Sofronas, T. (2008 Sep). Case 46: rotary screw compressor failure. *Hyd. Proc.* 87 (9): 169.

Sofronas, T. (2014 Jun). Case 79: integral compressor failures. *Hyd. Proc.* 93 (6): 37.

Sorokes, J.M. (2013 Jun). Selecting a centrifugal compressors. *Chem. Eng. Prog.* 109 (6): 44–51.

Stephenson, G. (2011 Jun). Integrate compressor performance maps into process simulation. *Chem. Eng. Prog.* 107 (6): 42–47.

Temur, T., Haktanir, M., Uzman, F. et al. (2013 May). Optimize vacuum ejectors operations. *Hyd. Proc.* 92 (5): 99–102.

Teng, F., Medina, P., and Heigold, M. (2014 Feb). Compressible fluid flow calculation methods. *Chem. Eng.* 93 (2): 32–41.

Torzewski, K. (2008 Jul). Vacuum pumps. *Chem. Eng.* 115 (7), p 31.

Vibert, P. (2004 Oct). Mechanical pumps for vacuum processing. *Chem. Eng.* 111 (11): 44–51.

Wilcox, E. (2007 Aug). Performance testing guidelines for centrifugal compressors. *Hyd. Proc.* 86 (8): 59–69.

Willetts, I. and Nair, A. (2010 Apr). Using high-fidelity dynamic simulation to model compressor systems. *Chem. Eng. Prog.* 106 (4): 44–48.

Yu, F.C. (1999 May). Compressible fluid pressure drop calculation – isothermal versus adiabatic. *Hyd. Proc.* 78 (5): 89–95.

14

Formation and In-Plant Transfer of Solids

Unlike liquids and gases, solids have shape as part of their description. The shape may be a three-dimensional pellet or powder, a two-dimensional film, or a one-dimensional filament. A particular solid may be produced from the reduction in size from larger particles, from the removal of solvent, through coating and agglomeration, or by molding during solidification from a melt.

The operations presented in this chapter are focused solely on solids in the form of pellets, powders, or other small entities. Sections 14.1–14.3 deal with the solids themselves, treating size reduction, size agglomeration, measurement, and characterization. Sections 14.4–14.6 deal with the transport of pellets and powders within a plant.

14.1 Solid's Size Reduction

The objective, here, is to use size reduction to produce the desired pellet or powder, starting with some other solid that may already be a pellet or may be in some other form. There is no agreed-upon dividing line in size between pellets and powder – usually, it is quite obvious. The term "granule", coming from grain, encompasses both pellets and powders. "Particle" is also used to denote both the in-process and final products.

Several methods of size reduction are in use, and new embodiments of these methods are constantly being developed, with the goal of

- being able to handle wider classes of materials,
- producing tighter distributions of size and shape, and
- requiring less energy.

With regard to materials, one device – the cutting mill – is used for relatively soft materials and for materials near the top of the hardness scale. Some other devices are suitable only for harder materials. Their capability is often matched against the scale of hardness proposed by Friedrich Mohs (Table 14.1).

With regard to distribution, every collection of solid pellets has some variation of size, as illustrated in Figure 14.1. Most often it is desired that this graph be as close to the ideal line as possible – i.e., that the pellets be uniform. A long tail on the left-hand side extension of the graph means there is a fraction of small particles that could end up as dust. The fineness of cut is often quantified by $\mathbf{d_{97}}$, which is the diameter below which 97% of the weight of the particles falls.

With regard to shape, uniformity is usually desired. The shape is commonly a sphere or an ovoid. Also common is the right-angle cylinder (length/diameter = 1, generally) obtained by continuous cutting of a moving strand of material.

Practical Process Design for Chemical Engineers, First Edition. Keith Marchildon and David Mody.

Table 14.1 Mohs scale of hardness.

Mohs hardness value	Material
1	Soapstone and talc
2	Gypsum, rock salt, and soft coal
3	Calcite, marble, soft limestone, and chalk
4	Fluorite, soft phosphate, and limestone
5	Apatite, bauxite, and hard limestone
6	Feldspar, ilmenite, and orthoclase
7	Quartz and granite
8	Topaz
9	Corundum, sapphire, and emery
10	Diamond

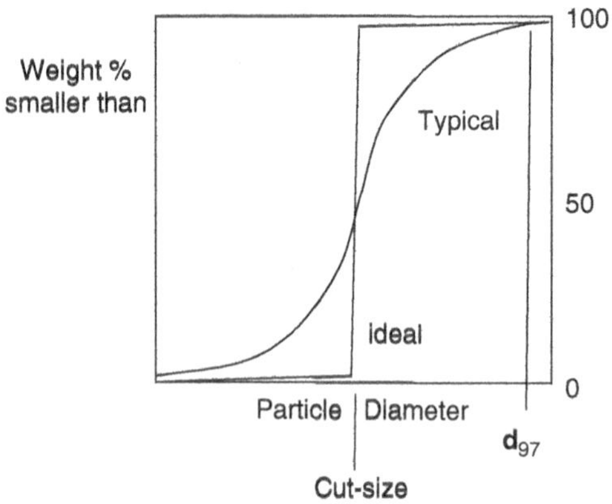

Figure 14.1 Particulate size distributions.

With regard to energy consumption, there are three different types of expressions:

$$\text{Rittinger } E = C_R\left(d_p^{-1} - d_f^{-1}\right) \tag{14.1}$$

$$\text{Kick } E = C_K \ln(d_f/d_p) \tag{14.2}$$

$$\text{Bond } E = C_B\left[\left(100\, d_p^{-1}\right)^{1/2} - \left(100\, d_f^{-1}\right)^{1/2}\right] \tag{14.3}$$

where E is the energy in kW-h ton^{-1}, the coefficients C are specific to the expression and the device, and d_f and d_p are the diameters of the feed and final product, respectively.

Even if the values of the coefficients cannot be ascertained for a particular device, the expressions allow comparisons for various feed and product diameters.

The methods of size reduction can be divided into two categories:

passive – where the material is simply cut or crushed;

impact – where the material is moved about at high speed and participates in its own reduction.

The impact category itself can also be divided into two subcategories:

- without media and
- with media, where an extraneous and harder set of pellets or objects is mixed with the process material to help with its reduction.

Silverberg et al. (1998) reviewed the family of size-reduction methods at that time, and much of their information is included here.

14.2 Cutting Mills

A cutting mill (a passive device from the point of view of the process material) uses rotating and stationary knives to cut materials on the shearing edges. Cutting mills are often used with soft, medium-hard, fibrous, tough materials, such as vegetable matter, plastics, and paper, and industrial scrap. The feed is often not a pellet but rather a continuous strand or a sheet. The end product size is a pellet typically 1–6 mm in size.

A different type of cutter is used to convert strands of polymer into pellets. See Figure 14.2.

14.2.1 Crushers

Compressive force between two solid surfaces causes a size reduction. Crushers are often used in the mining industry to reduce the size of minerals, gravel, and ores. The final particle size is 1–100 mm usually. Crushers may be used in series in order to achieve a low enough final particle size. A crusher may be used to prepare a feed for another, more sophisticated method of size reduction. Kumana (2015) recommended practices for optimizing crushing and explained the size distribution – which is Gaussian.

14.2.2 High Compression Roller Mills

These mills use compressive forces of one particle on another by way of a stationary roll or surface pressed against a moveable rotating roll to reduce the size of the material (Figure 14.3). This process again is passive to the process material. They exert at least 10-fold more pressure on the particles than a coarse crusher. The material in the mill may be air classified (elutriation), or classified upon exiting the mill. Alternatively, a roll press mill will have exit material in the form of a slab, and a hammer mill or other universal mill is used to reduce it to its component particles, followed by air

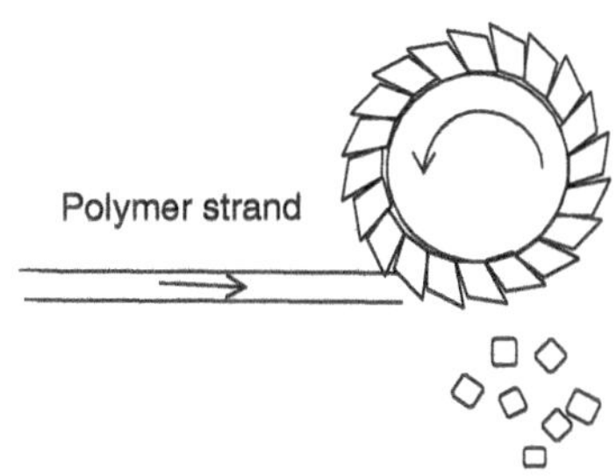

Figure 14.2 Strand pelletizing.

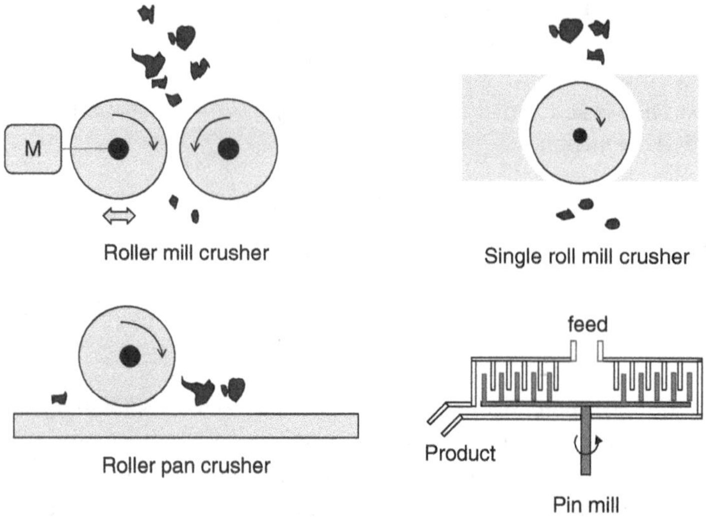

Figure 14.3 Roller and pin mills.

classification. Coarse particles are returned to the roller mill. The roller mill can produce powders from 5 to 45 μm, and with material which has a Mohs hardness up to 10. Compared to ball mills (see Section 14.2.7), the power usage can be 50% lower.

14.2.3 Universal and Pin Mills

"Universal" comprises a whole class of mechanically driven impact devices. These are also referred to as fine-grinding impact mills and make use of the principle of impact against both the grinding surfaces and impact from interparticle collision and attrition. An example of this type of mill is the pin mill which uses a set of rotating pins and a set of stationary pins. The particle size is determined by the feed rate and the rotational rate. Other mills are equipped with hammers and screens, but in all situations, the material hardness is limited to 3 Mohs because of internal (e.g. pin) materials. The applications of this type of mill are wide and varied and include milling of pharmaceuticals, food confectionaries, animal feeds, fertilizers, paints, pesticides, etc. If a problem with heat rise is foreseen, then the mixing chamber is designed to be larger. Universal mills can produce particles as coarse as $\mathbf{d}_{97}$ of 200 and as fine as $\mathbf{d}_{97}$ of 10 μm.

14.2.4 Hammer Mills

The material enters a chamber where it is struck repeatedly by hammers attached to a shaft that is rotating at high speed. The material is crushed or shattered by the hammer impacts as well as by impacts with the wall and with itself. Perforated screens or grates at the exit allow particles to leave when small enough.

It is simple to stop the mill and change a screen, so the hammer mill is ideal when a wide range of different particle sizes or products are to be produced. Product dimensions are quoted as between the 90 and 850 μm range. Cryogenic processing can be used which allows soft materials to be made brittle for easier processing.

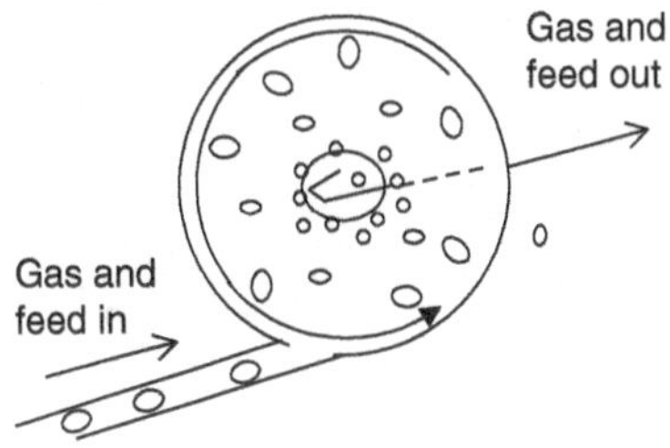

Figure 14.4 Simple jet mill.

14.2.5 Hammer Mill with Classifiers

Often, a hammer mill will be fitted with a classifier which entrains particles in a gas stream where a classifier wheel (see Chapter 17, Mechanical separations) is used to reject and return oversized material back to the mill. The classification system provides a tighter particle distribution in the final product. The constant air flow reduces temperature and is thus beneficial to a material which is heat sensitive.

14.2.6 Jet Mills

The original spiral and loop jet mills can produce very fine particles but only on materials with a hardness up to 3 Mohs. The principle of operation of the devices is to reduce the particle size by impact and to combine some degree of classification. The particle size is centered around the 10-μm region (Figure 14.4).

The fluidized bed jet mill (FBJM) incorporates more than one gas inlet and operates internally with gas at supersonic velocities. Particles are swept upward into an air (or gas) classifier which separates product from oversized granules, which it drops back into the high-speed flow. With this arrangement, there is greater separation by sizes than with the simple jet mill. The superiority of the FBJM is that it can handle materials with hardness of 10 Mohs and produce a final particle size of 2 μm.

Miranda (2017) reported the use of superheated steam, instead of air, as the motive gas in jet mills. So long as the process material is able to tolerate the temperature, there are advantages in fineness and size uniformity.

14.2.7 Dry Media Mills

The media is the *grinding media*, a set of spheres or rollers which accompany the material being reduced and which aids in the reduction. Typical of these mills is the horizontal dry-ball mill. These units along with a classifier can produce powders in the 10–40 μm size. A variety of different configurations and materials of construction are possible to provide an iron-free material or for products with a hardness above 4 Mohs. The media diameter is typically between 0.2 and 5 mm: The fineness of the product is directly related to the size of the media. Although the mill itself is fairly simple, the classification system can become sophisticated to ensure a narrow range in particle distribution.

14.2.8 Wet Media Mills

These mills are very similar to their dry counterparts but provide finer particulates, e.g., 1 μm in size, at lower energy costs than a dry mill. Wet media milling can produce particles less than

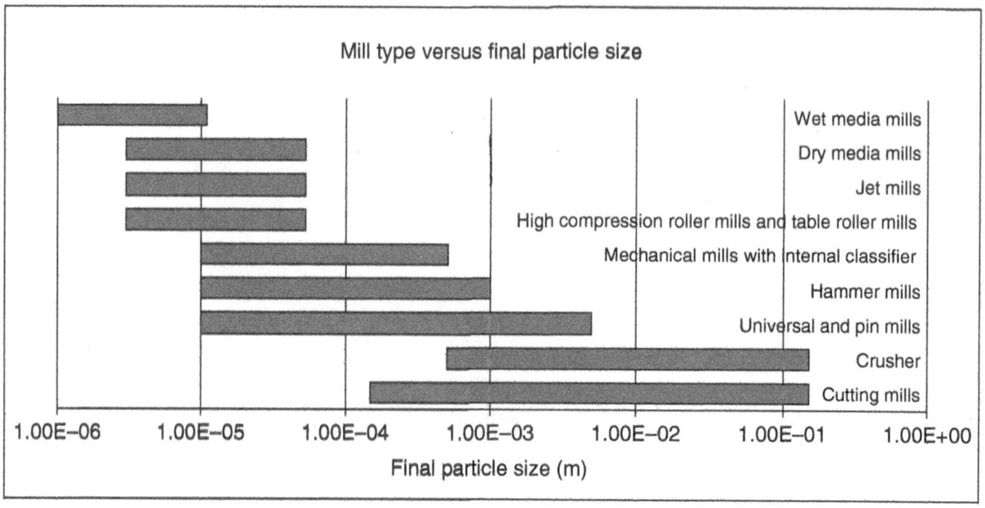

Figure 14.5 Size reduction by process (adapted from Clement & Purutyan 2002).

50 nm, which is the upper limit for an object to be classed as a *nanoparticle*. The liquid phase can be a solvent, and the mill can produce paints and inks. Way (2007) explained the technology. Wet processing can produce a particle size distribution that is not possible or not economical in dry processing. However, more recent developments in dry processing have narrowed the gap.

The various processes are compared in Figure 14.5 for the sizes of their end product.

For background on choosing a size-reduction device, the following references are useful: Jenkins (2012), Clement & Purutyan (2002), Johanson (2013), Dhodapkar & Theuerkauf (2011), Miranda (2017), and Silverberg et al. (1998).

14.3 Formation of Granules

Having dealt with the reduction in particle size, it is now important to look at growth. The objective is to cause individual particles to accrete together to form larger granules. The reasons for so doing are

- ease of handling, i.e., elimination of flow problems that occur with finer particles,
- requirement for larger granules in the end product,
- reduction of dust, and
- control of solubility.

It is conceivable that the results of a size-reduction step might then be subjected to a rebuilding step in order to narrow the distribution or swallow up troublesome low-size particles.

All processors are termed *granulators* because the end product is a granule. There are two groups of processes: *agglomeration* and *pressure compacting*. In all cases, success depends on the particles actually holding together. Mehos & Kozicki (2017) listed the possible binding forces, which may be summarized as follows:

- solid binders that adhere to the particles,
- liquid bridges that fill the interspaces and develop capillary force,
- Van der Waals forces, valence forces, and hydrogen bonding, and
- solid bridges resulting from sintering and from partial melting and recrystallization.

Mehos & Kozicki listed a number of potential binders, 10 of which are inorganic and 11 organic.

Ennis (2010A) listed the ingredients that may be used in wet granulators. They are the solid itself, water or other solvent, binders, diluents, flow-aids, surfactants, wetting agents, lubricants, fillers, and end-use aids.

The ratio of water to solid in wet grinding is 40–90% of the water which could be held in the particle-to-particle matrix. Any excessive water must subsequently be removed, typically by vacuum drying or in a fluidized bed.

In wet processes, the mechanism of granule building comprises three parallel steps:

- wetting and nucleation, where two or more particles become attached,
- coalescence and growth, and
- consolidation, where an equilibrium is attained between growth and attrition.

The following descriptions are of four wet agglomeration processes.

14.3.1 Tumbling Granulators

Disc or drum tumbling granulators are suitable for generating solids in the 1–20 mm range. Generally, they use a continuous feed of solids and wetting liquid to slowly enlarge the solid size. In the case of a drum granulator, a large (up to 4 m in diameter) inclined (0 to 10°) cylinder rotates at 7–17 rpm, while the material flows in almost plug flow (first in, first out) manner. Scrapers are sometimes employed to prevent buildup on the inside of the drum.

Disc granulators operate on the same agglomeration principle but with a more sophisticated geometry. A revolving disc is mounted at an angle closer to vertical than horizontal. Solid and liquid are continually fed onto the disc, and its turning movement brings solids up to the top where they roll down and pick up other solids. As granules reach a certain size, they become too heavy for this lift and they collect at the bottom where they are removed. There is enough classification that there is no need for recycle, in contrast to the drum process which does use recycle because of its broad range of product sizes in its output.

14.3.2 Mixer Granulators

Material 0.1–2 mm in size can be produced using mixer granulators. These devices use an agitator to mix particles and liquid to create the granulation. They are particularly adept at handling sticky, plastic type materials and are less sensitive to operating conditions than tumbling granulators.

14.3.3 Fluidized Granulators

Fluidized bed and spouted beds are suitable for producing solids in the range of 0.1–2 mm. The particles can exhibit two distinct characteristics: They may become highly porous due to the agglomeration of powders or evolve into high-strength layered granules as a result of seed particles being coated by liquid feeds. Fluidized bed granulators provide simultaneous drying and particle size growth.

14.3.4 Centrifugal Granulators

Primarily used in the pharmaceutical industry to generate particles of size 0.3–3 mm. In all of the various designs, the feed enters a horizontal disk rotating at high speed to form a rotating rope-like solid. The particles created are spherical and can be coated if a liquid feed is provided. The particles created are usually denser than those made in a granulator. Capital costs are higher than most other types of granulators. Rotary fluidized bed designs are possible which provide simultaneous drying.

14.3.5 Spray Methods (Spray Drying and Prilling)

Spray processes include spray dryers, prilling towers, and flash dryers. A variety of different feeds (solution gel, paste, emulsion, slurry, and melt) can be accommodated. For a spray process, the particle size is typically small (0.05–0.5 mm), which is dependent on the spray droplet size. The solvent which is evaporated must be dealt with.

Prilling also uses a spray, but the purpose of the spray is simply to act as a heat-transfer medium, to cool and solidify pellets of molten material dropped down a tower.

In the situation where fines are an issue, a fluidized system or hybrid fluid-bed spray dryer can be used.

14.3.6 Pressure Compaction

Pressure compaction uses an external force to join particles which then maintain shape through interparticle bonds. There are confined-pressure devices such as pistons, tableting, and roll presses which can exert the necessary forces; alternatively, extrusion devices such as pellet mills and screw extruders can be used.

14.3.7 Thermal Sintering

Solids may be combined through the use of elevated or reduced temperatures. This phenomenon may be applied in conjunction with other size enlargement processes or by itself. Metal powders are often joined with this technology (sometimes with a binder). High temperatures with associated operating costs are required (Figure 14.6).

Further information on these technologies is obtained in the following references:

Ennis (2010A, 2010B), Mehos & Kozicki (2011), Mehos & Kozicki (2017), and Pietsch (2007).

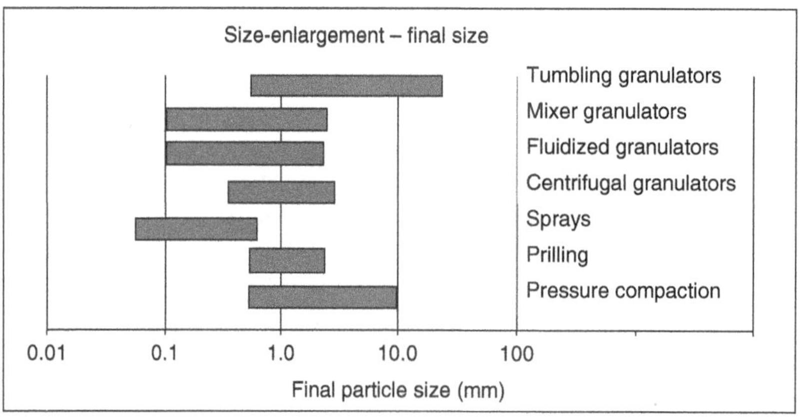

Figure 14.6 Product sizes from methods of enlargement

14.4 Measurement and Classification

Along with the processes that produce particles, another process is sometimes carried out, namely *classification*, the sorting of a stream of particles into parallel streams of different size ranges. Controlling all of these processes requires feedback, which requires the analysis of the final products: determination of size and size distribution as well as particle shape. The methods may take place offline or, more modernly, online where control can be done in real time.

Trottier & Dhodapkar (2014) listed 11 methods of particle sizing, the ranges of which are shown graphically in Figure 14.7. They describe each method in some detail. Some of the older techniques, such as *sieving* and *sedimentation*, are done offline. Others require high skill levels and are also performed in the lab.

At the moment, the most common method in the industry of determining size and distribution has been *laser diffraction*, a technique which covers a wide range of sizes and which can support online feedback control. Online systems are discussed by DeNigris & Ferrari (2012), Pugh (2007), and Pugh & Blasco (2005) based on laser diffraction. This technique is based on the extent of diffraction being dependent on the particle size. Light is shone through a stream of particles in liquid, and the diffracted beam is passed through a Fourier lens which directs the extents of diffraction to a number of sensors. Jenkins & Duffy (2016) singled out four common techniques: sieving, laser diffraction, imaging, and dynamic light scattering.

A number is needed to denote the average diameter of the particles. There is more than one way to define this number, that is, there are different diameters for the same collection of particles,

$$\text{Number–average diameter} = [1{,}0] = \sum n_\text{i}\, d_\text{i} \Big/ \sum n_\text{i} \tag{14.4}$$

where n_i is the number of particles in the size range around d_i. This definition of average counts every particle of whatever size as contributing equally to the average.

The denotation [1,0] means that d_i appears to the first power in the numerator and to the zeroth power in the denominator.

$$\text{Area–average diameter} = [2{,}0] = \left(\sum n_\text{i}\, d_\text{i}^{\,2} \Big/ \sum n_\text{i}\right)^{1/2} \tag{14.5}$$

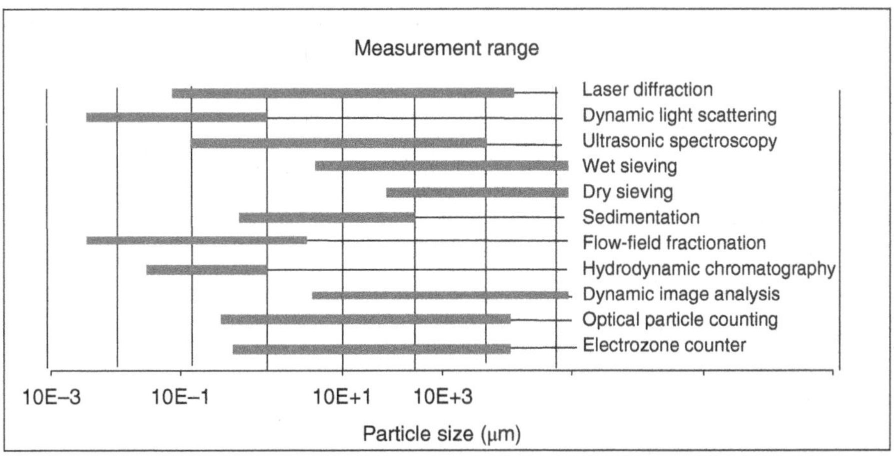

Figure 14.7 Range of particle size measurement.

In this average, particles with larger area count for a more significant part of the average,

$$\text{Volume–average diameter} = [3,0] = \left(\sum n_i\, d_i^{\,3} \Big/ \sum n_i\right)^{1/3} \tag{14.6}$$

In this definition, the higher-volume particles exert a greater contribution to the average.

The area-average is often suitable because it is at the surface of particles that mass transfer or reaction or catalysis occurs. However, there is a fourth definition which is even more suitable,

$$\text{Sauter mean diameter} = [3,2] = \sum n_i\, d_i^{\,3} \Big/ \sum n_i\, d_i^{\,2} \tag{14.7}$$

This expresses the ratio of volume to surface area.

To compare these expressions, consider a mixture of 3 particles of diameter 2, 5 of diameter 6, and 4 of diameter 9,

$[1,0] = (3\times 2+5\times 6+4\times 9)/(3+5+4) = 6.0$

$[2,0] = [(3\times 2\times 2+5\times 6\times 6+4\times 9\times 9)/12]^{1/2} = 6.6$

$[3,0] = [(3\times 2\times 2\times 2+5\times 6\times 6\times 6+4\times 9\times 9\times 9)/12]^{1/3} = 6.94$

$[3,2] = (3\times 2\times 2\times 2+5\times 6\times 6\times 6+4\times 9\times 9\times 9)/(3\times 2\times 2+5\times 6\times 6+4\times 9\times 9) = 7.8.$

The order of these numbers is general for whatever distribution, not just this example.

Applying these definitions to an actual collection of particles requires representative sampling, a detection method that can sort the particles into size groups, and an algorithm to calculate these averages. Depending on the end use toward which the particles are heading, one or the other of the averages can be used as the control variable for the particle production or classification process.

A stream of particles may have too wide a size distribution for the intended application. For instance, the low end of the size range may cause a dusting problem. The high end may have oversized particles that disrupt the smoothness of paints and finishes. *Classification* is needed in order to narrow the distribution. Wet or dry screening can eliminate unwanted sizes. Sedimentation can weed out the oversized particles. Chapter 17, Section 2, although it is aimed at separating particles of different composition, is useful in discussing solid–solid separation in general. Crawley et al. (2002) compared three methods of air classification:

- elutriation, useful for fine particles, can be used as a preprocessor for vortices,
- free-vortex, similar to a cyclone, separates coarse particles from the rest, and
- forced vortex uses a vaned rotor to aid peripheral motion.

Some of the size-reduction processes incorporate a built-in classifier to eliminate dust and to separate large particles which are recycled to the reduction zone.

Powders may pose flow problems with caking, and this subject is taken up by Armstrong et al. (2014), Marinelli (2014), Mehos & Clement (2008), Mehos (2016), and Purutyan et al. (2005).

Powders or dust from pellets are a potential explosion hazard. Amyotte et al. (2003), Brazier & Rooker (2009), Dhodapkar et al. (2006), Dhodapkar et al. (2007), Ebadat (2003), Kessler (2014), and Zalosh et al. (2005) presented remedial measures for this very important problem.

14.5 In-Plant Transfer of Solids

Nearly every chemical plant must handle a solid at some point in the process, whether as a raw material, additive, or finished product. Often the handling equipment for solids falls into the domain of the mechanical engineering department, but it is not uncommon for chemical engineers to be involved also. Preferably gravity (via chutes usually) should be used since it is the most reliable. Where gravity cannot be used, two classes of solids conveying systems are used: mechanical conveyors and pneumatic conveying.

In first determining the means by which a solid should be transported, the physical properties of the material should be understood. Properties such as:

- is the material friable (breaks apart easily)?
- is the material sticky, hydroscopic?
- what is the particle size and is it dusty?

A mechanical conveyor system is suitable when:

- materials require gentle handling and
- a high capacity (material flow rate) is required.

A pneumatic conveying system is suitable when:

- flexible routing is required,
- sealing the system (i.e. conveying under nitrogen, dry air, or when high levels of dusts might be harmful), and
- there are multiple pickups or discharges.

Mechanical conveyors are summarized briefly:

- Belt
  - High capacity
  - Can handle large particles
  - Does not break up the material (low attrition)
- Screw (Rigid or flexible helix)
  - Screw conveyors can convey under dry air or nitrogen
  - Heating or Cooling can be done
  - Inclined and vertical orientation are possible
  - Can damage the material, but usually not very much
- Vibratory
  - Good for short distances
  - Slight inclines are allowed. Spiral vibratory conveyors for larger inclines
  - Can be sealed at both ends providing dust tight systems
  - Not good for sticky, damp powders
- En Masse
  - Drag disk and aeromechanical
  - Drag chain and Redler
  - Bucket elevator

Screw conveyors are discussed by Boger (2008) and by Podevyn (2009).

Pneumatic conveying utilizes the principal that energy in the form of air pressure and velocity will move solids through a pipe. The principal is commonly encountered in your household vacuum cleaner. Two things need to be present to make solids flow using air. They are as follows:

1. The velocity of the air must be greater than the pickup or "saltation" velocity of the solids. In other words, the material must become swept up into the air stream before it will move. The pickup velocity can be determined from a theoretical point of view using drag coefficient calculations, or you can measure it using a real system.

2. As the air and solids move down the line, particles impact the pipe, slow down, and then are accelerated again by the air. This absorbs energy and steadily reduces the pressure of the air. Thus, there must be enough initial pressure in the air to offset the energy losses or the line will stop moving and become plugged.

There are two basic types of conveying systems, *vacuum* and *pressure*. Pressure systems are then subdivided into *dilute-phase* (high velocity) and *dense-phase* (low velocity) systems.

Within the dense phase category, there are several different variations provided by a variety of different vendors (Figure 14.8).

Dense-phase conveying maintains the particles in a state of incipient fluidization, and it allows the use of lower velocities, which normally means the material has less attrition by the time it reaches the end of the conveying system. Dense-phase conveying systems are usually more expensive to purchase (partially due the need for pressure piping, flanges, and screw compressors or use of higher-pressure air), so the advantages offered by reduced velocities need to be justified. The sizing of dense-phase conveying systems is normally left to the vendor and usually vendors that can provide both dilute and dense-phase conveying should be consulted so as to seek out the most economical system that meets the needs of the process. It is not uncommon for the pneumatic conveying vendor to request a sample of the material for testing so that they can properly design the system and provide performance guarantees. Crouch (1998) compared the dilute and dense approaches.

Dilute-phase conveying can be provided in positive-pressure and negative-pressure systems. Negative-pressure systems are well suited to situations where there are multiple pickups (i.e. a central vacuum system in your house), and positive-pressure systems are well suited where there are multiple destinations. In both systems, the standard equipment used is either a positive-displacement rotary lobe type blower or in small systems, sometimes a regenerative type blower (Figures 14.9 and 14.10).

Dilute phase systems can be sized with reasonable accuracy using straightforward calculations that are based on an energy balance [Gerchow (1975)]. There is one empirical factor, the material sliding friction coefficient, which depends on material properties and should be based on actual experience with the material being conveyed.

The gas flow can be estimated early on from the phase density for a dilute phase conveying system which is approximately 20 lb of material per lb of air for pressure systems and 10 lb of material per pound of air for vacuum systems.

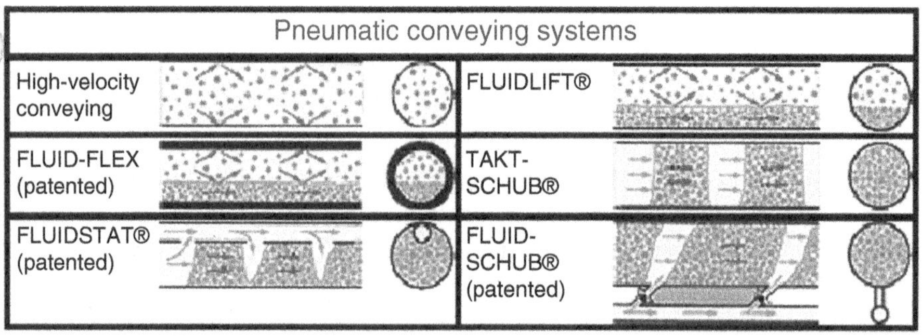

Figure 14.8 Apparatus for Dense-phase solids conveying. Source: Courtesy of Buhler Canada (www.Buhlergroup.com).

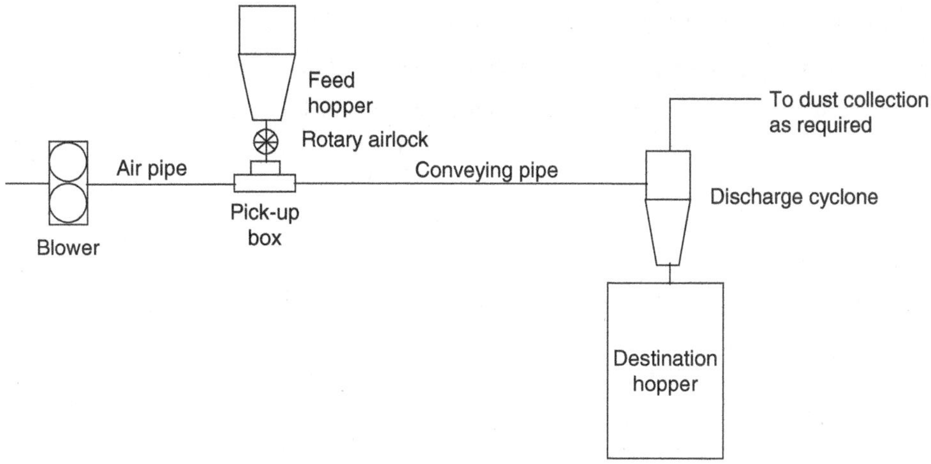

Figure 14.9 Typical dilute phase, positive-pressure system.

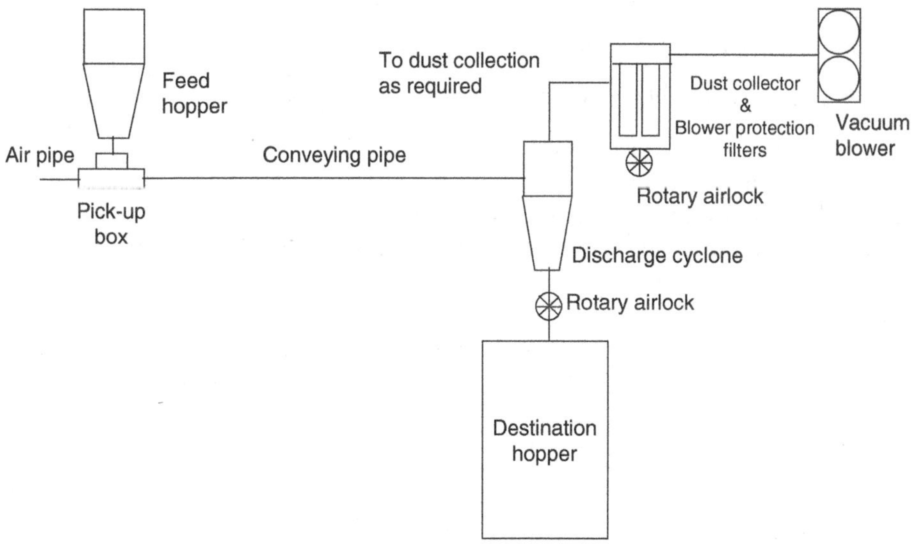

Figure 14.10 Typical dilute phase, vacuum system.

The pipe size is chosen to make the gas velocity high enough to pick up the solid particles from the bottom of the horizontal pipe. Cabrejos & Klinzing (1994) gave the following equation:

$$\text{Required pickup velocity (m s}^{-1}) = 0.042\,8 \times \text{pipe Reynolds number}$$
$$\times\ (\text{Pipe diameter/particle diameter})^{0.25}\ (\text{solid density/gas density})^{0.75}$$
$$\times\ \sqrt{(\text{acceleration due to gravity} \times \text{particle diameter})}$$

Knowing the gas volumetric flow, different pipe diameters can be chosen and tested with this equation until a match is achieved. A smaller diameter and a larger velocity will be chosen, but constraints are that (i) resistance to the air flow itself adds to the cost of operation and (ii) high velocities erode the pipe surface.

As the solids flow, they lose energy, which is calculated in the following four-part summation:

$$\begin{aligned}\text{TME} &= \text{Total material energy loss rate in foot–pounds force/second (or in watts)}\\ &= \text{acceleration} + \text{elevation} + \text{horizontal friction} + \text{elbow loss}\end{aligned} \tag{14.8}$$

where

acceleration loss = $0.5\,M \times V^2/g_c$,
elevation loss = $M \times H$,
horizontal loss = $M \times$ distance $\times$ Sfc,
elbow loss = number of elbows $\times [M \times V^2/(g_c \times R)] \times (2 \times \pi \times R/4) \times$ Sfc,

M is the solid mass flow, pounds per second (or kg s^{-1}),
V is the velocity of the gas–solid mixture, feet per second (or m s^{-1}),
g_c is the conversion – 32.2 lb mass × ft/(lb force × s^2) or 1 kg mass × m/(kg force × s^2),
H is the height of the vertical section (positive if upward), feet (or meters),
Sfc is the sliding friction coefficient, around 0.7 for plastic pellets, and
R is the radius of the elbow

(note that the radius of the elbow divides out in the Eq. 14.8).

Convert the total material energy loss rate into differential pressure:

Pressure drop due to solids (lbf/sq ft) = TME (ft-lb force/sec)/gas volume flow (cu ft/sec)

The material losses are not the only the pressure drop in a system. If the chosen line size is too small, the air losses through the system can be excessive and other items such as pickup boxes, cyclones, filters, and diverter valves should be added (Table 14.2). The air losses should be calculated using compressible flow calculations for the entire length of pipe (air only pipe + material conveying pipe),

Total system pressure drop = Solids ΔP + Air ΔP + Other Losses.

When the differential pressure has been calculated, the actual pickup air velocity should be calculated and compared to that which is required (for polymer pellets, the pickup velocity should be greater than 4200 ft min^{-1}).

The smallest line diameter should be selected that:

(1) ensures the pickup velocity for the material is met,
(2) the differential pressure of the blower is within practical values (for a rotary lobe blower, the max discharge pressure is usually about 14 psig and minimum vacuum pressure is −7 psig), and
(3) other considerations such as future expansion should be considered.

Table 14.2 Other typical losses in pneumatic conveying.

Item	Differential pressure (inches H_2O)	Differential pressure (lbf/ sq ft)
Pickup boxes/entry losses	1.55	0.29
Cyclones	3	0.58
Y-diverter valves	0.8″ @ 5000 ft min^{-1} (scale for other vel)	0.15
Filters	24	4.6

14.6 In-Transit Storage

A solids handling system will inevitably include one or more points where the material is accumulated or lagged. Quite often this point is just before a feeder which is being controlled to supply the next process step. It would seem that such a prosaic operation of material-holding could have little effect on the overall system. This is true if the holding vessel is properly designed. Otherwise there is the possibility of introducing nonuniformity into the particle stream and/or of causing the outflow to be erratic or even stopped. These phenomena have been much studied over time.

The vessel that holds the pellets or powders is generally vertical and cylindrical and is called a *bin* or *silo*. At the bottom is the *hopper*, a contraction down to an opening suitable for supplying the feeder. The hoped-for result is that material will flow uniformly down the vessel, i.e., first in is also first out. This is termed *mass flow*. Mass flow does not correct any time-wise variability in the incoming feed, but it does not introduce any new variability.

The alternate to mass flow is *funnel flow*, where substantial material settles on the sloping walls of the hopper. The steady flow occupies only the central part of the hopper, as shown in Figure 14.11, a phenomena known as "rat-holing". Parts of the held-up material may break off at random intervals, thus introducing a potential variability into the outflow. That material may, for instance, have become degraded or at least different.

A second difficulty occurs right at the point of discharge, where a collection of material, instead of exiting, forms a coherent bridge or *arch* across the opening. This obstruction may cause the outflow to be reduced, either steadily or randomly, either partially or totally, and can starve the feeder.

These two difficulties mostly affect powder flow, where the interparticle forces are greatest. Flow of pellets are less likely to suffer.

A third problem, which affects all sizes of particles, is *segregation* of sizes. If there is more than a 30% difference in the diameters among particles, then small and large particles have a tendency to separate. This may happen at the top of the pile, where larger particles roll across the surface toward the wall. It may also happen if a gas (e.g. air) is used to aerate the mass. Also, if particles are of different species and have different densities, then segregation can occur.

The measurement of solid properties is essential to answer these problems. Five properties are listed by Mehos et al. (2018):

- cohesive strength,
- internal friction,
- compressibility,

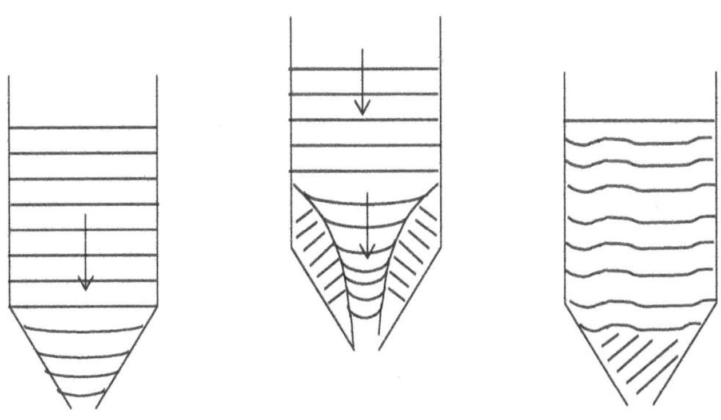

Figure 14.11 Bin modes: mass flow, funnel flow, and arching.

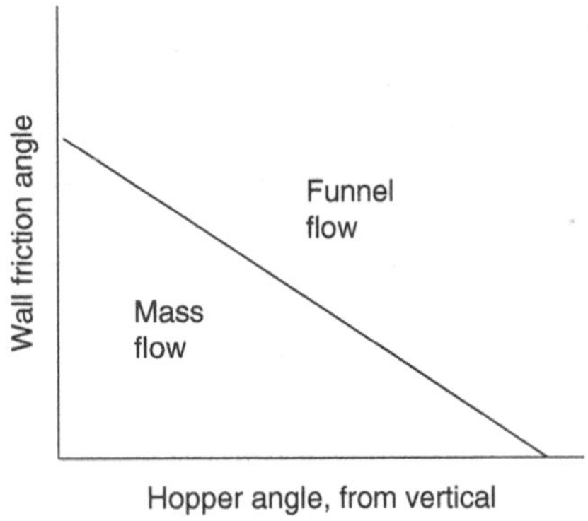

Figure 14.12 Allowable slope of hopper walls.

- wall friction, and
- permeability.

In the vertical part of the vessel, gravity is the dominant force, and friction exerts only a minor influence. However, in the hopper, gravity is reduced to a vertical component. Sometimes hoppers are made too short in order to save height and material. Friction and high cohesive strength combine together to plate the walls. Mehos et al. described test procedures (ASTM D-1628, D-6773) to measure these properties and how they are used to predict how steep the hopper walls must be in order to allow free flow and the desired *mass flow* situation. The tests produce a *wall friction angle* which, in combination with hopper angle, delineates the boundary between *mass flow* and *funnel flow*. See Figure 14.12.

Although the vertical part of the vessel may be cylindrical, the hopper can be transitioned into a wedge shape. As it turns out, such a configuration is less prone to arching, particularly if the bottom opening extends lengthwise across the bottom of the wedge. Wedges may also be allowed to be less steep that conical. In general, all hoppers should err on the steep side, even at the expense of requiring higher vessel height.

Mehos et al. also proposed criteria for arching.

Dhodapkar et al. (2016) and Carson et al. (2016) gave calculations for discharge rates from bins.

Maynard (2012) dealt very completely with the segregation of particles due to size or density. The mechanisms are described, and possible remedial steps are presented. These steps group into

- changes to the material,
- changes to the process, and
- changes to the equipment.

Other references on segregation are Bates et al. (2010), Carson (2004), Maynard (2008), Marinelli (2006), and Mehos & Maynard (2009).

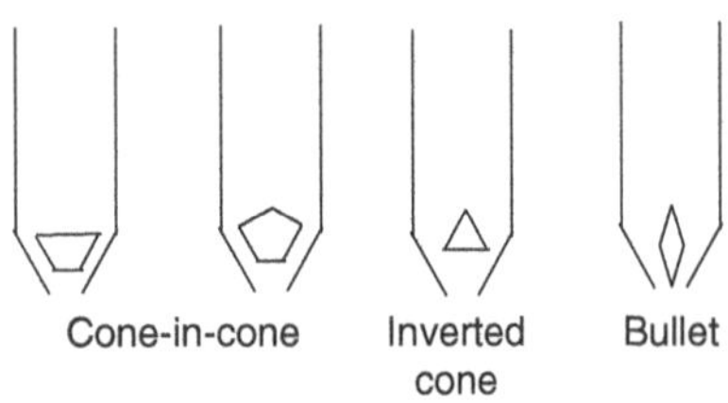

Figure 14.13 Hopper inserts.

With the exception of segregation-due-to-fluidization, a vessel in *mass flow* mode is better able to minimize the effects of segregation. Fluidization occurs when gases are added to a powder in order to help flowability [see Dhodapkar & Konanur (2005) and Purutyan et al. (2006)].

Bates et al. showed inserts for the transition zone of a bin which help to promote mass flow. See Figure 14.13. They list the specific problems which each style of insert helps ameliorate.

General references on bin flow are Carson et al. (2008), Johanson (2002), and Maynard (2013).

14.7 Solids Feeding

The end point of a solids system is delivery either into a package or into a processing vessel. The required consistency of feed depends on the nature of the destination. There are two types of feeder – volumetric and gravimetric – and the former typically provides accuracy of weight flow within 2–5%. Gravimetric is required for tighter control.

Marinelli & Miller (2017) listed seven types of volumetric feeders, of which the following are the most used: screw, belt, rotary valve, and vibratory.

Screw feeders consist of a rotating shaft with flights wrapped around it and turning in a close fitting trough. The flights push material along as they turn. The feeder is fed from a bin, and it is important that the bin has been designed to allow a steady flow to reach the screw. The screw may have a fairly sophisticated design along its length, with the flight depth being varied by changes to the diameter of the rotating shaft and with the pitch of the flights also changing. Some designs eliminate the shaft and simply have a spirally wound impeller turning in an enclosure.

Screw feeders can be oriented at any degree from the horizontal, from 0 to 90°. Roberts (2015) showed graphs of throughput versus speed for various angles of orientation. Bell et al. (2003) presented a guide to selection and use of screw feeders.

Belt feeders convey material on an endless belt. The rate depends on belt speed and belt loading. In this case, to control rate, the discharge from the bin needs either a vertical screw in the hopper or an adjustable gate.

Rotary valve feeders make use of rotating vanes attached in parallel to a horizontal turning shaft, as shown in Figure 14.14. The system is enclosed in a tight-fitting casing so that neither particles nor air can flow backward. The speed of the rotation dictates the material flow, providing the bin is designed to provide a uniform flow to the valve. A pipe section between the bin and valve helps to eliminate preferential flow from one side of the bin. This type of feeder is not suitable for sticky materials, nor should it handle abrasive solids which will damage the casing or the vanes and ruin the seal. Rotary valves are also used at pellet pickup points when the pneumatic transport is under pressure.

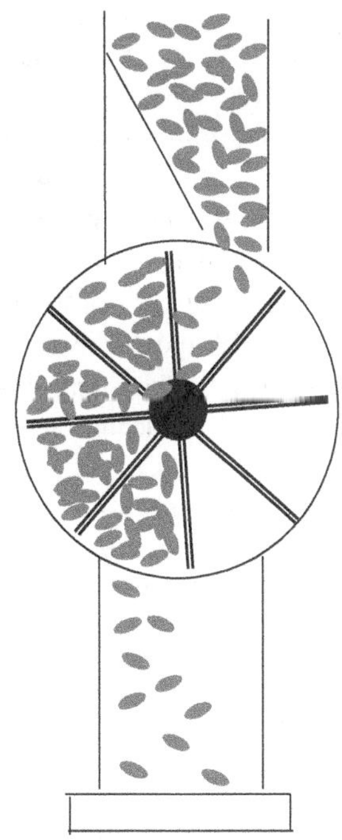

Figure 14.14 Rotary valve.

Vibratory feeders agitate the solid particles on a pan with a cyclic motion, which throws them up slightly and forward slightly. The rate is controlled by frequency. The extent to which a material responds to the vibration depends on the particle size, cohesive strength, and other properties. Key factors in design are the frequency range, amplitude, and throw distance. Yandrick (2009) discussed up to date developments.

Gravimetric feeders comprise weigh-belt feeders, loss-in weight (LIW) systems, and gain-in-weight (GIW) systems.

The weigh-belt feeder is like the volumetric belt feeder, but is mounted on load cells for continuous weighing. The material flow rate is calculated from the speed of the belt combined with the weight of the apparatus (minus the tare).

The LIW system precedes the feeder (e.g. a screw feeder) with a hopper mounted on load cells. The rate of weight loss, and hence the feed rate, is controlled by manipulating the speed of the feeder. Provision must be made for the short occasions when the hopper has to be replenished. Titmas & Carey (2007) discussed the choice of scales and load cells.

GIW applies only to batch operations, with the receiving vessel being mounted on load cells.

References

Amyotte, P.R., Khan, F.I., and Dastidar, A.G. (2003 Oct). Reduce dust explosions the inherently safer way. *Chem. Eng. Prog.* 99 (10): 36–43.

Armstrong, B., Brockbank, K., and Clayton, J. (2014 Oct). Understand the effects of moisture on powder behavior. *Chem. Eng. Prog.* 110 (10): 25–30.

Barrett, P. (2003 Aug). Selecting in-process particle-size analyzers. *Chem. Eng. Prog.* 99 (8): 26–32.

Bates, L., Dhodapkar, S., and Klinzing, G. (2010 Jul). Using inserts to address solids flow problems. *Chem. Eng.* 117 (7): 32–37.

Bell, T., Couch, S. W., Feise, H. J., Krieger, L. T. (2003 Feb), Screw feeders: A guide to selection and use, *Chemical Engineering Progress*, 99 (2), 44–51

Bharadwaj, R. (2012 Sep). Using DEM to solve bulk material handling problems. *Chem. Eng. Prog.* 91 (9): 54–58.

Boger, D. (2008 Apr). Move difficult-to-handle bulk materials with flexible screw conveyers. *Chem. Eng.* 115 (4): 36–39.

Brazier, G. and Rooker, M. (2009 Oct). Preventing dust explosions. *Chem. Eng.* 116 (10): 49–51.

Cabrejos, F. J. and Klinzing, G. E. (1994 May) Pickup and saltation mechanisms of solid particles in horizontal pneumatic transport, *Powder technology*, 79 (2), 173–186.

Carson, J.W. (2002 Aug). Equipment, modeling and testing bulk solids handling. *Chem. Eng.* 109 (8): 98–100.

Carson, J.W. (2004 Feb). Preventing particle segregation. *Chem. Eng.* 111 (2): 29–31.

Carson, J.W., Troxel, T.G., and Bengtson, K.E. (2008 Apr). Successfully scale up solids handling. *Chem. Eng. Prog.* 87 (4): 33–40.

Carson, J., Pittenger, B.H., and Marinelli, J. (2016 Apr). Characterize bulk solids to ensure smooth flow. *Chem. Eng.* 123 (4): 50–59.

Clement, S. and Purutyan, H. (2002 Jun). Narrowing down equipment choices for particle-size reduction. *Chem. Eng. Prog.* 98 (6): 50–54.

Crawley, G., Malcolmson, A., Crosley, I., and McLeish, A. (2002 Apr). Particle classification: making the grade. *Chem. Eng.* 109 (4): 54–60.

Crawley, G. and Malcolmson, A. (2004 Sep). Online particle sizing as a route to process optimization. *Chem. Eng.* 111 (9): 37–41.

Crouch, C. (1998 Aug). Conveying: how dilute phase stacks up against dense phase. *Chem. Eng. Prog.* 94 (8): 79–86.

DeNigris, J. and Ferrari, A. (2012 Feb). Control strategies based on realtime particle size analysis. *Chem. Eng.* 119 (2): 41–44.

Dhodapkar, S. and Konanur, M. (2005 Aug). Selection of discharge aids for bins and silos. *Chem. Eng.* 112 (8): 27–31. 2005 Oct, *112(10)*, 71–82.

Dhodapkar, S., Bates, L., and Klinzing, G. (2006 Aug). Don't fall for common misconceptions. *Chem. Eng.* 113 (8): 31–35.

Dhodapkar, S., Manjunath, K., and Jain, P. (2007 Jan). Design safer solids processing plants. *Chem. Eng.* 114 (1): 34–39.

Dhodapkar, S., Trottier, R., and Smith, B. (2009 Sep). Measuring dust and fines in polymer pellets. *Chem. Eng.* 116 (9): 24–29.

Dhodapkar, S. and Theuerkauf, J. (2011 Jun). Maximizing performance in size reduction. *Chem. Eng.* 118 (6): 45–48.

Dhodapkar, S., Jacob, K., and Kodam, M. (2016 May). Determining discharge rates of particulate solids. *Chem. Eng. Prog.* 12 (5): 50–60.

Ebadat, V. (2003 Oct). Is your dust collection system an explosion hazard? *Chem. Eng. Prog.* 99 (10): 44–48.

Ennis, B.J. (2010A Mar). Agglomeration technology. *Chem. Eng.* 117 (3): 34–39.

Ennis, B.J. (2010B May). Agglomeration technology: equipment selection. *Chem. Eng.* 117 (5): 50–54.

Fox, B. (2005 Mar). True grit: granulation & drying of delicate products. *Chem. Eng.* 112 (3): 35–38.

Freeman, T. (2011 Oct). Characterizing powder flow. *Chem. Eng.* 118 (10): 66–72.

Gerchow, Frank., J. (1975 Feb). How to Select a Pneumatic Conveying System. *Chemical Engineering*, 82 (4); 72–86.

Jenkins, S. (2011 Apr). Hopper inserts for improved solids flow. *Chem. Eng.* 118 (4): 26.

Jenkins, S. (2012 Nov). Particle-size reduction. *Chem. Eng.* 119 (12): 25.

Jenkins, S. and van Hulsel, H. (2013 Nov). Calculating volumes of bulk solids. *Chem. Eng.* 120 (11): 25.

Jenkins, S. and Duffy, J. (2016 Mar). Particle-sizing technology selection. *Chem. Eng.* 123 (3): 46.

Johanson, J.R. (2002 Apr). Troubleshooting bins, hoppers and feeders. *Chem. Eng. Prog.* 98 (4): 24–36.

Johanson, K. (2013 Nov). Selecting the proper mill for your product. *Chem. Eng.* 120 (11): 47–54.

Kessler, W.S. (2014 Jul). Dust control in the chemical processing industries. *Chem. Eng.* 121 (7): 59–63.

Klinzing, G.E., Rizk, F., Marcus, R., and Leung, L.S. (2010.). *Pneumatic Conveying of Solids*, 2nde, vol. 8. Particle Technology Series.

Kumana, J. (2015 Apr). Increase profits in size-reduction plants. *Chem. Eng.* 122 (4): 68–72.

LePree, J. (2015 May). Improved particle-size analysis boosts quality. *Chem. Eng.* 122 (5): 28–34.

LePree, J. (2020 Jan). Particle size matters. *Chem. Eng.* 127 (1): 12–15.

Marinelli, J. (2006 Apr). Will mass flow solve all your segregation problems? *Chem. Eng.* 113 (4): 40–43.

Marinelli, J. (2006 Nov). The do's and don't's of chute design. *Chem. Eng.* 113 (12): 63–64.

Marinelli, J. (2014 Apr). Overcoming solids caking with flow aids. *Chem. Eng.* 121 (4): 38–41.

Marinelli, J. and Miller, S. (2017 Aug). Feeder Design for Solids Handling, *Chemical Engineering*, 124 (10), 43.

Marshall, R. (2007 Mar). Pneumatic conveying systems. *Chem. Eng.* 114 (3): 39.

Maynard, E. (2008 May). Blender selection and avoidance of post-blender segregation. *Chem. Eng.* 115 (5): 67–71.

Maynard, E. (2012 Apr). Avoid bulk solids segregation problems. *Chem. Eng. Prog.* 108 (4): 35–39.

Maynard, E. (2013 Nov). Ten steps to an effective bin design. *Chem. Eng. Prog.* 109 (11): 25–32.

McGee, E. (2006 May). Predicting powder flow behavior – a new approach. *Chem. Eng.* 113 (5): 34–36.

McGregor, R. (2015 Sep). Solids discharge: characterizing powder and bulk solids behavior. *Chem. Eng.* 122 (9): 62–65.

Mehos, G. and Pittenger, B. (2007 Aug). Using bins and silos to heat or cool bulk solids. *Chem. Eng.* 114 (8): 57–62.

McGuire, K. (2007 Feb). Silo design and selection. *Chem. Eng.* 114 (2): 27–30.

Mehos, G. and Clement, S. (2008 Aug). Prevent caking and unintended agglomeration. *Chem. Eng.* 115 (8): 55–61.

Mehos, G. and Maynard, E. (2009 Sep). Handle bulk solids safely and effectively. *Chem. Eng. Prog.* 105 (9): 38–42.

Mehos, G. and Kozicki, C. (2011 Jan). Consider wet agglomeration to improve powder flow. *Chem. Eng.* 118 (1): 46–49.

Mehos, G. and Morgan, D. (2016 Jan). Hopper design principles. *Chem. Eng.* 123 (1): 58–63.

Mehos, G. (2016 Apr). Prevent caking of bulk solids. *Chem. Eng. Prog.* 112 (4): 48–55.

Mehos, G. and Kozicki, C. (2017-08 Oct). Choosing agglomeration equipment New York: Access Intelligence, LLC Chemical Engineering. 124 (10): p. 51.

Mehos, G., Eggleston, M., Grenier, S. et al. (2018 Apr). Designing hoppers, bins, and silos for reliable flow. *Chem. Eng. Prog.* 114 (4): 50–58.

Mills, D. (2000 Dec). Optimizing pneumatic conveying. *Chem. Eng.* 107 (12): 74–80.

Mills, D. (2005 May). Pneumatic conveying: know your options. *Chem. Eng.* 132 (5): 58–63.

Mills, D. (2005 Jun). Pneumatic conveying: more options plus guidelines. *Chem. Eng.* 132 (6): 46–51.

Mills, D. (2015 Nov 11). *Pneumatic Design Guide*, 3rde. Butterworth Heinemann.

Miranda S., The quest for nanotechnology and the evolution of wet and dry milling processes, Powder & Bulk Solids, 2017 Mar 30. Retrieved https://www.powderbulksolids.com/size-reduction/the-quest-for-nanotechnology-and-the-evolution-of-wet-and-dry-milling-processes.

Patel, C.M. (2019 Jul). Particle size characterization and analysis. *Chem. Eng.* 126 (7): 54–60.

Pawar, J. (2003 Sep). Wet grinding at its finest. *Chem. Eng.* 110 (9): 39–44.

Pietsch, W. (2007 Nov). Understanding agglomeration. *Chem. Eng. Prog.* 103 (11): 18–20.

Podevyn, M. (2009 Feb). Selecting a conveyor. *Chem. Eng.* 116 (2): 27–29.

Pugh, D. and Blasco, A. (2005 Nov). Online particle analysis in wet processes. *Chem. Eng.* 112 (12): 55–60.

Pugh, D. (2007 May). Sizing up online particle size analysis. *Chem. Eng. Prog.* 103 (5): 23.

Purutyan H., Troxel T. G., Cabrejos F., Propel your pneumatic conveying system to higher efficiency *Chem. Eng. Prog.* 2001 Apr, 97*(4)*, 42–55.

Purutyan H., Pittenger B. H., Tardos G. I., Prevent caking during solids handling, *Chem. Eng. Prog.* 2005 May, 101*(5)*, 22–28.

Purutyan H., Carson J. W., Troxel T. G., Increase powder flow by direct injection, *Chem. Eng. Prog.* 2006 Jul, 102*(7)*, 38–43.

Rawle, A.F. (2018 Apr). Taking representative samples in solids-handling processes. *Chem. Eng.* 125 (4): 38–46.

Rentz, J. and Churchman, C. (1998 May). Streamline predictions for pneumatic conveyors. *Chem. Eng. Prog.* 94 (5): 47–54.

Roberts, A. W. (2015 Feb). Bulk Solids: Optimizing Screw Conveyors. *Chemical Engineering*. 122 (2), 62–67.

Silverberg, P.M., Miranda, S., and Yaeger, S. (1998 Nov). Homing in on the best size reduction method: start with particle-size distributions and hardness to simplify the task. *Chem. Eng.* 105 (12): 102–118.

Solt, P.E. (2002 Jan). Solve the 5 most common pneumatic conveyor problems. *Chem. Eng. Prog.* 98 (1): 52–55.

Theuerkauf, J., Dhodapkar, S., and Jacob, K. (2007 Apr). Modeling granular flow. *Chem. Eng.* 114 (4): 39–46.

Titmas, R. and Carey, S. (2007 December). Weighing Your Options: The 10 Most Important Scale Considerations. *Chemical Engineering*, 114 (13), 61–65.

Trottier, R., Dhodapkar, S., and Wood, S. (2010 Apr). Particle sizing across the CPI. *Chem. Eng.* 117 (4): 59–65.

Trottier, R. and Dhodapkar, S. (2012 Apr). Sampling particulate materials the right way. *Chem. Eng.* 119 (4): 42–49.

Trottier, R. and Dhodapkar, S. (2014 Jul). A guide to characterizing particle size and shape. *Chem. Eng. Prog.* 110 (7): 36–46.

Way, H. (2007 Jul). Nanoparticles + mild dispersion. *Chem. Eng.* 114 (7): 44–48.

Yandrick, R. (2009 June), Vibratory Feeders And Conveyors: Useful Selection Tips, *Chemical Engineering*, 116 (6), 47–49.

Zalosh, R., Grossel, S.S., Kahn, R., and Sliva, D.E. (2005 Dec). Safely handle powdered solids. *Chem. Eng. Prog.* 101 (12): 23–30.

Zeeuwen, P. and Ebadat, V. (2011 Aug). Preventing self-heating and ignition in drying operations. *Chem. Eng.* 118 (8): 45–47.

Zenz, F.A. (1964 Feb). Conveyability of materials of mixed particle size. *Ind. Eng. Chem. Fundam.* 3 (1): 65–75.

15

Heating, Cooling, and Change of Phase

It is a rare process in the chemical industry that does not require some exchange of thermal energy. Examining many possible processes shows that 10 situations arise, as shown in Figure 15.1. For gases and liquids, the substance may be a process stream or it may be a heat transfer fluid (HTF) conveying heating or cooling to the process flow. The apparatus for any process must, among its other considerations, be chosen and designed to facilitate the required exchange of energy.

In Section 15.1, several means by which energy is exchanged are considered individually for ten situations. In subsequent sections, a number of ancillary topics are discussed: heat-transfer media, insulation, shell-and-tube heat exchangers, plate-and-frame heat exchangers, modeling and control, thermal integration and pinch technology, new developments in heat exchangers. A recent review by Kleijn (2020) deals with heat exchangers and with the causes of heat-transfer surface fouling.

15.1 Process Substances and Their Thermal Modifications

The three states of matter present different faces to the transfer of heat, made more interesting by the occasions where one state converts into another. Gases must often be raised to high temperatures and pressures if they are to react at a desired rate and equilibrium. Liquids, if they are to stay liquid, must operate within constraints. Solids are the state which has shape: powder, pellets, film, and thread-lines. Sometimes a mixture of a liquid and a solid – e.g., a paste or suspension – requires energy exchange.

15.1.1 Gas/Vapor Heating

The sensible heating of a gas or vapor is impeded by its relatively low coefficient of heat transfer. When the gaseous and liquid forms of the same substance are subjected to the standard Dittus–Boelter equation for flow in a pipe,

$$h\,D\,k^{-1} = 0.023\ \mathrm{Re}^{0.8}\ \mathrm{Pr}^{0.333} \tag{15.1}$$

the calculated "h" values (the heat transfer coefficients) can be an order of magnitude different.

However, a gas or vapor has the advantage of tolerating extremely hot heating surfaces. Heating from an electrical element is effective but expensive at large scale. Direct firing is cheaper. Cross (2002) discussed temperature gradients in fired heaters. Radiant heat for complex gases (unlike simpler gases like oxygen) works at high temperature, where gases are often raised

Practical Process Design for Chemical Engineers, First Edition. Keith Marchildon and David Mody.

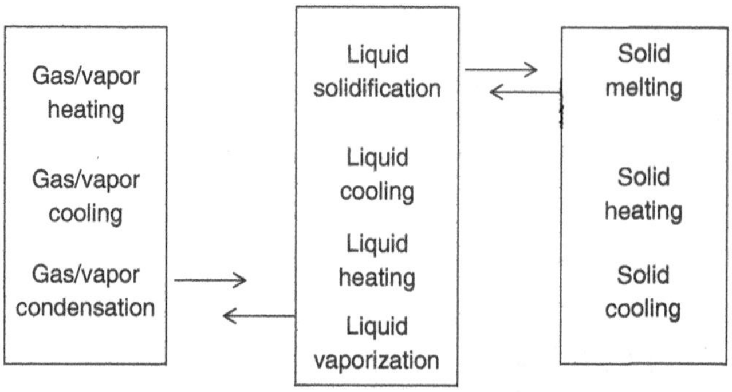

Figure 15.1 Varieties of thermal transfer for process materials.

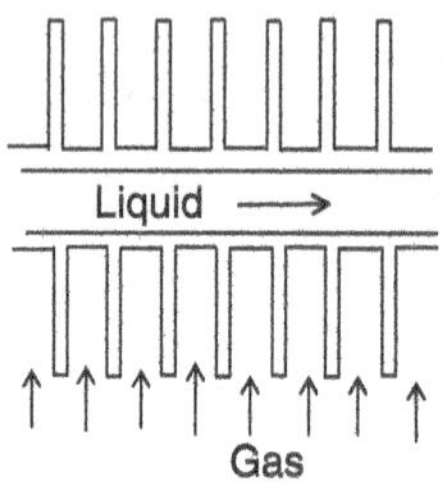

Figure 15.2 Enhanced heat exchange with fins.

for reaction. If a gas is combustible, then a small amount of air or oxygen can be injected to combust a small fraction of the flow and provide heat.

Because convection varies as $h \times$ area, then one way to speed it is to increase the area – that is, the area in contact with the gas. This is done by fitting the heat source with fins, over which the gas flows. See Figure 15.2. The inside of the heat source has good heat transfer, so the fins have high temperature. Ganapathy (2013) presented quantitative data on types and density of fins.

15.1.2 Gas–Vapor Cooling

Cooling of a gas or vapor presents the same problem of low heat transfer coefficients. As with heating, a heat extractor can be equipped with fins for effectiveness. Anecdotally, a study of a stream of water in a pipe being cooled by an external flow of air showed a threefold increase in heat transfer by the use of fins.

Water or other liquids can cool gases, either by spraying or by countercurrent flow in a packed column.

15.1.3 Vapor Condensation

A vapor may condense on the cold surface of a condenser vessel. If the heat transfer is too great, then a difficult-to-deal-with "fog" forms in the bulk of the vapor as well as at the cold surface. Fogs are assumed to have low gas-like heat transfer coefficients versus the higher heat transfer coefficients of condensing gases.

Condensation can take place into a falling film of the liquid phase. Also, a spray of liquid may be enough to turn most of the vapor into liquid, leaving the rest for a secondary condensation.

Non-condensable gases may likely be present and must be allowed access for their escape.

Figure 15.3 shows an efficient condenser for handling small flows. The slope in the piping inlet for the incoming process gas allows for non-condensable gases to leave when inlet gas velocities are low. High gas velocity on the process side may benefit from a non-condensable gas vent at the outlet of the exchanger, or at a high point, depending on the relative density of the gases, and where gas velocities have dropped.

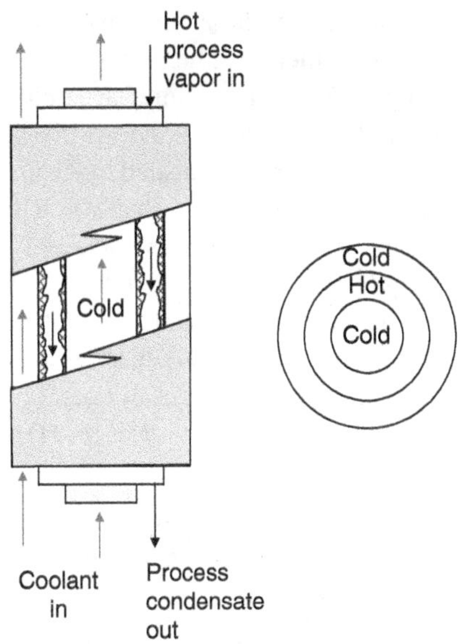

Figure 15.3 Narrow-gap condensation.

Vapors sometimes must be disposed of while under vacuum. A condenser may be used to turn most into liquid, leaving the rest to be handled by an ejector or mechanical producer of vacuum. Lines (2017) discussed this arrangement.

Wilhelmisson (2003) showed a compact condenser. Other references are Buecker (2003) and Durand et al. (2002)

15.1.4 Liquid Vaporization

Occasionally, a pure liquid must be turned into its vapor. Usually, the liquid is part of a solution, with the vaporization being done to separate components that are volatile. Two cases are as follows:

- reboiling in a distillation column and
- concentrating of components, as in the refining of sugar.

Reboiling is typically done in a kettle heated by internal piping carrying heat-transfer fluid (e.g. steam). A kettle is defined as a shell-and-tube exchanger with no baffles. Figure 15.4 shows a schematic.

Kettle operation is discussed by Das (2012), Hagan & Kruglov (2010), Kister & Chaves (2010), Ondrey (2011), and Tammami (2008).

Evaporation of a solvent is often done inside the pipes of a vertical shell-and-tube heat exchanger. When the motive force to keep the liquid and vapor mixture moving is the difference in density

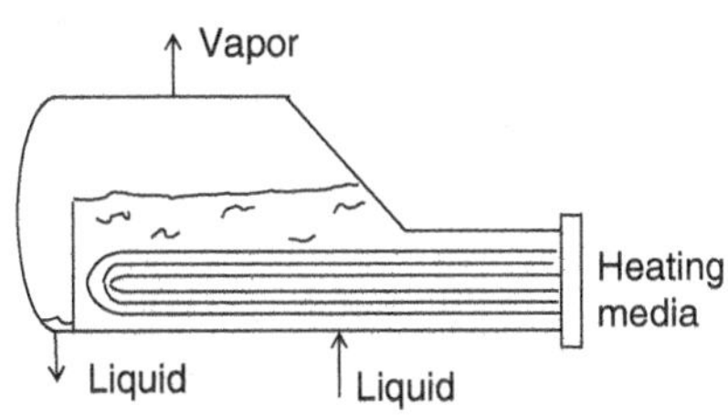

Figure 15.4 Distillation column reboiler.

between the two-phase mixture in the tubes and the liquid phase elsewhere, the exchanger is usually called a thermosiphon reboiler.

Evaporation is energy intensive unless heat can be recovered from the generated vapor. In some cases, if the evaporation is at high pressure, the vapor may be used for heating. If the vapor is steam and if the evaporator is heated by steam – a common combination – then vapor recompression can greatly cut the need for fresh steam. The concept is shown in Figure 15.5.

The benefits are shown in Table 15.1. Conditions are for a difference of 30 weight units between the liquid feed and liquid product. In the calculation, no account is taken of the dependence of latent heat on pressure and temperature.

Cooper & Lyon (2004) discussed applications of mechanical vapor recompression (MVR). A companion concept uses an ejector to add pressure to the recycle flow.

Gabelman (2020) and Glover (2004) discussed evaporators.

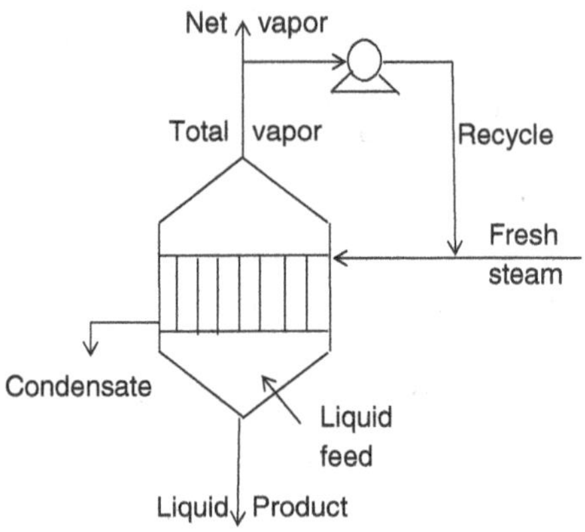

Figure 15.5 Continuous evaporation with mechanical vapor recompression (MVR).

Table 15.1 MVR for evaporation.

Recycle flow	Total vapor	Net vapor	Condensate	Fresh steam required
0	30	30	30	30
12	30	18	30	18
20	30	10	30	10
25	30	5	30	5

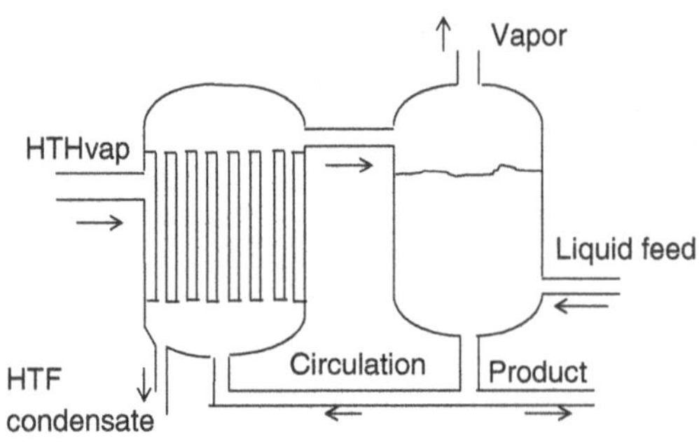

Figure 15.6 Thermosiphon evaporator.

Thermosiphon evaporators are highly efficient vaporizers. They comprise two parts:

- a vertical tube bundle inside a heated shell and
- a holding tank to provide head for the circulatory motion of the mixture (Figure 15.6).

The source of heat is the condensation of a heat-transfer fluid vapor: It could be steam if there is sufficient temperature across the tubes. Technology of these evaporators is referenced in Kister & Van Doorn (2019), Zadakbar et al. (2008), and Zygula & Barkat (2000).

Liquid can also be vaporized by being sprayed into a hot atmosphere.

Figure 15.7 shows a vaporizer that would be useful at lab or pilot-plant scale.

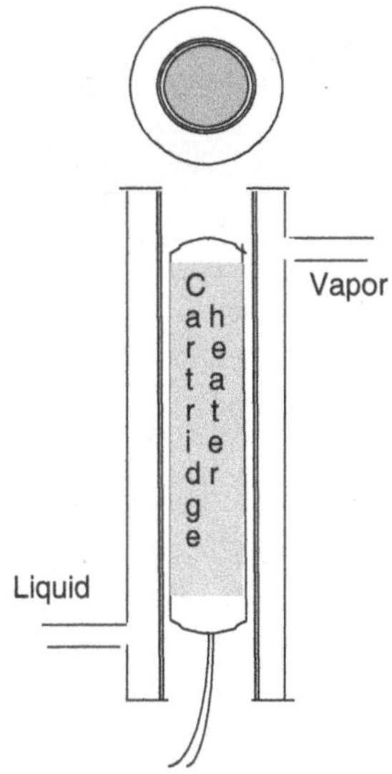

Figure 15.7 Electrical cartridge heater and vaporizer.

15.1.5 Liquid Heating

A liquid can be heated by electrical dissipation, by infra-red radiation, and, if the liquid possesses an adequate dielectric constant, by microwaves. However, the most common approach is to use another fluid of higher temperature and to extract from it latent and/or sensible heat. A simple example is the double-pipe heat exchanger, shown in Figure 15.8.

Double pipes are not common, but they do provide a gateway into the most common type of exchangers, the *shell-and-tube* (Figure 15.9), where one or more tubes pass through a heated shell. The exit temperatures – from fluid leaving the tubes and from fluid leaving the shell-side – are calculated as in Figure 15.8, but then a correction factor is applied to account for deviation from the pure countercurrency of the double-pipe configuration. These factors are given in Perry's handbook for various types of arrangements. It is recommended that no design be based on a value of this if the factor is less than 0.8.

There are many combinations of dimensions, number of tubes, number of baffles, and number of tube passes. More information and some references on this important device are provided in Section 15.4.

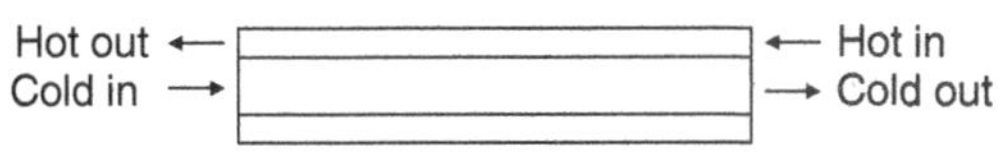

$$Q_{\text{hot}} = M_{\text{flow hot}}\,Cp_{\text{hot}}\,(T_{\text{hot in}} - T_{\text{hot out}})$$

$$Q_{\text{cold}} = M_{\text{flow cold}}\,Cp_{\text{cold}}\,(T_{\text{cold out}} - T_{\text{cold in}})$$

$$Q_{\text{transferred}} = \text{U Area LMTD}$$

$$\text{LMTD} = ((T_{\text{hot in}} - T_{\text{cold out}}) - (T_{\text{hot out}} - T_{\text{cold in}}))/\ln((T_{\text{hot in}} - T_{\text{cold out}}) / (T_{\text{hot out}} - T_{\text{cold in}}))$$

$$Q_{\text{hot}} = Q_{\text{transferred}} = Q_{\text{cold}}$$

$$M_{\text{flow hot}}\,Cp_{\text{hot}}\,(T_{\text{hot in}} - T_{\text{hot out}}) = \text{U Area LMTD} = M_{\text{flow cold}}\,Cp_{\text{cold}}\,(T_{\text{cold out}} - T_{\text{cold in}})$$

Figure 15.8 Countercurrent double-pipe heat exchanger.

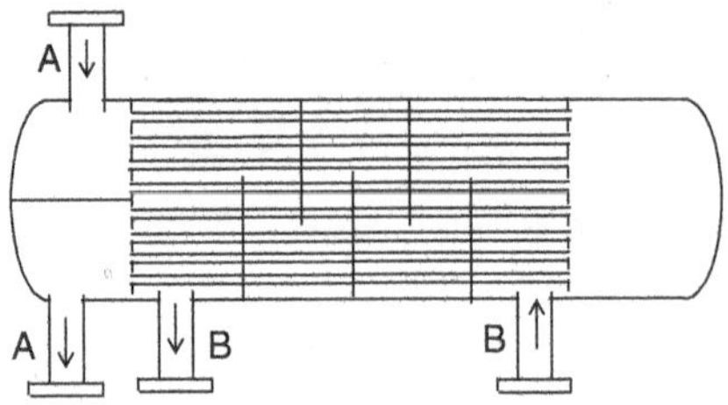

Figure 15.9 Shell-and-tube heat exchanger – typical.

Liquid heating (and cooling) is often done in vessels, which is discussed by Garvin (2001, 2005) and Gentilcore (2000). A coil carrying hot heat-transfer fluid, vapor or liquid, is a good addition to heating through the walls.

Bagley (2002) proposed submerged combustion for heating water.

An alternative to shell-and-tube is provided by the plate-and-frame exchanger, which is discussed in Section 15.1.6.

15.1.6 Liquid Cooling

If the liquid to be cooled is water, then there are traditional methods: cooling towers and spray ponds. Large cylindrical cooling towers are ubiquitous, sited around power-generating plants where hot water is a by-product. Hot water is fed around the periphery at the top, and air (at less than saturation) is fed from the bottom. Dendy (2008) and Goyal (2012) presented suggestions for design and upgrading of cooling towers.

A shell-and-tube can certainly be used to cool liquids and is the choice at high pressures. Another device is the *plate-and-frame* heat exchanger. It consists of vertical plates, sealed around the edges, with the space between them alternately being traversed by one or the other of the two fluids. Figure 15.10 illustrates the principle. A frame is used to support and to align the plates for close leak-free fit. One of the great advantages is that more plates can be added at will. All plates are supported on a frame. Plate and frame exchangers are often installed to recover heat from the cooling of liquids.

15.1.7 Liquid Solidification

Processing may require a material to be in liquid form, but the ultimate product may be a solid. Pellets, film, and strands are common.

If a liquid forms itself into a ball after passing through a multi-hole plate, and then drops into a pool of water, a relatively uniform pellet results. Subsequently, the pellets must be separated from the water and dried.

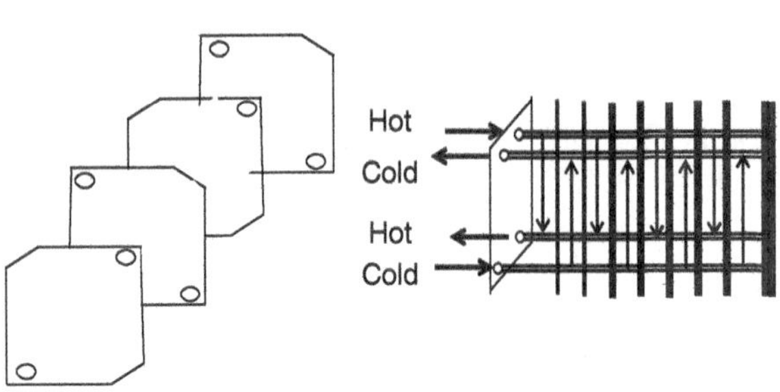

Figure 15.10 Plate-and-frame heat exchanger – conceptual.

Figure 15.11 Thermal screw.

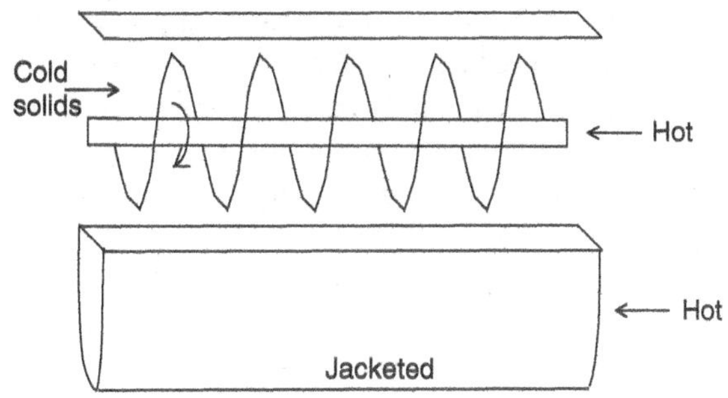

The Sandvik company market their Rotoform® process, which turns a liquid stream into liquid pellets, which are then cooled and solidified on a moving belt. The belt is sprayed with cooling water on its underside. The process produces highly uniform pellets and is used in the food industry and the polymer industry.

Watery solutions of wood (and other) fibers are processed over a series of felts and heated roles to remove water and produce paper.

Molten polymer is "spun" as slender streams from multi-hole "spinnerets" and solidified by cool air into solid fibers.

15.1.8 Solid Melting

Solids are frequently melted on a heated plate.

They may simply be added continuously to an inventory pool of already molten material.

They may be crushed and heated in the screws of an extruder.

15.1.9 Solid Heating

A solid can be heated in a column by a countercurrent flow of hot air or other gas. It can also be heated by a hot gas in a fluidized or spouted bed, provided it is not friable.

Some solids with the right electrical properties are susceptible to microwaves.

A gentler heating is obtained in a thermal screw, provided by a number of manufacturers and shown schematically in Figure 15.11.

15.1.10 Solid Cooling

The cooling of solid particles can also be accomplished in a thermal screw or in packed or fluidized beds. Adham (2000) described different types of equipment for extracting heat from hot solids: stationery coolers, rotary coolers, fluidized beds, and conveyor coolers. Mehos (2014) examined different types of bed cooling. Solids on trays are cooled by passage of countercurrent air, nitrogen, or other gas.

15.2 Heat Transfer Media

The term "media" encompasses all the fluids that aid in the heating, cooling, and transformation of process materials. Thus, water, steam, air, and even liquid nitrogen qualify, along with the specialty materials that we usually think of as HTFs.

Beteta & Ivanova (2015) discussed an application where liquid nitrogen was the coolant.

Chu (2005) gave heat transfer predictions for air-cooled exchangers.

Durand et al. (2012) and Winter & Peress (2003) examined cooling water practices.

Risko (2004) discussed the control of steam heating during fluctuating demand.

HTFs are used when steam or water is inadequate. For instance, steam at 257.1 °C has a condensing pressure of 40 atmospheres or bars, which would require impractically heavy-walled piping and vessels to contain it. Instead, a well-known HTF, Dowtherm A®, a eutectic mixture of diphenyl and diphenyl oxide (DP:DPO) is used which condenses at one bar. Even at 400 °C, its maximum useable temperature in practice, its pressure is only 10.6 bars. This material can serve as both liquid media and condensing-vapor media. As a liquid, it serves in cases where it may have to control temperature by supplying or absorbing heat. The disadvantage (of any HTF used as a liquid) is that its temperature changes as heat is given up, so large amounts may need to flow. As a vapor, it supplies its latent heat (although only about 1/5 as much as steam) and its flow path must be designed to avoid flooding and vapor lock. A schematic of a liquid system is shown in Figure 15.12 and of a vapor system in Figure 15.13.

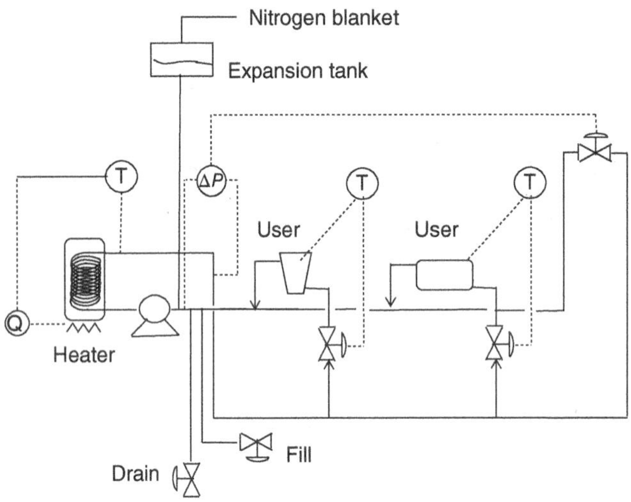

Figure 15.12 Heat transfer fluid – liquid system.

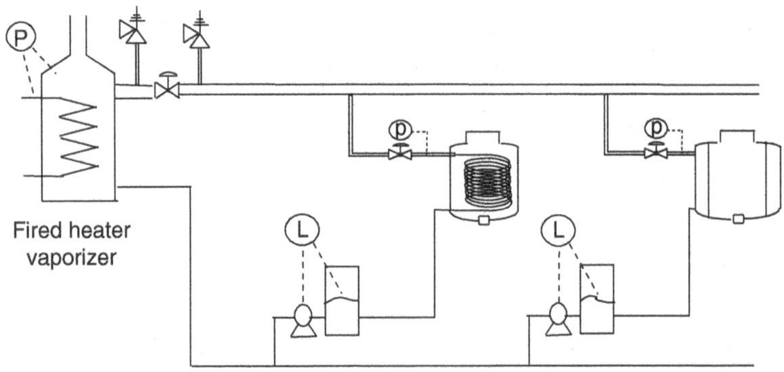

Figure 15.13 Heat transfer fluid – vapor system.

Dowtherm A, along with other Dowtherm's, is sold by the Dow Chemical Company. The same composition is sold as Therminol 66 by the Eastman Chemical Company, again as part of a stable of HTFs.

Hudson (2011), Jenkins (2009, 2012), and Sahasranaman (2005) assisted in designing HTF systems. Gamble (2006) spoke of cost management. Arseneault (2009) and Armstrong & Arseneault (2015) discussed degradation of HTFs, which is to be anticipated because of the high temperatures (in spite of the relative stability of the compounds). Gamble & Schopf (2010) and Jenkins (2013) dealt with leaks, and Oetinger (2002) discussed fires. Oetinger (2011) troubleshoot HTF systems.

15.3 Insulation, Tracing, and Fouling

One of the persistent problems in heat transfer is the gradual accumulation of degraded material on heat transfer surfaces, which reduces the overall heat transfer coefficient. Kleijn (2020) distinguished four sources of fouling:

- chemical, where the chemistry in the process produces a by-product that adheres,
- biological, where organisms in the process fluid grow and adhere to surfaces,
- deposition, where particles in the process fluid settle out, and
- corrosion, where the metal surfaces produce a layer of low conductivity.

While every process substance has its own propensity for fouling, some general guidelines can be followed:

- minimize the temperature of surfaces in contact with the process material,
- avoid regions of stagnation,
- maximize the velocity of the process flow,
- provide and maintain smooth surfaces of metal surfaces,
- avoid high fluxes that bring on film boiling, and
- make sure the material of the heat-transfer surface is compatible with the process.

An anonymous article (2006) and Panchal (2004) have suggestions regarding fouling. Neste (2004) discussed tube velocities to minimize fouling rates and therefore factors and costs. Cross (2009-11) discussed fouling for fired heaters.

Insulation is everywhere in a chemical plant, and it should logically be included in the design. It is an initial expense, and it can be an ongoing financial drain if it is not adequate or if it needs to be frequently replaced. Durand et al. (2015), Hart & Yarbrough (2010), Hart (2014), Jenkins (2016), and Melton & Bittner (2019) provided information on this subject.

If piping is kept warm only by insulation, then the process fluid will lose heat at some small rate. Heat transfer tracing may be needed. Electrical heating tape is common. Jenkins (2010, 2015) spoke about steam tracing and later about hot-oil tracing.

15.4 Shell-and-tube Heat Exchangers

Any serious proposal to design a shell-and-tube exchanger needs to start with Tubular Exchanger Manufacturers Association (TEMA), the TEMA. This group has established standards for construction, has defined the various combinations of shell and tube, and has developed software for flow-induced vibration and expansion joints.

Shell-and-tube exchangers have received much study over the years. Several authors explained the basic design: Bennett et al. (2007), Bhattacharyya & Mukherjee (2016), Greene (1999), Jones (2002), Poddar & Polley (2000), Raza (2013), and Singh (2015).

Some specific aspects of design are as follows:

- selecting baffles, Bouhairie (2012),
- designing for flexibility including phase change, Butterworth (2004),
- calculation of pressure drop, Gulley (2004),
- choice of TEMA exchanger category, Mukherjee (2004), and
- allowing for two-phase flow in design, Polley et al. (2012C).

Enhancement of tubes is discussed by Nasr & Polley (2002), Ploix & Lang (2012), and Shilling (2012).

Shell-and-tube heat exchangers are vulnerable to any action that weakens the sealing of tubes at the headers, or tubesheet. One mechanism that makes this happen is the vibration caused by the lateral passage of the shell-side fluid across the tubes. Specifically, the wake behind a tube releases as a regular string of vortices, which can interact with the natural frequency of the tube. This may lead to tube leakage, with mixing of the two fluids – with more or less serious consequences. A number of workers have written about vibration: Babakr et al. (2010), Dole et al. (2015), Polley et al. (2012B), Sofronas (2007), and Sutar (2016). Yokell (2005, 2007) has written on the general topic of mechanical integrity and maintenance. A second mechanism that can cause the tube to tubesheet connection to fail is differential thermal expansion of the tubes and shell. This can occur when there are different materials used between the tube and shell, when there are significant differences of temperature between the tube and shell, and when fixed tube sheet exchangers are utilized. Consideration and checking for thermal expansion should be given whenever the shell and tube process temperatures at either end differ by more than 93 °C (Ludwig 2002).

15.5 Plate-and-frame and Other Heat Exchangers

Design of plate-and-frame exchangers is treated by Haslego & Polley (2002) and, in more detail, by Srinaphasawadi & Tanthapanichakoon (2014). Kerner (2009) cautioned against misconceptions. Polley & Haslego (2002A, 2002B) examined one use of these exchangers, namely the recovery of heat from process streams. Broad & Kauders (2019) described welded-plate exchangers.

An anonymous article (2011) presented two other high-efficiency devices: the *fluted-plate block heat exchanger* and the *spiral heat exchanger*. LePree (2013), Moretta (2010), and Wajciechowski (2011) also described spiral exchangers. Gunnarsson et al. (2009), Polley (2002), and Wadekar (2000) presented compact heat exchangers, some of them cast as a unit and some welded together in order to avoid leakage at gaskets which can happen with standard plate-and-frame units.

15.6 Modeling, Control, and Design Tools

Every chemical process benefits from being accompanied by a mathematical model of its workings. Models come in two forms:

- statistical, based on observations of behavior over a range of conditions, and
- fundamental, based on physical and chemical principles.

The former type of model is useful on a day-to-day basis, especially if it incorporates a large range of operating conditions. It generally becomes less trustworthy if conditions depart from its data base.

The fundamental model, assuming the principles are correctly interpreted, is trustworthy over a larger range. Its precision improves if checked against ranges of operating conditions, which allows it to be "massaged". Beck et al. (2014) and Casenave (2012) expounded on the benefits of rigorous modeling. Models can be steady state: Bai et al. (2019), or they can be unsteady state: Perez & Lenferink (2015).

Albers (1999) presented a model-based control scheme for a heat exchanger, the calculation incorporating a fundamental dynamic (unsteady-state) mathematical model.

Commercial heat exchanger design software can be obtained from Aspentech and the HTRI consortium. Either way, training is recommended.

15.7 Thermal Integration and Pinch Technology

Energy utilities, whether for heating or for cooling, are a significant cost factor in running a plant, so there is a strong incentive to use the energy in a stream that needs to be cooled to help raise the enthalpy in a stream that needs to be heated or vaporized. A plant that is thus *thermally integrated* can save a lot of ongoing money. Unfortunately, the equipment to fully realize this benefit constitutes a capital cost, so generally there is some optimum compromise in the integration.

To achieve heat transfer between streams requires a temperature difference. In the following discussions, this minimum approach of temperatures between hot and cold will be taken as 10 °C.

One of the first principles is that a design that keeps this difference of temperature as low as possible (in this case 10°) is a design that minimizes utility costs (while it increases capital costs because of the need for more heat transfer area).

In Figure 15.14, the positioning of the exchanger in the left-hand configuration means that the flows of both the cooling and heating medium can be less than for the coolant and heat in the right-hand configuration – thus saving energy costs for these two utility streams.

The heat recovery around the above reactor is straight forward but the situation rapidly becomes more complicated when more than two streams are considered. Be aware that the exchangers below are assumed to all be countercurrent flow, which may not be practical in some situations. One of the first treatments of multi-stream exchange is Linnhoff & Flower (1978). They provided a four-stream example which we will explore (Table 15.2).

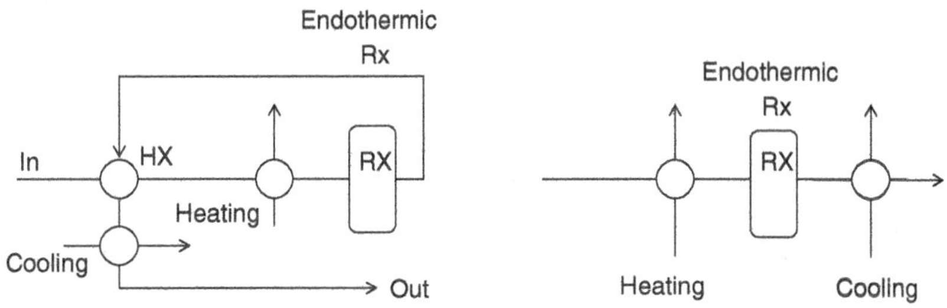

Figure 15.14 Heat recovery around a reactor.

Table 15.2 Heat-transfer network case study by Linnhoff and Flower.

Stream number	Flow × specific heat (kW °C^{-1})	Entering temperature (°C)	Desired exit temperature (°C)
1 (cold)	3.0	60	180
2 (hot)	2.0	180	40
3 (cold)	2.6	30	105
4 (hot)	4.0	150	40

There are a few variations on the original method, but the first step, as various authors agree on, is to plot the streams as arrows on a temperature graph, as shown in Figure 15.15.

It is noticed that the scale for the heating steams is displaced upwards by 10° from the scale for the cooling streams. The displacement results from the agreed-upon minimum difference of 10° between the streams in a heat exchanger. The graph provides guidance as to how the steams can be matched up for exchange.

The graph is divided into five segments, in each of which particular streams appear. The enthalpy (kW) of each stream is noted within each segment. These enthalpies are calculated as flow × heat capacity × temperature change. Starting at the top, the optimum design for each segment is to see that each stream matches the above enthalpy at its bottom, whether by transfer of heat between itself and another stream or by an externally provided infusion or a withdrawal of heat (Table 15.3).

The savings in coolant are 62% and in heating 81%. Linnhoff and Flower showed two possible networks, one of which is shown in Figure 15.16.

It is likely that some shortcuts would be taken, with cuts in the number of utility exchangers and increases in the input of the remaining exchangers. For instance, what would be the result if all intermediate exchangers were eliminated, and only the final heating or cooling of the four streams to meet their target temperatures were left in place? Most likely the system, interiorly, would reach a "pinch point", where temperatures between hot and cold streams become too low for effective heat transfer. Simulations of Heat Exchanger Networks (HENs) detect such a situation.

By finding the pinch point, the minimum number of exchangers can be determined assuming temperature crosses would not occur (which is not a valid assumption in all cases). The *in situ* enthalpy content of each of the five zones is given Figure 15.15. The pinch temperature occurs at the point where the enthalpy switches sign, that is between zone 3 and zone 4. The significance is in the application of utilities: Only heat should be applied above the pinch point and only cooling below. The theory behind this procedure is best explained in the Linnhoff papers.

The minimum number of exchangers is 6, and the energy consumed is predicted to be 60 kW of heating and 145 kW of cooling (Figure 15.17). However, if such a network is created (Figure 15.18), a temperature cross occurs cooling stream 4 with stream 1, which can be solved by decreasing the heat that can be recovered. The result is an exchanger network that still requires only 6 exchanges, and has 60 kW of heating, and 225 kW of cooling. It is advisable to review the assumption that true countercurrent exchangers can be used in the application since this will impact the LMTD correction factor and makes some designs infeasible. Three of the exchangers (H1, H3, and H5) in this example would have low F factors if they were two-pass exchangers.

Stream ->		2	4		3	1			Externally provided duty in region kW		Pinch determination
m * Cp	kW (C)	2	4		2.6	3					
	Hot stream temp, C			Region			Cold stream temp (C)	delta T			sum of req'd duty
	190						180				
				A // 1		30 kW		10	30	heat req	30
	180						170				
		−60 kW		B // 2		90		30	30	heat req	60
	150						140				pinch at Δ sign
		−70	−140	C // 3		105		35	−105	cool req	−45
	115						105				
		−90	−180	D // 4	117	135		45	−18	cool req	−63
	70						60				
		−60	−120	E // 5	78			30	−102	cool req	−165
	40						30				
	Totals, kW	−280	−440		195	360	External (cooling/heating)		−225 // 60		

Base case utility costs

Utility	Duty kW	Cost $/GJ	Cost $/yr
cld water	−280	0.5	$ 4 032
cld water	−440	0.5	$ 6 336
hot Utility	195	5	$ 28 080
hot Utility	360	5	$ 51 840
			$ –
			$ –
			$ 90 288
		Possible 10 yr savings	$ 722 304

Utility Costs with Maximum Heat Recovery

type	duty (kW)	cost (GJ)	cost $/yr
Cooling	225	0.5	$ 3 240
Hot Utility	60	5	$ 8 640
		Total	$ 11 880
		10 yr costs	$ 118 800

Figure 15.15 Comparison of cooling and heating streams.

Table 15.3 Utilities with and without recovery.

		No recovery		With recovery	
Stream	**Enthalpy change (kW)**	**Total coolant (kW)**	**Total heat (kW)**	**Total coolant**	**Total heat**
2	280				
4	440	720			
1	360				
3	195		555		
	Total	720	555	272	107

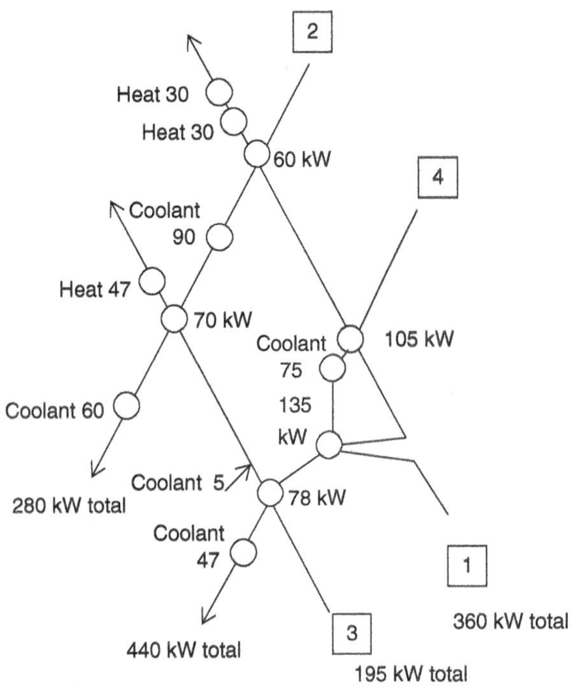

Figure 15.16 Linnhoff and Flower's example.

References on thermal integration include Linnhoff & Flower (1978) as well as Linnhoff & Hindmarsh (1983), Beaman & Rerse (2011), El-Temtamy & Gabr (2011), Gill (2005), Milosevic et al. (2013), Polley & Heggs (1999), Raza & Hussain (2019), Rikhtehgar (2011), Rossiter (2010), Sparrow (2000), and Thubaiti (2008). Polley et al. (2012A) wrote about thermal integration of reboilers.

References on cogeneration of steam and electricity are Buecker (2013) and Rutheal (2013).

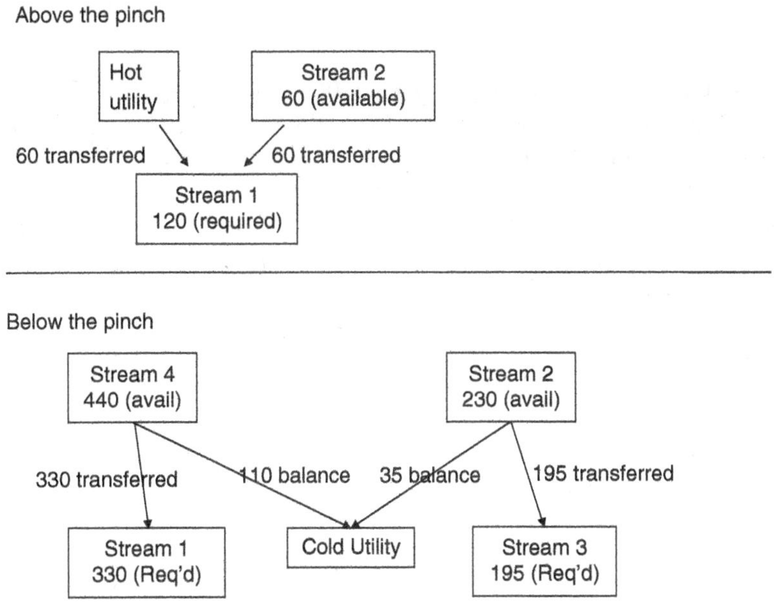

Figure 15.17 Minimum number of exchangers.

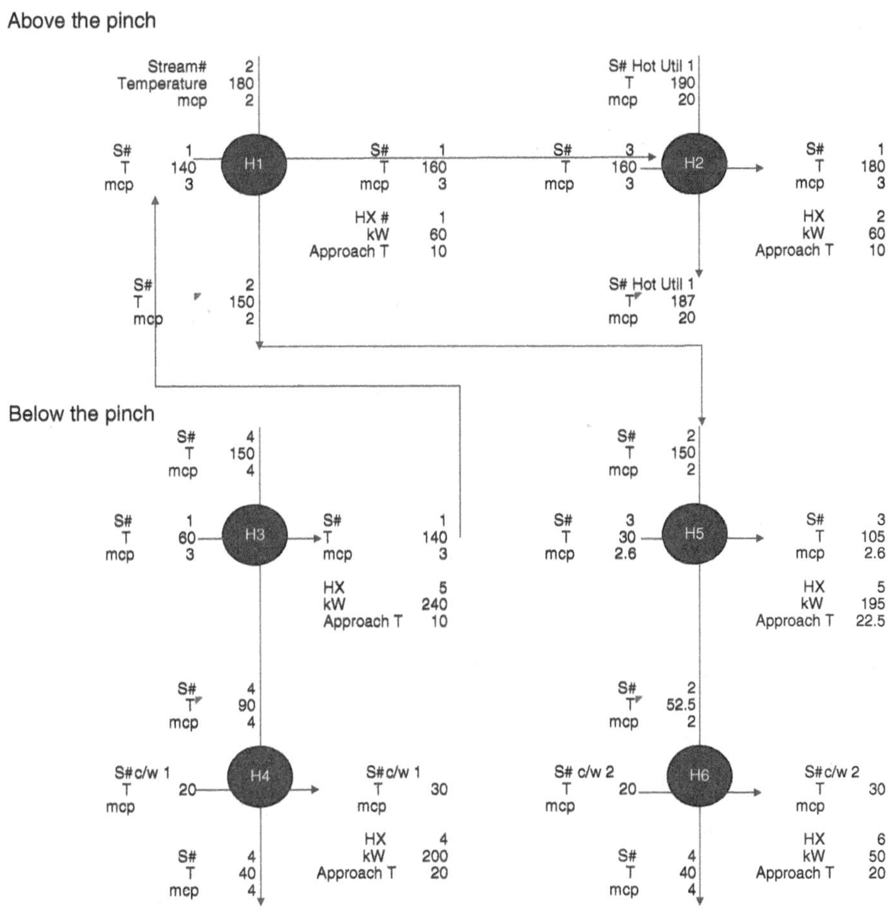

Figure 15.18 Exchanger network V2.

References

Adham, K. (2000). Recover heat from hot solids. *Chem. Eng. Prog.* 96 (8): 37–46.

Anonymous (2011 Jun). High efficiency CHES. *Chem. Eng.* 118 (6): 36–37.

Anonymous (2006 Feb). Heat exchanger performance. *Chem. Eng. Prog.* 102 (2): 14–15.

Albers, J.E. (1999 Nov). Nonlinear model-based control of a heat exchanger. *Hyd. Proc.* 78 (11): 101–106.

Armstrong, C. and Arseneault, G. (2015 Jan). Minimize degradation of heat-transfer fluids. *Chem. Eng. Prog.* 111 (1): 58–62.

Arseneault, G. (2009 Dec). Maximizing heat-transfer fluid longevity. *Chem. Eng.* 116 (13): 32–39.

Babakr, A., Ashiru, T., and Westhuizen, C. (2010 May). Flow-induced fatigue failure in tubular heat exchangers. *Hyd. Proc.* 89 (5): 27–33.

Bagley, J.L. (2002 Mar). Consider submerged combustion for hot water production. *Chem. Eng. Prog.* 98 (3): 48–51.

Bai, Y., Si, H., and Wang, X. (2019). Steady-state simulation of internal heat-transfer characteristics in a double tube reactor. *Chem. Eng. Process. Process Intensif.* 144: 1–9.

Beaman, R. and Rerse, C. (2011 Nov). Energy optimization using pinch analysis – part 2. *Chem. Eng.* 118 (12): 36–41.

Beck, R., Ralston, T., and Hegy, S. (2014 Jul). Incorporate rigorous heat exchanger models in simulation. *Chem. Eng. Prog.* 110 (7): 21–26.

Bennett, C.A., Kistler, R.S., Lestina, T.G., and King, D.C. (2007 Apr). Improving heat exchanger designs. *Chem. Eng. Prog.* 103 (4): 40–45.

Beteta, O. and Ivanova, S. (2015 Sep). Cool down with liquid nitrogen. *Chem. Eng. Prog.* 111 (9): 30–35.

Bhattacharyya, S. and Mukherjee, S. (2016 Mar). Shell-and-tube heat exchangers: the design cycle. *Chem. Eng.* 123 (3): 61–67.

Bouhairie, S. (2012 Feb). Selecting baffles for shell-and-tube heat exchangers. *Chem. Eng. Prog.* 108 (2): 27–33.

Broad, R. and Kauders, P. (2019 Jun). Welded plate heat exchangers cut refinery process costs – part 2. *Hyd. Proc.* 98 (6): 55–58, 60.

Buecker, B. (2013 Apr). Combined heat and power for the CPI: modern concepts. *Chem. Eng.* 120 (4): 50–53.

Buecker, B. (2003 Feb). Track condenser performance. *Chem. Eng. Prog.* 99 (2): 40–43.

Butterworth, D. (2004 Dec). Heat exchanger duty: going for gold. *Chem. Eng.* 111 (13): 44–49.

Casenave, J. (2012 Sep). What are the benefits of rigorous modeling in heat exchanger design? *Hyd. Proc.* 91 (9): 94–100.

Chu, C. (2005 Nov). Improved heat transfer predictions for air-cooled heat exchangers. *Chem. Eng. Prog.* 101 (11): 46–48.

Cooper, A. and Lyon, M. (2004 Jul). Make the most of MVR. *Chem. Eng.* 111 (7) 44–47.

Cross, A. (2002 Jun). Evaluate temperature gradients in fired heaters. *Chem. Eng. Prog.* 98 (6): 42–46.

Cross A., Direct-fired heaters: evaluation thermal performance and the effects of fouling, *Chem. Eng.*, Nov 2009-11, 116 (12): 47–51.

D'Aquino, R. (2003 Sep). Improve heat flux control with piped reactor jackets. *Chem. Eng. Prog.* 99 (9): 14–15.

Das, T. (2012 Mar). Achieve optimal heat recovery in a kettle exchanger. *Hyd. Proc.* 91 (3): 87–88.

Dendy, T. (2008 May). Evaluating cooling tower upgrades. *Chem. Eng. Prog.* 104 (5): 44–48.

Dole, R.H., Vivekanand, S., and Sridhar, S. (2015 Dec). Mitigate vibration issues in shell-and-tube heat exchangers. *Hyd. Proc.* 94 (12): 57–60.

Dukler, A.E. (1959 Oct). Dynamics of vertical falling film systems. *Chem. Eng. Prog.* 55 (10): 62–67.

Durand, A.A., Guerrero, C.A.A., and Ronces, E.A. (2002 Apr). Gravity effects in horizontal condensers. *Chem. Eng.* 109 (4): 91–94.
Durand, A.A., Suarez, S.R.G., Roman, F.J.P. et al. (2012 Sep). Cooling water outlet temperature: evaluating the best maximum value. *Chem. Eng.* 119 (9): 46–50.
Durand, A.A., Maldonado, A.O., Holly, A.A. et al. (2015 Oct). Pipe insulation: finding the optimal thickness. *Chem. Eng.* 122 (10): 61–65.
El-Temtamy, S. and Gabr, E.M. (2011 Apr). Flexible heat exchanger networks. *Chem. Eng.* 118 (4): 32–38.
Gabelman, A. (2020 Jan). Evaporators: design concepts and equipment selection. *Chem. Eng.* 127 (1): 27–38.
Gamble, C.E. (2006 Jul). Cost management in heat-transfer systems. *Chem. Eng. Prog.* 102 (7): 22–26.
Gamble, C.E. and Schopf, M. (2010 Dec). Heat transfer fluid leaks: break the fire triangle. *Chem. Eng.* 117 (13): 26–33.
Ganapathy, V. (2013 Sep). Understanding finned heat exchangers. *Chem. Eng.* 120 (9): 62–65.
Garvin, J. (2001 Apr). Estimate heat transfer and friction in dimple jackets. *Chem. Eng. Prog.* 97 (4): 73–75.
Garvin, J. (2005 Aug). Evaluate flow and heat transfer in agitated jackets. *Chem. Eng. Prog.* 101 (8): 39–41.
Gentilcore, M.J. (2000 Mar). Estimate heating and cooling times for batch reactors. *Chem. Eng. Prog.* 96 (3): 41–45.
Gill, B.S. (2005 May). Selecting streams in pinch systems [1]. *Chem. Eng.* 112 (5): 55–57.
Glover, W.B. (2004 Dec). Selecting evaporators for process applications. *Chem. Eng. Prog.* 100 (12): 26–33.
Goyal, J. (2012 Feb). Effective thermal design of cooling towers. *Chem. Eng.* 119 (2): 24–29.
Greene, B. (1999 Jan). A practical guide to shell-and-tube heat exchangers. *Hyd. Proc.* 78 (1): 79–86.
Gulley, D. (2004 Jul). More accurate exchanger shell-side pressure drop calculations. *Hyd. Proc.* 83 (7): 71–76.
Gunnars, on, J., Sinclair, I., and Alanis, F.J. (2009 Feb). Compact heat exchangers: improving heat recovery, part 2. *Chem. Eng.* 116 (2): 44–47.
Hagan, M.D. and Kruglov, V.N. (2010 Nov). Understand heat flux limitations on reboiler design. *Chem. Eng. Prog.* 106 (11): 24–31.
Hart, G.H. and Yarbrough, D.W. (2010 Oct). Industrial insulation systems: material selection factors. *Chem. Eng.* 117 (10): 34–40.
Hart, G.H. (2014 Jun). Selecting thermal insulation. *Chem. Eng. Prog.* 110 (6): 22–27.
Haslego, C. and Polley, G. (2002 Sep). Designing plate-and-frame heat exchangers. *Chem. Eng. Prog.* 98 (9): 32–37.
Hudson, J. (2011 Dec). Thermal fluid systems: design considerations, part 1. *Chem. Eng.* 118 (13): 30–33.
Jenkins, S. (2009 Sep). Heat transfer system design II. *Chem. Eng.* 116 (9): 31B.
Jenkins, S. (2010 Mar). Steam tracer lines. *Chem. Eng.* 117 (3): 24A.
Jenkins, S. (2012 Aug). Heat-transfer expansion tank design. *Chem. Eng.* 119 (8): 32A.
Jenkins, S. (2013 Aug). Detection and prevention of DP:DPO fluid leaks. *Chem. Eng.* 120 (8): 25.
Jenkins, S. (2016 May). Insulating heat-transfer-fluid piping. *Chem. Eng.* 123 (5): 38.
Jenkins, S. and Gamble, C. (2015 Aug). Hot-oil heat-tracing systems. *Chem. Eng.* 122 (8): 36A.
Jones, A.E. (2002 Mar). Thermal design of the shell-and-tube. *Chem. Eng.* 109 (3): 60–65.
Kerner, J. (2009 Feb). Plate heat exchangers: avoiding common misconceptions, part 1. *Chem. Eng.* 116 (2): 40–43.
Kister, H.Z. and Chaves, M.A. (2010 Feb). Kettle troubleshooting. *Chem. Eng.* 117 (2): 26–33.

Kister, H.Z. and Van Doorn, E. (2019 Feb). Stable thermosiphon reboiler operation: with or without a valve in the inlet piping? *Hyd. Proc.* 98 (2): 73–76.

Kleijn, A.B. (2020 Feb). Heat exchanger overview. *Chem. Eng. Prog.* 116 (2): 34–38.

Klein, R. (2006 Jan). Immersion heaters: selection & implementation. *Chem. Eng.* 113 (1): 44–50.

LePree, J. (2013 Nov). The new face of heat exchangers. *Chem. Eng.* 120 (11): 20–22.

Lestina, T.G. (2011 Jun). Selecting a heat exchanger shell. *Chem. Eng. Prog.* 107 (6): 34–38.

Lines, J. (2017 Jan). Key considerations for vacuum process condensers. *Hyd. Proc.* 96 (1): 77–81.

Linnhoff, B. and Flower, J.R. (1978 Jul). Synthesis of heat exchanger networks. *AICHE J.* 24 (4): 633–642.

Linnhoff, B. and Hindmarsh, E. (1983). The pinch design method for heat exchanger networks. *Chem. Eng. Sci.* 38 (5): 745–763.

Ludwig, E., *Applied Process Design for Chemical & Petrochemical Plants*, 2002, Vol 3, Chap 10, Gulf Publishing.

Mandal, M. (2010 Dec). Optimize heat exchanger installations. *Hyd. Proc.* 89 (12): 53–56.

Mehos, G. (2014 Nov). Operating direct-contact heat exchangers. *Chem. Eng.* 121 (11): 58–64.

Melton, K. and Bittner, J. (2019 May). Introduction to industrial insulation. *Chem. Eng. Prog.* 115 (5): 23–29.

Milosevic, Z., Rudman, A., and Brown, R. (2013 Jun). Are you using pinch technology effectively in your daily operations? Part 2. *Hyd. Proc.* 92 (6): 99–102. Jul, *92(7)*, 77–81.

Moretta, A.A. (2010 May). Spiral plate heat exchangers: sizing units for cooling non-newtonian slurries. *Chem. Eng.* 117 (5): 44–48.

Mukherjee, R. (2004 Apr). Does your application call for an F-shell heat exchanger? *Chem. Eng. Prog.* 100 (4): 40–45.

Nasr, M.R.J. and Polley, G.T. (2002 Apr). Should you used enhanced tubes? *Chem. Eng. Prog.* 98 (4): 44–50.

Neste, J., Reduce fouling in shell-and-tube heat exchangers, *Hyd. Proc.*, Jul, 2004, 83 (7): 77–82.

Oetinger, J. (2002 Jan). Prevent fires in thermal fluid systems. *Chem. Eng. Prog.* 98 (1): 46–48.

Oetinger, J. (2011 Dec). Troubleshooting heat-transfer fluid systems, part 2. *Chem. Eng.* 118 (13): 34–36.

Ondrey, G. (2011 Jan). Reboiler circuits for trayed columns. *Chem. Eng.* 118 (1): 26–35.

Panchal, C.B. (2004 Oct). Poorly performing heat transfer equipment. *Chem. Eng. Prog.* 100 (10): 18.

Pelini, R.G. (2008 Dec). Heat flux and film temperature in fired, thermal-fluid heaters, part 1. *Chem. Eng.* 115 (13): 34–40.

Perez, R.Y.U. and Lenferink, J.E.O. (2015 Oct). Design pressure reduction in high-pressure heat exchangers with dynamic simulation. *Hyd. Proc.* 94 (10): 37–41.

Perry R., Green D. W., *Perry's Chemical Engineers' Handbook*, n.d. 7th Edition, Chapter 11, 6–7.

Ploix, B. and Lang, T. (2012 Jan). Apply new enhanced tubes to optimize heat transfer in LNG trains. *Hyd. Proc.* 91 (1): 55–60.

Poddar, T.K. and Polley, G.T. (2000 Sep). Optimize shell-and-tube heat exchanger design. *Chem. Eng. Prog.* 96 (9): 41–46.

Polley, G.T. and Heggs, P.J. (1999 Dec). Don't let the pinch pinch you. *Chem. Eng. Prog.* 95 (12): 27–36.

Polley, G. and Haslego, C. (2002A Oct). Using plate exchangers in heat recovery networks, part 2. *Chem. Eng. Prog.* 98 (10): 48–51.

Polley, G. and Haslego, C. (2002B Nov). Retrofit heat recovery networks with plate exchangers, part 3. *Chem. Eng. Prog.* 98 (11): 46–51.

Polley, G.T. (2002 Mar). Cast heat exchangers in a leading role. *Chem. Eng.* 109 (3): 56–58.

Polley, G.T., Vasquez-Ramirez, E.E., Avila, M., and Jaramillo, D.J. (2012A Apr). Thermal integration of reboilers. *Chem. Eng.* 119 (4): 50–53.

Polley, G.T., Farfan, M.A.V., and Pugh, S.J. (2012B Jan). Designing shell & tube heat exchangers: avoid vibration from the start. *Chem. Eng.* 119 (1): 30–34.

Polley, G.T., Vasquez-Ramirez, E.E., and Avila, M.R. (2012C Jan). Designing shell and tube heat exchangers: consider two-phase flow. *Chem. Eng.* 119 (1): 36–39.

Rutheal, R. (2013 Apr). Boosting energy efficiency in chemical manufacturing: a look at cogeneration. *Chem. Eng.* 120 (4): 46–49.

Raza, A. (2013 May). Specifying shell-and-tube heat exchangers. *Chem. Eng.* 120 (5): 47–53.

Raza, S.A. and Hussain, I. (2019 Oct). Pinch analysis: a retrofit approach. *Hyd. Proc.* 98 (10).

Rikhtehgar, F. (2011 May). Design an efficient exchanger network. *Hyd. Proc.* 90 (5): 113–116.

Risko, J.R. (2004 Nov). Steam heat exchangers are underworked and over-surfaced. *Chem. Eng.* 111 (12): 58–62.

Rossiter, A.P. (2010 Dec). Improve energy efficiency via heat integration. *Chem. Eng. Prog.* 106 (12): 33–42.

Sahasranaman, K. (2005 Mar). Get the most from high-temperature heat-transfer-fluid systems. *Chem. Eng.* 112 (3): 46–50.

Shilling, R.L. (2012 Sep). Selecting tube inserts for shell-and-tube heat exchangers. *Chem. Eng. Prog.* 108 (9): 19–25.

Singh, S.K. (2015 Feb). Thermal design guidelines for optimizing shell-and-tube heat exchangers. *Chem. Eng.* 122 (2): 54–57.

Sofranos, T. (2007 Jul). Vibration failure of a heat exchanger tube. *Hyd. Proc.* 86 (7): 118.

Sparrow, E.M. (2000 Feb). Follow this strategy for better thermal design. *Chem. Eng. Prog.* 96 (2): 53–59.

Srinaphasawadi, G. and Tanthapanichakoon, W. (2014 May). Unlocking the secrets of plate-and-frame heat exchngers. *Chem. Eng.* 121 (5): 52–57.

Sutar, S.M. (2016 Jul). Experimental study of flow-induced pipe vibrations with guided end conditions. *Hyd. Proc.* 95 (7): 67–72.

Sofronas, T. (2007 Jul). Case39: vibration failure of a heat exchanger tube. *Hyd. Proc.* 86 (7): 118.

Taborek, J. (2002 Aug). Much evolution; challenges remain process heat transfer. *Chem. Eng.* 109 (8): 104–105.

Tammami, B. (2008 Mar). How to select the best reboiler for your processing operation. *Hyd. Proc.* 87 (3): 91–94.

Thubaiti, M. (2008 Mar). Optimize heat transfer networks. *Hyd. Proc.* 87 (3): 109–115.

Turton, R. et al. (2018). *Analysis, Synthesis, and Design of Chemical Processes*, 5the. Prentice Hall.

Wadekar, V.V. (2000 Dec). Compact heat exchangers. *Chem. Eng. Prog.* 96 (12): 38–49.

Wajciechowski, C. (2011 May). Spiral heat exchanger in desalter service solves fouling issues. *Hyd. Proc.* 90 (5): 75–76.

Wilhelmisson, B. (2003 Jul). Compact condensers offer sizable payback. *Chem. Eng.* 110 (7): 60–65.

Winter, S. and Peress, J. (2003 Nov). Tempering loops for temperature control. *Chem. Eng.* 110 (12): 68–74.

Yokell, S. (2007 Mar). Heat exchangers: tube-to-tubesheet joint tightness. *Chem. Eng.* 114 (3): 46–51.

Yokell, S. (2005 Jan). Get more life out of heat exchangers. *Chem. Eng.* 112 (1): 54–67.

Zadakbar, O., Vatani, A., and Karimpour, K. (2008 Nov). Get more from vertical thermosiphon reboilers. *Chem. Eng.* 115 (12): 52–54.

Zygula, T.M. and Barkat, O. (2000 Jul). Troubleshoot vertical thermosiphon reboilers. *Chem. Eng. Prog.* 96 (7): 71–75.

16

Mixing and Agitation

In the production and processing of materials, it is very common to have to bring components together into a mixture of some sort. The operation may be batch or continuous. We identify, and will look at, six different combinations of components that may have to be mixed:

1. miscible liquids,
2. immiscible liquids,
3. gas and liquid,
4. solid particles in liquid,
5. solid particles with each other, and
6. solid particles and gas.

The process designer should follow these steps:

(1) specify the desired results of the mixing, including uniformity of the mixture both at the fine scale and across the total mass being mixed,
(2) estimate or measure the properties of the components and of the final mixture,
(3) estimate the difficulty of mixing,
(4) choose and size the mixing device, and
(5) calculate the required power.

Mixing devices are and have been the subject of much development, all to the benefit of the processes of which they are a part. For the first four of the above combinations, that is the ones involving a liquid, and if the process materials are of low viscosity, then the common choice for a device is the *turbine agitator*. It is a ubiquitous item that has received much study, and it is discussed at the start of Section 16.1, Blending of Miscible Liquids.

General references are textbooks and/or compilations by Nagata (1975), Uhl & Gray (1966, 1967, 1981), Harnby et al. (1985), Tatterson (1991), Hemrajani (1995), and Paul et al. (2003).

16.1 Mixing or Blending of Miscible Liquids

The requirement is to bring two or more liquids together and form a single phase. The mixture is intended to be permanent, unlike other mixing situations where the phases are combined into a mechanical mixture, which can subsequently separate or be separated.

Practical Process Design for Chemical Engineers, First Edition. Keith Marchildon and David Mody.

Table 16.1 Mixing situations for miscible liquids.

Type of mixing required			
Full: axial plus radial		**Radial only**	
Stirred vessel		**In-line mixer**	
Turbulent liquid	Viscous liquid		
Turbine or propeller	Wall wiping	Static	Dynamic

There are two general cases:

1. the entering liquids are different from one another, and the resulting mixture is different from any of the feeds:
2. the liquid is a single stream which may have fluctuations in its properties, and it is the intent of the mixing to attenuate this variability (see Section 16.2.3).

For miscible liquids, the choices can be considered according to the following scheme in Table 16.1.

16.1.1 Turbine Mixers

The mixer consists of a rotating shaft upon which is mounted one or more impellers which disturb the liquid mixture. In a vessel, the shaft is generally, but not always, vertically and centrally mounted. Impellers have various forms according to the effects they are intended to achieve, and if the mixer has more than one impeller, the impellers need not necessarily be the same. Two of the most common styles are shown in Figure 16.1, and the flow patterns which they generate are shown in Figure 16.2.

Fasano (2015) showed many different styles of impellers, and D'Aquino (2004) showed some novel types. Wyczalkowski (2004) showed an unusual case of counterflow impellers for a specific task. Benz (2010) showed hydrofoil-shaped blades which are useful for slurries of fibrous solids.

The outer reaches of an impeller have a diameter, D, which is set as some fraction of tank diameter T, this fraction being usually between 0.2 and 0.6. The rotational speed of the mixer is given by Ns.

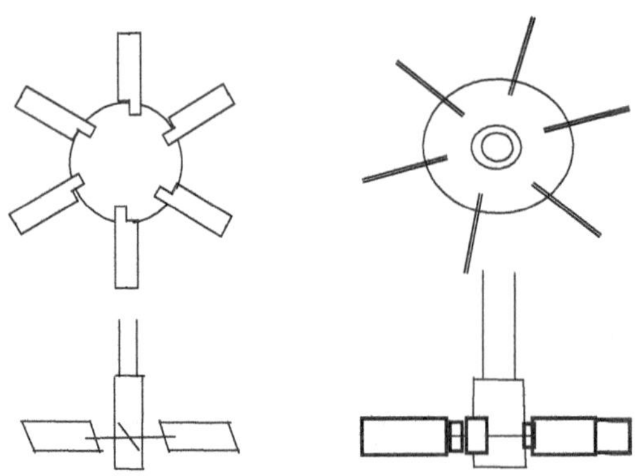

Figure 16.1 Pitched-blade and straight-blade turbine impellers.

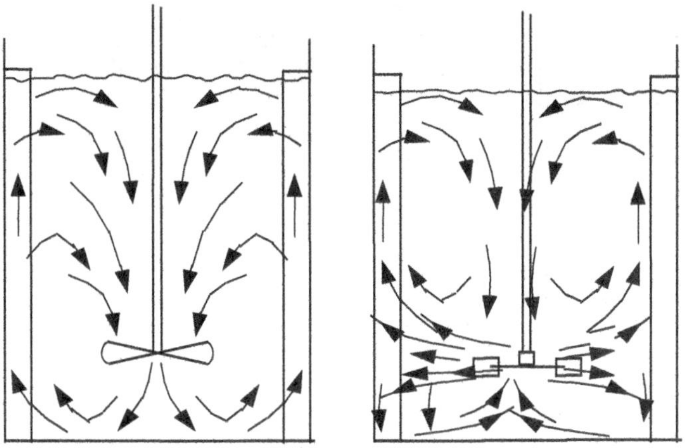

Figure 16.2 Flow patterns for pitched and straight impellers.

The definition of Reynolds number, an important parameter of the impeller, is

$$\mathrm{Re} = \rho\, \mathrm{Ns}\, D^2\, \mu^{-1} \tag{16.1}$$

where ρ and μ are liquid mixture density and viscosity, respectively. Another nondimensional parameter is the pumping number (also called the flow number),

$$N_\mathrm{Q} = Q\, \mathrm{Ns}^{-1}\, D^{-3} \tag{16.2}$$

where Q is the vertical volumetric flow of fluid, up and down, caused by the impeller. The value of this flow in a particular case will be seen to arise from the design procedure. N_Q is related to Re, as shown in Figure 16.3.

Finally, the power consumption, P, of the agitator is included in N_P, the power number,

$$N_\mathrm{P} = P\, D^{-5}\, \rho^{-1}\, \mathrm{Ns}^{-3} \tag{16.3}$$

This dimensionless number results from the hydrodynamic fact that power dissipation by an impeller is proportional to (i) the fifth power of impeller diameter, (ii) the density of the mixture, and (iii) the cube of the agitator rotational speed.

The power is that of a single impeller, so a multi-impeller shaft will add the powers together. For multiple impellers, an effort should be made to optimize (equalize) the power consumptions – see Benz (2012A).

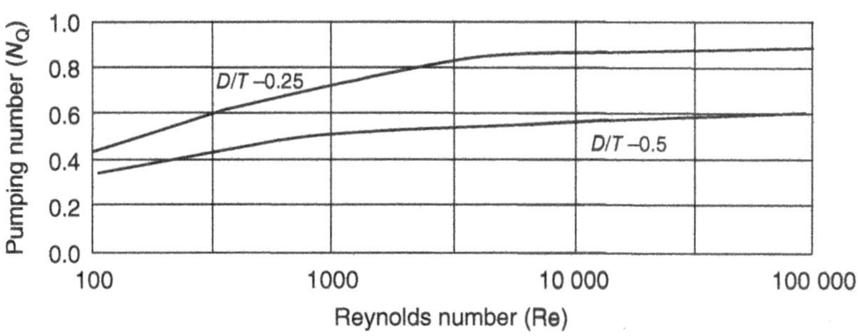

Figure 16.3 Typical dependence of the power number on the Reynolds number and on the ratio of impeller diameter to tank diameter.

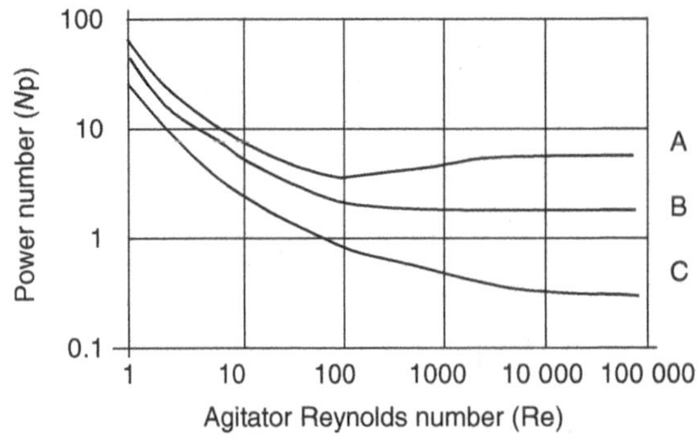

Figure 16.4 Typical power numbers for the *A* straight-blade impeller, *B* pitched-blade impeller, and *C* impeller with no baffles.

Both the pumping number and the power number are functions of Reynolds number, and typical values are shown in Figures 16.3 and 16.4.

Hicks et al. (1976) provided tabular data showing available agitator speeds and corresponding agitator power for various viscosities, equivalent volumes (liquid volume × specific gravity), and severity of agitation (SA).

All of these impellers require baffles along the wall, typically four equally spaced and of standard width and position: baffle width typically 1/12 of the vessel diameter and allowing a gap of 1/72 of the vessel diameter between the baffle and wall to avoid stagnation. Without baffles, the liquid simply swirls with little mixing. Myers et al. (2002) discussed baffles, and Pogal & Kehn (2018) provided a very comprehensive treatment. Baffles and their design can play a significant role in

Table 16.2 *Chemineer* series on turbine mixing.

Authors	Date	Title	Pages
Gates L. E., Fenic J. G., Henley T. L.	1975 Dec. 8	How to select the optimum turbine	110–114
Dickey D. S., and Fenic	1976 Jan.	Dimensional analysis for fluid agitation systems	139–145
Dickey and Hicks R. W.	1976 Feb. 2	Fundamentals of agitation	93–100
Hicks, Morton J. R., and Fenic	1976 Apr. 26	How to design agitators for desired process response	102–110
Gates, Morton, and Fondy P. L.	1976 May 24	Selecting agitator systems to suspend solids in liquids	144–150
Hicks and Gates	1976 Jul. 19	How to select turbine agitators for dispersing gas into liquids	141–148
Hill R. S. and Kime D. L.	1976 Aug. 2	How to specify drive trains for turbine agitators	89–94
Ramsey W. D. and Zoller G. C.	1976 Aug. 30	How the design of shafts, seals and impellers affects agitator performance	101–108
Meyer W. S. and Kime	1976 Sept. 27	Cost estimation for turbine agitators	109–112
Rautzen R. R., Corpstein R. R., and Dickey	1976 Oct. 25	How to use scale-up methods for turbine agitators	119–126
Hicks and Dickey	1976 Nov. 8	Applications analysis for turbine agitators	127–133
Gates, Hicks, and Dickey	1976 Dec. 6	Application guidelines for turbine agitators	165–170

mixing and power consumption. There is obviously difficulty in baffling a glass-lined vessel, and Dickey et al. (2004) illustrated how this is dealt with.

Benz (2004, 2012B) talked about the interplay between choices of impeller diameter as it affects the speed and power. Dickey & Fasano (2004) espoused the use of CFD for the study of flow patterns, including the identification of dead spots.

16.1.2 Chemineer Series on Turbine Mixing

In the mid-1970s, staff at Chemineer, Inc., published a 12-part series of articles in the Chemical Engineering magazine, dealing with agitation of low viscosity liquids using turbine mixers. These articles presented some very useful detailed design procedures which are still referenced after several decades (Table 16.2). The set are still available in reprint form from the magazine. The design procedures described in the first sections of this chapter are based largely on those articles.

16.2 Blending Calculation

If the viscosities of the components and mixture are moderate, then a turbine agitator is used, with pitched-blade impellers. Drury & Gates (2001) presented a sequence of calculations to determine vessel dimensions, impeller diameter, power requirement, and other quantities of interest. The calculations presented here already assume the dimensions of the tank and the impeller and are directed at determining the required speed and power.

Step 1: Decide on the *SA* on a scale of 1–10. This severity depends mainly on the disparities (i) between feed liquid viscosities and/or (ii) between feed liquid densities, as shown in Table 16.3. Choose SA to satisfy the more extreme of the two conditions.

Step 2: Note the required bulk fluid velocity, γ_B, corresponding to SA. This is a measure of the vertical velocity, up and down, and indicates the rate with which the mixture must be turned over by the agitator.

Table 16.3 Mixing intensities.

Blending and motion: assessing the task			
Ratio of viscosities	**Difference of specific gravities**	**Severity of agitation (SA)**	**Required bulk fluid velocity (ft min^{-1}) γ_B**
	<0.1	1	6
0.01–100		2	12
		3	18
	<0.6	4	24
0.0001–10 000		5	30
		6	36
		7	42
	<1.0	8	48
10^{-5}–10^{+5}		9	54
		10	60

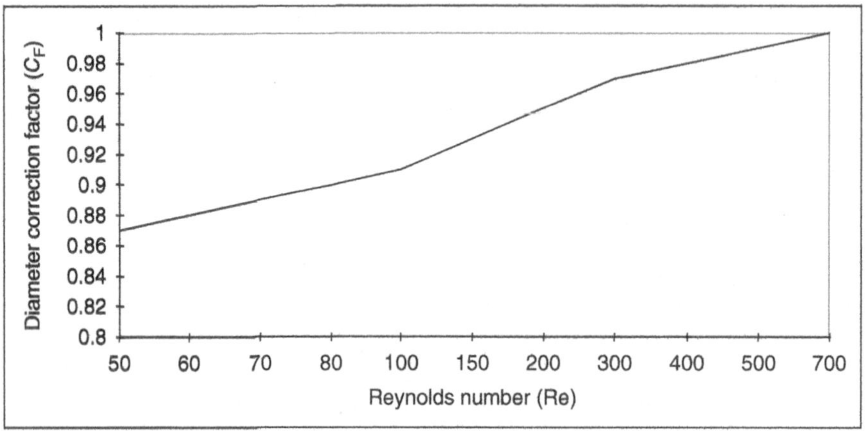

Figure 16.5 Correction to the impeller diameter for low Reynolds number.

Step 3: Calculate effective pumping capacity, $Q = \gamma_B \times A$, where A is the cross-sectional area of the tank.

(Calculations must be consistent with the units being used.)

Step 4: Select a trial impeller diameter, D, between 0.2 and 0.6 times the tank diameter.

Step 5: Initially assuming the motion to be turbulent (i.e. high Reynolds number), read the pumping number, N_Q, from the graph in Figure 16.3. This graph is a generic one, so if a more specific relation is available, then it should be used.

Step 6: From Eq. (16.2), calculate agitator speed NS, and from Eq. (16.1), calculate Reynolds number Re.

Step 7: If the Reynolds number is less than 700, the indication is that a turbine agitator is becoming unsuited for this high-viscosity application. If using a turbine agitator, then obtain the diameter correction factor, C_F, from the graph in Figure 16.5 and calculate the corrected agitator diameter. Repeat steps 4, 5, and 6.

Step 8: Choose the number, n, of agitator impeller elements to mount on the shaft, from Table 16.4.

Step 9: Obtain the power number, N_P, for each impeller.

Figure 16.4 shows the general trend of power numbers with Reynolds number. For the agitators described in this section, a power number of 1.6 is a good estimate. However, the designer will seek more clarity on this important coefficient because the estimation of power is directly proportionate to N_P.

Table 16.4 Single- and dual-impeller elements.

Viscosity (cp)	Liquid height/tank diameter (Z/T)	Recommended number of impellers (N)	Recommended distance from tank bottom: lower impeller	Recommended distance from tank bottom: upper impeller
Up to 25,000	Up to 1.4	1	Z / 3	
	Up to 2.1	2	T_S / 3	(2/3) Z
Over 25,000	Up to 0.8	1	Z / 3	
	Up to 1.6	2	T_S / 3	(2/3) Z

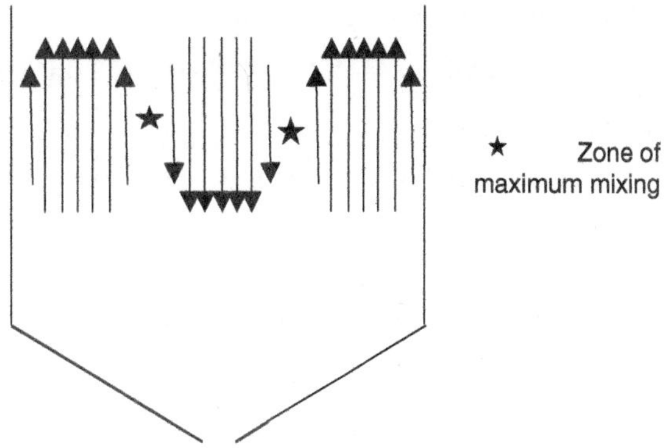

Figure 16.6 Schematic of circulating-flow blending.

Calculate the agitator power (watts) as

$$P = N_P\, D_S^{\,5}\, n\, \rho\, N_S^{\,3} \tag{16.4}$$

With all quantities in SI units – D in meters, density in kg m^{-3}, and N_S in s^{-1} – the result is power in watts.

Pitched-blade impellers cause internal countercurrent liquid motion, producing zones of high shear at the interface between upflow and downflow (see Figure 16.6). It is likely that most of the intermolecular mixing takes place in this zone of high strain.

16.2.1 Wall-Wiping Impellers

For viscous liquids, a completely different type of agitator is used. In the absence of turbulence, it is necessary to move the liquid directly instead of relying on momentum imparted just at the center of the vessel. Two styles are shown below in Figure 16.7.

Bakker & Gates (1995) provided the following procedures for calculating the agitator power.

For *helical-ribbon* agitators, calculate the agitator Reynolds number in the usual manner.

If Re is less than or equal to 100, power number N_P equals (350/Re) $\times f_{geo}$.

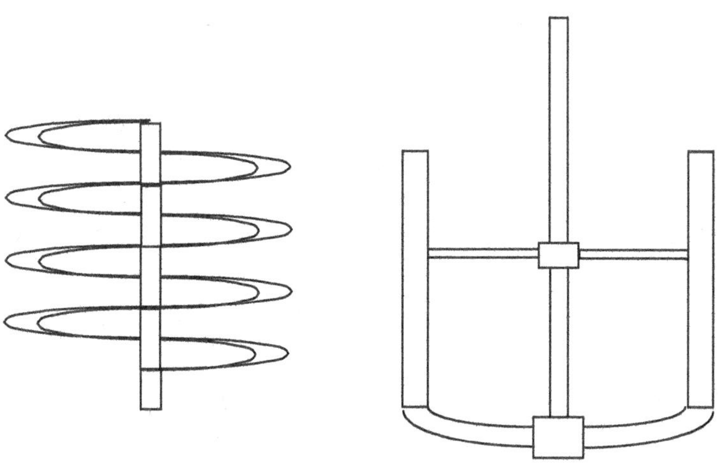

Figure 16.7 Helical-ribbon agitator and anchor agitator.

If Re is greater than 100, N_P equals

$$1.12 \times Exp\,[6.15 - 1.37\ \ln(Re) + 0.0613\,(\ln(Re))^2] \times f_{geo}$$

where f_{geo} equals

$$(D/\text{pitch})^{1/2}\,(\text{height}/D)\,(\text{width}/D/0.1)^{0.16}\,(D/24/(T - D))^{1/2}\,(n_f/2)^{1/2}$$

with D being the outer diameter of the agitator,
pitch being the vertical distance for one full turn of the spiral,
height being the height of the agitator,
width being the width of the blade,
T being the diameter of the tank, and
n_f being the number of flights for the helix.

For anchor agitators, if Re is less than or equal to 10, N_P equals $(400/\text{Re}) \times f_{geo}$.
If Re lies between 10 and 10 000, N_P equals

$$1.05 \times Exp\,[5.64 - 0.783\ \ln(Re) - 0.0523\,(\ln(Re))^2 + 0.00674\,(\ln(Re))^3] \times f_{geo}$$

where f_{geo} equals $(\text{height}/D)\,(W/D/0.1)^{0.16}\,(D/49/(T\text{–D}))^{1/2}$.

We have experimented recently with a new design of mixer for viscous liquids, built from a circular rod or tubing in the form of a helix or spiral. Generally, we have used a double helix. The outer diameter of the spiral may be as large as the tank diameter itself (minus a small clearance), which helps keep the walls well wiped to prevent fouling and possibly enhance heat transfer. The spiral arms are attached to a central drive shaft, which may enter the vessel from either the top or bottom. The rod or tube diameter is chosen so that the force which it encounters does not cause excessive deflection of the spiral. We have found it practical to design these agitators with no cross bracing even at tank diameters up to 5 ft. The advantages are simplicity, low cost, elimination of points of stagnation, and clear space within the spiral arms which allows for heating/cooling coils, instrumentation, etc.

References are Bakker & Gates (1995) and Gibson (1999). Benz (2017) discussed heat effects in high-viscosity mixing.

16.2.2 Attenuation of Variation

The previous discussion concerned the blending of two somewhat different streams. A second objective of blending can be to attenuate the fluctuations in properties of a single stream, simply by providing a zone of full back-mixing.

Examining this second case, it can be shown mathematically that if an entering stream with a regular fluctuation of frequency f in some property is blended in a fully mixed vessel of residence time Φ, then the amplitude of the fluctuation is attenuated by the factor $\frac{1}{\sqrt{[1+(2\pi)^2(f\times\Phi)^2]}}$. For example, see Table 16.5.

This calculation relies on the assumption of *ideal* "back-mixing". If liquids bypass the main flow or if there are zones of stagnation, then the attenuation will not be as great.

16.2.3 In-Line Mixers

In-line pipeline mixers are dedicated to the task of ensuring uniformity across the cross section of the flow. Typically, they are used after two streams have been brought together. The mixer may be *dynamic*, with a rotating element within the flow or it may be *static* or *motionless*, making use of fixed deflectors to bring about interchange of streamlines. Some types of dynamic mixers

Table 16.5 Smoothing effects of total back-mixing.

$f \times \Phi$	Attenuation factor
0	1.0
0.1	0.85
0.2	0.62
0.5	0.30
1.0	0.16
2.0	0.08
5.0	0.03

are described in Perry & Green (1997). Static mixers add extra pressure drop to the line in which they are located, so they are using the energy of the stream itself to bring about mixing. Several varieties of static mixers are on the market.

Obviously, there would be a great advantage in eliminating the moving parts associated with a mechanical agitator. Several companies have created designs of static or motionless mixers that can be built into pipes to bring about radial homogenization. One of the simplest is the Kenics mixer (Chemineer, Inc.), which is a sequence of successive elements consisting of metal strips 1.5 × pipe ID long, each twisted through 180°, with the twist of successive elements being in opposite directions. The start of each strip is at 90° to the end of the strip that it follows.

Static mixers are able to convert a highly nonuniform stream into a stream of any desired uniformity if adequate length is provided. The effectiveness of various mixers is evaluated in terms of

(1) the length required to achieve a given reduction (e.g. 95%) of incoming nonuniformity and
(2) the pressure drop per unit length of mixer or, more specifically, the ratio of mixer pressure gradient to the pressure gradient of a mixer-less pipe.

Godfrey (1985) compared the pressure drop and the effectiveness of various offerings. One can choose between a device with high effectiveness and high-pressure gradient or a device with less effectiveness per unit of length but lower-pressure gradient.

The two criteria can be multiplied into a single ratio, which is desired to be as low as possible. There is a great range among commercial mixers, and there are some brands that we would stay away from (See Table 16.6). Looking at the two Sulzer mixers, we might choose a long length of the high-pressure-gradient SMX type if complete mixing were critical and adequate pressure were available to us. In any case, the best mixers make the most effective use of pressure dissipation in achieving uniformity. This is the indicator in the last column.

References are Bayer et al. (2003), Godfrey (1985), Myers et al. (1997), Perry & Green, Ch 18, p. 20–21 (1997), Crisafulli (2012)

16.3 Immiscible Liquids

Immiscible liquids are brought together for one of the four reasons:

1. to extract a component out of one phase into the other phase,
2. to react components contained separately in different phases,
3. to cool a phase by direct contact with another phase, and

Table 16.6 Static mixer characteristics and comparison.

	L/D for 95% uniformization $(L/D)_{95\%}$	**Ratio of pressure gradient relative to empty pipe (P/P_o)**	**Inverse of effectiveness (low number is best) $(L/D)_{95\%} \times P/P_o$**
Best	9		
Best		6	
Best			190
Worst	320		
Worst		300	
Worst			5000
Kenics	29	6.9	200
Sulzer SMXL	26	7.7	200
Sulzer SMX	9	39	350

4. to create a permanent emulsion of one phase in another.

In cases 1, 2 and 3, the phases are subsequently separated from one another: Care may have to be taken not to disperse so finely that a permanent emulsion results. In case 4, separation is undesirable.

Turbine agitators are again suitable. For dispersing one liquid in another, the impeller blades are in their straight form, as shown in Figure 16.1. The objective is to convert power of rotation into comminution of droplets.

The drop size created by an agitator of specific size, shape, and speed depends on the properties of the liquids. The coalescence of small droplets may be taking place simultaneously, so agitation must be carried on continuously while mass transfer and/or chemical reaction is taking place. In-line mixers are not suitable for this application because the residence time in the mixer is too short.

If the objective is to create a permanent emulsion, then an in-line device is suitable. Several designs of equipment are on the market, generally under the heading of *homogenizers*.

Figure 16.8 shows a continuous multistage agitated column for liquid–liquid extraction, where an immiscible phase is alternately dispersed in a second phase to achieve mass transfer, then allowed to settle out in order to flow separately to the next stage. Benz (2014) discussed such a column.

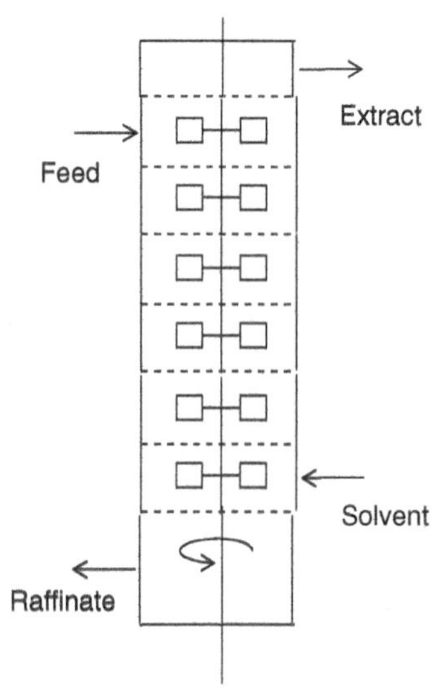

Figure 16.8 Multistage stirred extraction column.

References are Cohen (1998) and Pandolfe (2000).

Permanent mixtures of immiscible liquids, with droplets of one liquid in the continuous phase of the other liquid, constitute the class of *emulsions*. An analogous type of mixture with solid particles in a liquid is called a *suspension*. The creation of these substances is described in Section 16.2.3

16.4 Gas and Liquid

Gases are dispersed as bubbles in liquids and liquids are dispersed as sprays in gases, for such purposes as cooling, mass transfer, and chemical reaction.

An adequate dispersion of gas in liquid may sometimes be attained simply by sparging the gas into the liquid. If a finer dispersion is desired (e.g. to obtain greater interfacial surface area), then a rotating agitator may be used (Figure 16.9). The same style of flat-bladed turbine impeller is used as for dispersing immiscible liquids. A problem arises in that the cloud of bubbles may overwhelm the action of the agitator and make it ineffective. Hicks & Gates (1976) and also Deutsch & Van Essen (1998) provided predictive equations for this effect and for how it may be overcome. The following calculation follows that of Hicks and Gates.

Step 1: Assess the required SA from Table 16.7.

Step 2: Choose an agitator diameter, D. For this application, typically D is set at 0.15–0.5 times tank diameter T.

Step 3: Calculate the superficial velocity of gas, U_S, equal to the volumetric flow of gas evaluated at the point of entry, divided by the horizontal cross-sectional area of vessel.

Power for the impeller is the maximum of

- a rising function of the SA, regardless of the amount of gas, and
- a rising function of the amount of gas to be dispersed.

In Figure 16.10, the top line shows the first effect and the close-to-vertical lines of superficial velocity show the second effect.

Step 4: Calculate parameter $Y = \text{SA} \times (D/T)^{1.2}$.

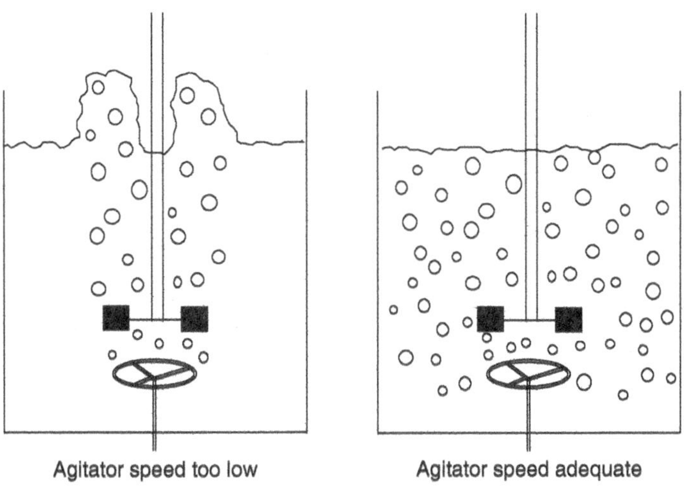

Figure 16.9 Sparging calculation.

Table 16.7 Dispersion of gases: assessing the task.

Desired condition	Severity of agitation
For noncritical degree of gas dispersion; provide non-flooded impeller conditions for a coarse dispersion of gas in liquid; suitable for applications that are not mass-transfer limited	1–2
For moderate degree of gas dispersion; drives fine bubbles completely to the vessel wall; provide recirculation of dispersed bubbles back into the impeller	3–5
For critical gas–liquid reactors where rapid mass transfer is required	6–10

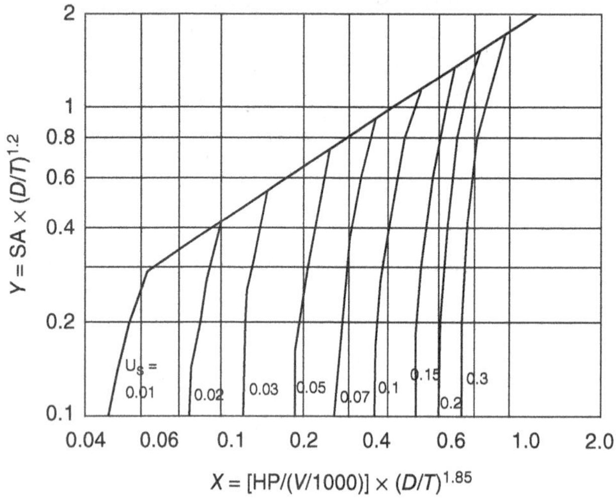

Figure 16.10 Agitator power for adequate sparging.

Step 5: From the following graph, determine the parameter

$$X = [P/(V/1000)]\,(D/T)^{1.85}$$

where the Y line intersects both the top line and the line specific to the superficial velocity, the higher value of X should be used.

Step 6: Calculate power required to achieve dispersion,

$$P_1 = (V/1000) \times X \times (D/T)^{-1.85}$$

The units are specifically horsepower for P and gallons of liquid for V.

Divide by 0.001 341 to convert horsepower to watts.

As opposed to the blending calculation of 16.1.2, where we set D and Ns and calculated power, in the present calculation, we first determine the required power and then determine the speed to give that power, thus steps 7–10.

Step 7: Make an estimate of the impeller rotational speed, N_S, and then use Eq. (16.4) to calculate the power (watts) that the agitator would require if operating at this speed in an un-gassed liquid,

$$P_{2U} = N_P\, D^5\, \rho\, N_S^{\,3}$$

where 4.5 is good estimate for the power number for this style of agitator.

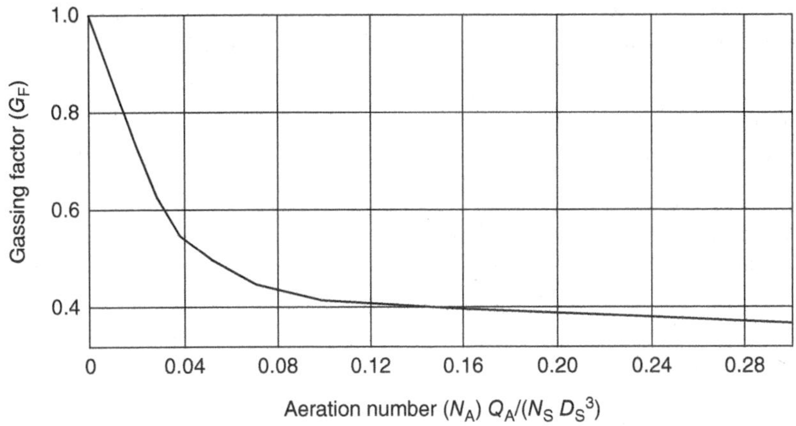

Figure 16.11 Gassing factor for agitator flooding.

Step 8: Calculate the aeration factor $N_A = Q_A/(N_S\, D^3)$,
where QA is the total volumetric flow rate of the gas.

Step 9: From Figure 16.11, determine a value of gassing factor G_F based on N_A.

Step 10: From Figure 16.11, obtain a "gassing factor" and apply the gassing factor to the power calculated in Step 7. This accounts for the reduction, due to the interference of the gas, in power transmitted to the fluid by the agitator,

$$P_2 = G_F \times P_{2U}$$

Step 11: Compare the value of P_2 with P_1. If they are not sufficiently close, then return to Step 7 and choose a new value of agitator speed. The objective is that the true rate of dissipation of energy between the impeller and fluid is in fact P_1.

Step 12: Choose the number of agitator impellers according to the following rule:

for Z/T less than or equal to 1.0, one impeller situated at a distance of $T/6$ from the floor of the vessel,

for Z/T up to 1.8, two impellers, at $T/6$ and at $(2/3)\, Z$ above the floor,

with Z and T being the height of liquid and the tank diameter, respectively.

References are Hicks & Gates (1976), Deutsch & Von Essen (1998), Lee & Tsui (1999), and Himmelsbach et al. (2007).

The other type of gas–liquid mixing is the dispersion of a liquid into a gas, generally by means of atomization. The main example is *spray drying*. Other applications are cooling of gases, scrubbing of particulates, humidification, absorption of vapors, and chemical reaction. All of these phenomena rely on a high ratio of surface-to-volume.

References are Ondrey (1995) and Welander & Vincent (2001). Chapters 17 and 18 also have references.

16.5 Solid Particles in Liquid

Some situations where solid particles need to be suspended in a liquid are

- heterogeneous catalysis,
- dissolution of solids, and
- preparation of a permanent slurry or emulsion.

Assuming a turbulent liquid, the agitator of choice is a turbine with blades pitched to produce a component of axial flow. The flow pattern is shown in Section 16.1.1. The speed and power of the agitator are a function of

- the terminal velocity of the particles and
- the degree of suspension required.

Obviously, the greater the terminal velocity, the harder it is to keep the particles suspended. Particles that are small and/or similar in density to that of the liquid are easier to keep in suspension.

The degree of suspension refers to the uniformity of solids concentration in the vessel: Is it enough just to sweep particles periodically off the bottom of the vessel or must there be uniformity almost right to the top. If a slurry is being prepared, then the height to which uniformity must extend is dictated by the point of slurry draw-off.

16.5.1 Suspension Calculations

Step 1: Once again, the calculation begins with selecting a SA. The following Table 16.8 gives the criteria.

Step 2: Select impeller diameter, D, between 0.2 and 0.6 times the tank diameter, T, with 0.3 being a commonly used ratio for this application.

Step 3: Determine the terminal settling velocity of a single solid particle, U_t.

Step 4: Correct the calculated terminal velocity to take account of interparticle interactions based on the correction factor f_w from Figure 16.12,

Step 5: Based on D/T and on the chosen value of SA, obtain the value of the parameter φ from Figure 16.13.

Step 6: Calculate the required agitator speed as $N_S = (\varphi\ U_d/D^{2.81})^{1/3.75}$, where the units are specifically

rpm for N_S
ft min^{-1} for U_d
inches for D.

Step 7: Choose the number, n, of agitator elements (impellers) to mount on the shaft as follows:
for Z/T up to 1.2, one impeller situated $Z/4$ above the bottom of the tank;
for Z/T up to 1.8, two impellers situated $T/4$ and (2/3) Z above the bottom.

Table 16.8 Suspension of solids: assessing the task.

Desired condition	Severity of agitation (SA)
Produce some motion of all solids; suitable for periodic lifting of solids from bottom	1–2
Provide uniformity of slurry to a height of at least 1/3 of the total liquid height; suitable for slurry draw-off at low exit nozzle elevations	3–5
Provide uniformity of solids concentration up to 95% of the total liquid height; suitable for slurry draw-off up to 80% of liquid height	6–8
Provide uniformity of solids concentration up to 98% of total liquid height; suitable for slurry draw-off by overflow	9–10

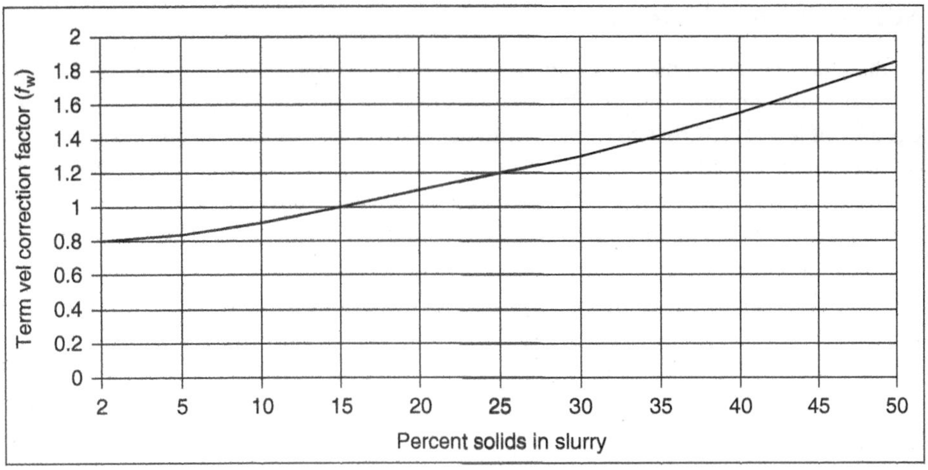

Figure 16.12 Terminal velocity correction.

$$U_d = U_t \times f_w$$

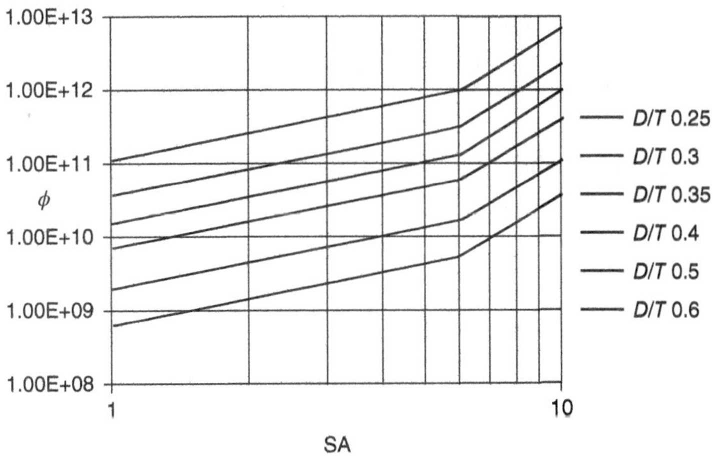

Figure 16.13 Solid's suspension factor.

Step 8: Calculate agitator power (watts) where we are back to SI units for D (m), ρ (kg m^{-3}) and N_S (s^{-1})

$$P = N_P\, D_S{}^5\, n\, \rho\, N_S{}^3$$

where N_P of 1.6 is inferred from the power formula in the original calculation:

$$D = 394\ \left(\text{horsepower}\ n^{-1}\ N_S{}^{-3}\ \text{specific gravity of slurry}^{-1}\right)$$

with impeller speed in revolutions per minute.

The power number can be checked against the curves in Figure 16.4.

References are as follows: Ditl & Rieger (2006), Gates et al. (1976A), Heywood (1999), Ozcan-Taskin (2008), Russell-Hill (2004), Shaw (1992), Von Essen (1999), and Langhorn and Banaszek (2009).

16.5.2 Problem Mixing and Extreme Mixing

Assumed in the above descriptions and calculations is that all liquids behave in the Newtonian fashion, i.e., with strain and stress linear with each other starting at zero. A shear-thinning liquid may create a zone of lesser disturbance in the vicinity of the agitator, which lessens the hoped-for transfer of momentum from agitator to process. Other varieties of fluid peculiarities can cause unexpected effects. Several authors addressed this problem, including Dickey (2000, 2015), Wilkens et al. (2003), and Wyczalkowski (2004) with his counterflow impeller.

Turbine mixers and wall-wiping mixers are unable to generate and mix the fine droplets and fine particles that make up emulsions and permanent suspensions. High-speed high-shear mixers take on this task. This is a field of current development, with articles including Myers et al. (1993), Shelley (2004, 2007), Cohen (2005), Dickey (2015), and Banaszek (2016, 2018).

16.6 Solid Particles with Solid Particles

Probably, most cases of particle mixing have the objective of removing or blending out nonuniformities. The operation may be batch or continuous. Particle blenders come in different styles.

The *tumbling* blender (Figure 16.14) rotates, causing particles to tumbling over one another, hopefully in a random manner.

Convective blenders are stationary and have an internal rotating system of paddles or a screw to impart motion to the particles. *Fluidized beds* use a gas to move particles relative to one another. A *recirculating* vessel with multiple staggered downcomers can rearrange particles (Figure 16.15).

Some considerations in particle mixing:

1. extra effort is required to mix particles that have a tendency to agglomerate,
2. tumble mixing typically takes 10–15 minutes,
3. a fluidized bed gives the fastest mixing,
4. particles of different size or density have a tendency to settle relative to one another. The mixer must overcome this action, and
5. the speed of mixing needs to be restrained if particles are fragile. Also, rapid mechanical agitation may raise the temperature.

Clump, "Mixing of Solids", in Uhl & Gray (1967), Goris & Van der Wel (2003),

Harnby, "The selection of powder mixers", in Harnby et al. (1985), Maynard (2008), Williams, "Mixing of Particulate Solids", in Uhl & Gray, vol. III (1986).

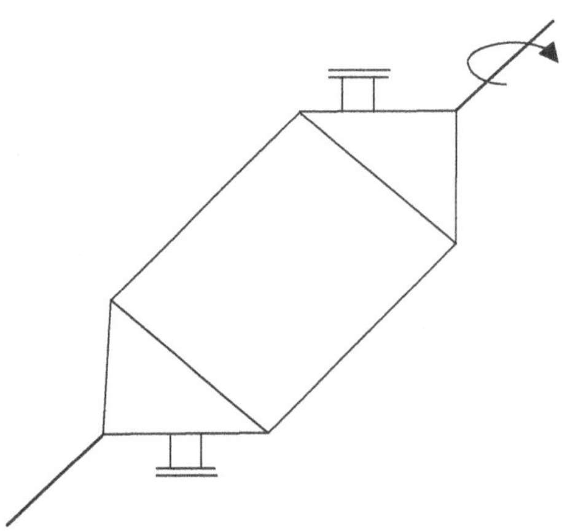

Figure 16.14 Rotating solids blender.

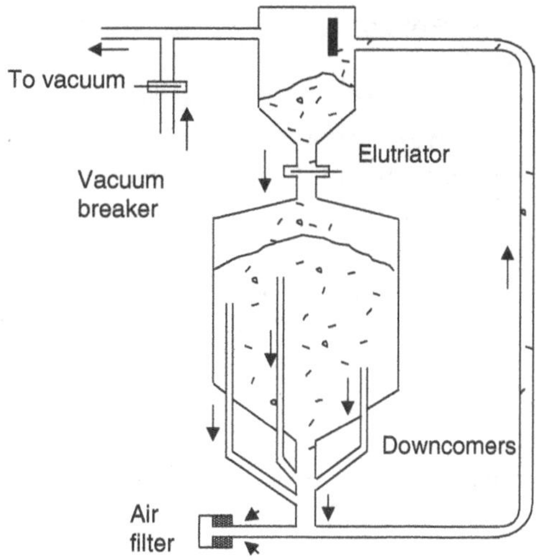

Figure 16.15 Recirculating blender for solids.

16.7 Solid Particles and Gas

Some reasons why a solid may be mixed with a gas are

- to be conveyed,
- to be reacted,
- to be dried or otherwise treated, and
- to catalyze a reaction.

Conveying (e.g. by air) is carried out in one of two modes: *dilute phase* or *dense phase*. As indicated by the nomenclature, the ratio of solids to gas is quite different in the two modes. In the dilute phase, the particles are mixed into the gas stream simply by being dropped in; a rotary feed valve is needed if the gas system is under pressure. In the dense phase, the solids are assembled and then gas is introduced to start moving them.

If the objective is mass transfer or chemical reaction, then intimate mixing can be done in a fluidized bed. The bed is designed in accordance with well-established fluidization technology. High rates of surface transfer are achieved relative to say a packed or slowly moving bed (Figure 16.16).

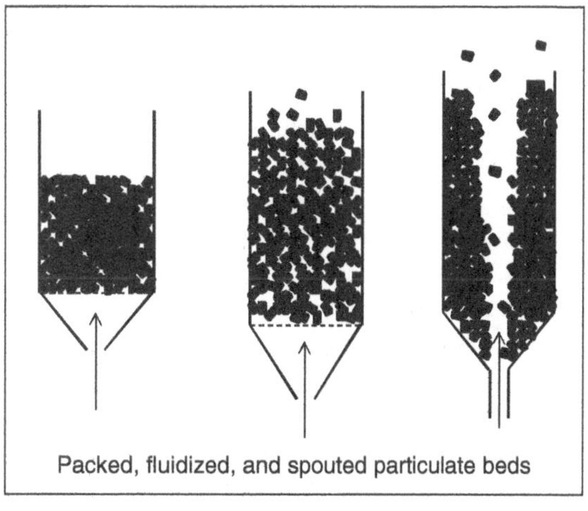

Figure 16.16 Particulate behaviors in solid beds.

References

Bakker, A. and Gates, L.E. (1995 Dec). Properly choose mechanical agitators for viscous liquids. *Chem. Eng. Prog.* 91 (12): 25–34.

Banaszek, C. (2016 May). Mixers: four innovations worth a closer look. *Chem. Eng.* 123 (5): 42–48.

Banaszek, C. (2019 Aug). Beyond simple mixing. *Chem. Eng.* 126 (8): 36–40.

Bayer, T., Himmler, K., and Hessel, V. (2003 May). Don't be baffled by static mixers. *Chem. Eng.* 110 (5): 50–57.

Benz, G.T. (2004 Feb). Sizing impellers for agitated aerobic fermenters. *Chem. Eng. Prog.* 100 (2): 18S–24S.

Benz, G.T. (2007 Mar). The G-value for agitator design: time to retire it. *Chem. Eng. Prog.* 103 (3): 43–47.

Benz, G.T. (2010 Jun). Agitating fibrous materials. *Chem. Eng. Prog.* 106 (6): 28–32.

Benz, G.T. (2012A Feb). Determining torque split for multiple impellers in slurry mixing. *Chem. Eng. Prog.* 108 (2): 45–48.

Benz, G.T. (2012B Nov). Cut agitator power costs. *Chem. Eng. Prog.* 108 (11): 40–43.

Benz, G.T. (2014 Jan). Designing multistage agitated reactors. *Chem. Eng. Prog.* 110 (1): 30–36.

Benz, G.T. (2017 Dec). Optimize heat transfer of viscous fluids in agitated vessels. *Chem. Eng. Prog.* 113 (12): 24–27.

Calabrese, R.V., Kresta, S.M., and Liu, M. (2014 Jan). Recognizing the 21 most influential contributions to mixing research. *Chem. Eng. Prog.* 110 (1): 2–29.

Cohen, D. (1998 Aug). How to select rotor-stator mixers. *Chem. Eng.* 105 (8): 76–79.

Cohen, D. (2005 Apr). High-shear mixing: don't fall victim to common misconceptions. *Chem. Eng.* 112 (4): 46–51.

Clump C. W., *Mixing of solids*, in Uhl & Gray 1967.

Crisafulli, K. (2012 Jan). Inline blending can help process plants cut costs and reduce quality give-away. *Hyd. Proc.* 91 (1): 19.

Deutsch, D.J. and Von Essen, J. (1998 Aug). Gas-liquid mixer avoids correlation flooding: scaleup from laboratory data can now yield gas-liquid agitators that are neither too big nor too small. *Chem. Eng.* 105 (8): 80–82.

D'Aquino, R. (2004 Sep). Impellers for niche mixing needs. *Chem. Eng. Prog.* 100 (9): 10–12.

Dickey, D.S. and Fenic, J.G. (1976 Jan). Dimensional analysis for fluid agitation systems. *Chem. Eng.* 83 (1): 7–13. or 139–145.

Dickey, D.S. and Hicks, R.W. (1976 Feb). Fundamentals of agitation. *Chem. Eng.* 83 (2): 14–21. or 9–100.

Dickey, D.S. (2000 May). Facing the challenge of mixing problem fluids. *Chem. Eng.* 107 (5): 68–75.

Dickey, D.S. (2001 Nov). Liquid mixing. *Chem. Eng. Prog.* 97 (11): 74.

Dickey, D.S. (2002 Aug). Equipment, technology, people – mixing. *Chem. Eng.* 109 (8): 88–89.

Dickey, D.S. and Fasano, J.B. (2004 Feb). How geometry and viscosity influence mixing. *Chem. Eng.* 111 (2): 42–46.

Dickey, D.S., Bittorf, K.J., Ramsey, C.J., and Johnson, K.E. (2004 Nov). Ynderstand flow patterns in glass-lined reactors. *Chem. Eng. Prog.* 100 (11): 21–25.

Dickey, D.S. and Patterson, G.K. (2008 Oct). Fluid mixing success through failures, part 1. *Chem. Eng. Prog.* 104 (10): 40–46.

Dickey, D.S. (2013 Aug). Ten things you may not know about liquid mixing scaleup. *Chem. Eng.* 120 (8): 32–39.

Dickey, D.S. (2015 Aug). Tackling difficult mixing problems. *Chem. Eng. Prog.* 111 (8): 35–42.

Dickey, D.S. (2015 Feb). Fundamentals of high-shear dispersers. *Chem. Eng.* 122 (2): 40–47.

Dickey, D.S. (2019 Aug). Liquid mixing in stirred tanks. *Chem. Eng.* 126 (8): 24–34.

Dietsche, W. (1998 Aug). Mix or match: choose the best mixers every time. *Chem. Eng.* 105 (8): 70–75.

Ditl, P. and Rieger, F. (2006 Jan). Designing suspension mixing systems. *Chem. Eng. Prog.* 102 (1): 22–30.

Drury, S.F. and Gates, L.E. (2001 Feb). Mixer specification à la spreadsheet. *Chem. Eng.* 108 (2): 62–69.

Fasano, J.B. (2015 Jun). Select the right impeller. *Chem. Eng. Prog.* 111 (6): 30–36.

Gates, L.E., Fenic, J.G., and Henley, L. (1975 Dec). How to select the optimum turbine. *Chem. Eng.* 82 (12): 2–6. or 110–114.

Gates, L.E., Morton, J.R., and Fondy, P.L. (1976A May). Selecting agitator systems to suspend solids in liquids. *Chem. Eng.* 83 (5): 31–37. or 144–150.

Gates, L.E., Hicks, R.W., and Dickey, D.S. (1976B Dec). Application guidelines for turbine agitators. *Chem. Eng.* 83 (12): 79–84. or 165–170.

Gibson, W.D. (1999 Apr). Impelling mixers to move viscous fluids. *Chem. Eng.* 106 (4): 49–57.

Godfrey J. C., Static mixers, in Harnby et al., *Mixing in the Process Industries*, 1985.

Goris, D. and Van der Wel, P. (2003 Mar). Pick the right solids mixer. *Chem. Eng.* 110 (3): 39–46.

Harnby N., The selection of powder mixers, in Harnby et al., *Mixing in the Process Industries*, 1985.

Harnby, N., Edwards, M.F., and Nienow, A.W. (1985). *Mixing in the Process Industries*. London: Butterworths.

Hemrajani R. R., Mixing and blending, in *Of Kirk-Othmer Encyclopedia of Chemical Technology, Fourth edition*, vol 6. 1995, John Wiley & Sons, New York.

Heywood, N.I. (1999 Sep). Stop your slurries from stirring up trouble. *Chem. Eng. Prog.* 95 (9): 21–41.

Hicks, R.W., Morton, J.R., and Fenic, J.G. (1976 Apr). How to design agitators for desired process response. *Chem. Eng.* 83 (4): 22–30. or 102–110.

Hicks, R.W. and Gates, L.E. (1976 Jul). How to select turbine agitators for dispersing gas into liquids. *Chem. Eng.* 83 (7): 38–45. or 141–148.

Hicks, R.W. and Dickey, D.S. (1976 Nov). Applications analysis for turbine agitators. *Chem. Eng.* 83 (11): 72–78. or 127–133.

Hill, R.S. and Kime, D.L. (1976 Aug). How to specify drive trains for turbine agitators. *Chem. Eng.* 83 (8): 46–51. 89–94.

Himmelsbach, W., Houlton, D., Keller, W., and Lovallo, M. (2006 Apr). Mixing systems: design and scale up. *Chem. Eng.* 113 (4): 46–53.

Himmelsbach, W., Keller, W., Lavallo, M. et al. (2007 Oct). Increase productivity through gas-liquid mixing. *Chem. Eng.* 114 (10): 5–58.

Jenkins, S. (2012 Sep). Gas sparging. *Chem. Eng.* 119 (9): 33.

Langhorn K., Banaszek C., Disperse difficult solids, *Chem. Eng.* 2009 Jul, 116*(7)*, 40–43.

Langhorn K., The unexpected rewards of testing a mixer, *Chem. Eng.* 2012 Jul, 119*(7)*, 30–33.

Lee, S.-Y. and Tsui, Y.P. (1999 Jul). Succeed at gas-liquid contacting. *Chem. Eng. Prog.* 95 (7): 23–24.

Machado, M.B. and Kresta, S.M. (2015 Jul). When mixing matters: choose impellers based on process requirements. *Chem. Eng. Prog.* 111 (7): 27–33.

Makwana, V.D., Sivalingam, G., and Vakil, S.M. (2015 Aug). Improving mass-transfer-limited reactions. *Chem. Eng.* 122 (8): 67–71.

Maynard, E. (2008 May). Blender selection and avoidance of post-blender segregation. *Chem. Eng.* 115 (5): 67–72.

Meyer, W.S. and Kime, D.L. (1976 Sep). Cost estimation for turbine agitators. *Chem. Eng.* 83 (9): 60–63. or 109–112.

Myers, K.J., Bakker, A., and Ryan, D. (1997 Jun). Avoid stagnation by selecting static mixers. *Chem. Eng. Prog.* 93 (6): 28–38.

Myers, K.J., Reeder, M.F., Ryan, D., and Daly, G. (1999 Nov). Get a fix on high-shear mixing. *Chem. Eng. Prog.* 95 (11): 42.

Myers, K.J., Reeder, M.F., and Fasano, J.B. (2002 Feb). Optimize mixing by using the proper baffles. *Chem. Eng. Prog.* 98 (2): 42–47.

Nagata, S. (1975). *Mixing Principles and Applications.* New York: John Wiley & Sons.

Noble, T. (2000 Nov). Mystery moves out of mixing. *Chem. Eng. Prog.* 96 (11): 11–13.

Oldshue, J.Y. (1997 Mar). Heed this advice in scaling up mixers. *Chem. Eng. Prog.* 93 (3): 70–73.

Ondrey, G. (1995 Dec). Models demystify spray drying. *Chem. Eng.* 102 (12–13): 30–31.

Ozcan-Taskin, G. (2008 Jul). Incorporate floating solids into liquids. *Chem. Eng.* 115 (7): 43–45.

Pandolfe, W.D. (2000 Aug). Milking more from homogenizers. *Chem. Eng.* 107 (8): 119–122.

Paul, E.L., Atiemo-Obeng, V.A., and Kresta, S.M. (2003). *Handbook of Industrial Science and Practice.* New York: John Wiley & Sons.

Perry, R. and Green, D.W. (1997). *Perry's Chemical Engineers' Handbook*, Seventhe, vol. Chapter 18, 20–21.

Pietsch, W. (2007 Nov). Understanding agglomeration. *Chem. Eng. Prog.* 103 (11): 18–20.

Pogal, G. and Kehn, R.O. (2018 Jan). Mixers: don't let baffles baffle you. *Chem. Process.* 80 (1): 1–4.

Post, T. (2010 Mar). Understand the real world of mixing. *Chem. Eng. Prog.* 106 (3): 25–32.

Ramsey, W.D. and Zoller, G.C. (1976 Aug). How the design of shafts, seals and impellers affects agitator performance. *Chem. Eng.* 83 (8): 52–59. or 101–108.

Rautzen, R.R., Corpstein, R.R., and Dickey, D.S. (1976 Oct). How to use scale-up methods for turbine agitators. *Chem. Eng.* 83 (10): 64–71. 119–126.

Russell-Hill, S. (2004 Nov). The better way to mix solids with liquids. *Chem. Eng.* 111 (11–12): 52–56.

Shaw, J.A. (1992 May). Succeed at solids suspension. *Chem. Eng. Prog.* 88 (5): 34–41.

Shelley, S. (2004 Apr). Taking high-shear mixing to the next level. *Chem. Eng.* 111 (4): 24–26.

Shelley, S. (2007 Nov). High-shear mixers: still widely misunderstood. *Chem. Eng. Prog.* 103 (11): 7–11.

Stadtaus, M., Weiss, H.-J., Himmelsbach, W., and Smith, J. (2010 Apr). High-performance agitated reactors. *Chem. Eng.* 117 (4): 38–47.

Tatterson, G.B. (1991). *Fluid mixing and Gas Dispersion in Agitated Tanks.* New York: McGraw-Hill Inc.

Uhl V.W., Gray J.B., eds, *Mixing Theory and Practice* 1966, 1967, 1981, Academic Press, New York.

Von Essen, J.A. (1999 Nov). Design agitated slurry storage tanks to minimize costs. *Chem. Eng. Prog.* 95 (11): 51–55.

Wienberg, S. (2006 Nov). Making emulsions. *Chem. Eng.* 113 (12): 60–62.

Welander, P. and Vincent, T.L. (2001 Jun). Select the right spray nozzle. *Chem. Eng. Prog.* 97 (6): 75–79.

Wilkens, R.J., Henry, C., and Gates, L.E. (2003 May). How to scale up mixing processes in non-newtonian fluids. *Chem. Eng. Prog.* 99 (5): 44–52.

Williams J. C., Mixing of particulate solids, in Uhl & Gray, *Mixing Theory and Practice*, vol. III, 1986.

Wyczalkowski, W. (2004 Jun). How fluid properties affect mixer and impeller design. *Chem. Eng.* 111 (6): 43–47.

17

Mechanical Separations

Mechanical separations comprise the operations in which different phases are parted from one another. There are five general situations, namely where the phases are

1. liquid and liquid (immiscible or only slightly miscible in each other),
2. solid and solid,
3. gas and liquid,
4. gas and solid, and
5. liquid and solid.

By contrast with *molecular separations*, which are discussed in the Chapter 18, mechanical separations are distinguished by the absence of component transfer between phases.

The need for a mechanical separation may arise in dealing with mixtures of phases found in natural raw materials or in purifying process materials where extraneous phases have infiltrated. A mechanical separation also may be the final step in a molecular separation: Two phases may have been brought into intimate contact in order to promote component molecular transfer between the phases; afterward the phases have to be separated. In another situation, a fluid (e.g. air) may have been used to convey solid pellets and must then, at the end of the travel, be separated from both the pellets and from any fines that were generated. A mechanical and a molecular separation may be used in sequence: In a familiar domestic operation such as drying a load of washing, the clothes are first wrung or spun to remove the bulk of the rinse water (a mechanical operation) and then hung or tumbled to diffuse out the rest of the water – a molecular operation.

An excellent reference for many of the operations discussed in this section is Volume 1, Chapter 4, "Mechanical Separations" in Ludwig (1999) (Figure 17.1, Table 17.1).

17.1 Liquid–Liquid Separations

There are many situations where a process ends up with a mixture of two (or more) liquid phases. The liquids may both be organics, or, very often, one is organic and the other is largely water. Some situations are as follows:

- liquid–liquid extraction,
- a chemical reaction that generates an immiscible product or by-product (e.g. water),
- a feed stream containing water,

Practical Process Design for Chemical Engineers, First Edition. Keith Marchildon and David Mody.

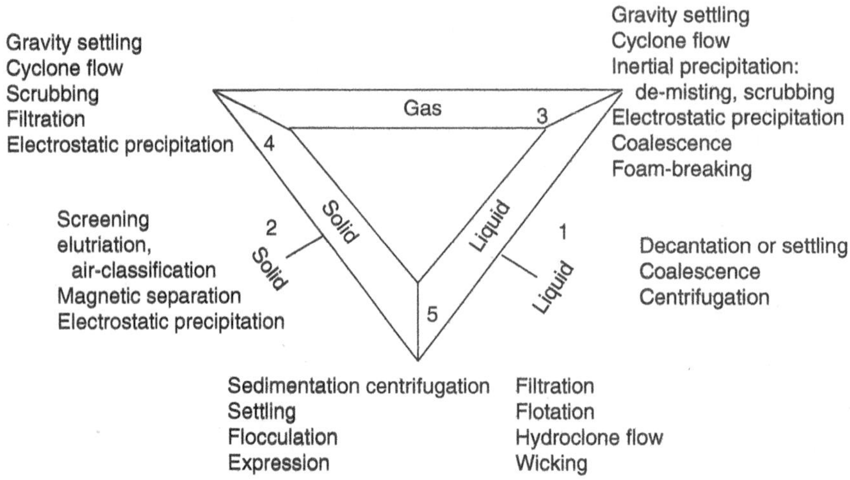

Figure 17.1 Varieties of the separation challenge.

- cooled condensate from azeotropic distillation,
- cooled condensate from steam stripping, and
- aqueous solution used to wash or otherwise treat an organic liquid.

A general reference on liquid–liquid separations is Cusack (2009).

The three common methods of separation rely on a difference in the density of the two phases.

Table 17.1 Summary of mechanical separations.

Method	Liquid–liquid	Solid–solid	Gas–liquid	Gas–solid	Liquid–solid
1. Decantation	Yes				
2. Coalescence	Yes		Yes		
3. Centrifugation	Yes	Yes			Yes
4. Screening		Yes			
5. Elutriation and classification		Yes			
6. Magnetic attraction		Yes			
7. Cyclone flow		Yes	Yes	Yes	
8. Settling, differential settling		Yes	Yes	Yes	Yes
9. Flotation		Yes			Yes
10. Inertial precipitation: Demisting and scrubbing			Yes	Yes	
11. Foam-breaking			Yes		
12. Electrostatic precipitation			Yes	Yes	
13. Filtration				Yes	Yes
14. Flocculation					Yes
15. Hydroclone flow					Yes
16. Wicking and expression					Yes

17.1.1 Decantation or Settling

If the density difference is sufficient, the simplest method is to use gravity to pull the phases apart. The device or vessel is called a *decanter* if it operates partially full and a *settler* if it operates full. If operating continuously, the mixture enters at one end and soon separates into a light top layer and a heavy bottom layer. Part of the light phase remains in dispersed form in the bottom layer, and part of the heavy phase remains in dispersed form in the top layer. Typically, a droplet size of 150 μm (a micron being 1/1000th of a millimeter) is used for calculation purposes. In the rest of the operation, the droplets rise or fall to join their appropriate phase and the vessel is sized to allow this to happen. Sigales recommended that, regardless of the calculated value of droplet fall or rise velocity, the velocity should never be assumed to be greater than 10 inches per minute. Ludwig (1999) quoted one hour as a typical time for separation. Provision is made at the end of the vessel to draw off the two phases. This is done in a controlled manner to regulate the level of the interface between phases and, in the case of a decanter, to regulate the overall height in the vessel (Figure 17.2).

References are Schultz (2005) and Sigales (1975).

17.1.2 Coalescence

To *coalesce* is to bring together and unite. In the present case, the coalescence is of small liquid droplets dispersed in another liquid. Typically, the droplets are in the size range of 0.1–50 μm, which is too small to separate under the influence of gravity. (This is because the terminal velocity of small droplets varies as the square of diameter.) In coalescence, the two-phase mixture is passed through a bed of fibers or particles onto which the dispersed liquid attaches (possibly adsorbs) and builds up into larger droplets (e.g. 500–5000 μm), which can then be separated by gravity. The effectiveness of coalescence in liquid–liquid separation requires some degree of interfacial tension (IFT) between the two liquids. The lower limit is about 20 dynes cm^{-1} when glass fibers are used. More recently, polymeric fibers, sometimes with a surface coating, have allowed coalescence down to IFT of 0.5 dynes cm^{-1}.

References are Brown & Wines (1993), Wines & Brown (1993), Bloch (2000), Wines (2000), Katrona et al. (2001), and Wines (2011).

17.1.3 Centrifugation

Centrifugation makes use of rotation to create large centrifugal forces, hundreds or thousands of times greater than gravity. Droplets that move very slowly under gravity can be significantly accelerated in a centrifuge. The subject of centrifugation is discussed in more detail in Section 17.5.1 on liquid–solid separation. For liquid–liquid separation, a *tubular bowl* type of centrifuge is generally used. The tube rotates about a vertical axis and achieves forces typically up to 62 000 times that of gravity in laboratory centrifuges and 16 000 in industrial units. The liquid mixture enters at the bottom of the tube and leaves in two phases through concentric outlets at the top.

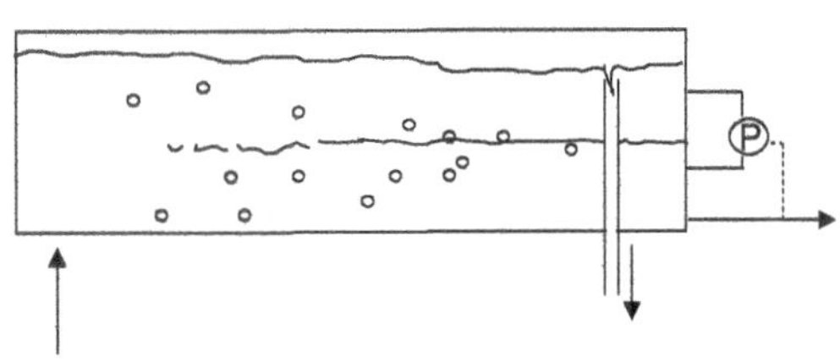

Figure 17.2 Liquid–liquid decanter.

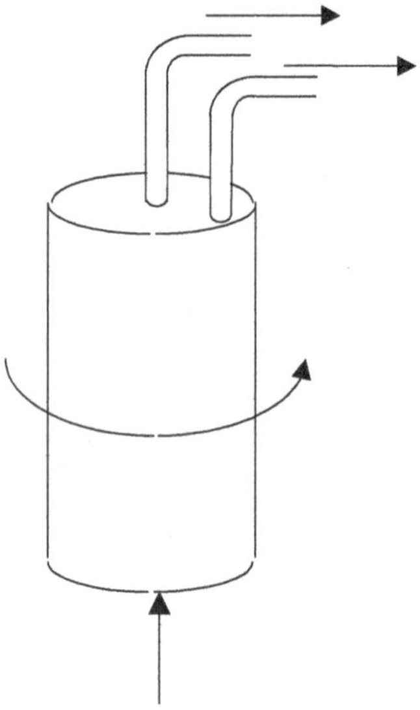

Figure 17.3 Tubular bowl liquid–liquid separator.

A reference for liquid–liquid centrifugation is Hertle (2012). See Section 17.5.1 for further references on centrifugation (Figure 17.3).

17.2 Solid–Solid Separations

Solid particles or pellets of different types often occur as mixtures. The difference in type may be simply size, or it may be the chemical nature of the particles. Some separation situations are as follows:

- remove oversized particles that would ruin a surface finish or cause an electronic product to fail,
- remove fine particles that would cause a dusting problem or that would get rubbed off in a molded or coated object,
- sort out and recycle oversized particles in a comminution operation,
- prepare feed for a mass-transfer operation or a chemical reaction, where small size processes faster, and
- remove "tramp" metal from a solid product.

Some separation mechanisms depend on the differential behavior in a separation system for gases and solids or liquids and solids, examples being *centrifugation*, *cyclone flow*, *hydroclone flow*, *differential settling*, and *flotation*. These operations are discussed subsequently. It should be noted that, in these cases, the behavior of the solid particles may depend on a combination of their size, shape, and density, so the separation with respect to the characteristic of interest may not be as distinct as desired.

Three methods specific to solid–solid separation are *screening*, *air classification*, and *separation by magnetic or electrostatic forces* and are discussed in this section.

Two terms that describe the nature and effectiveness of a size-separation operation are *cut size* and *size selectivity*. Cut size is the size (diameter) of particle such that 50% is sorted into the "large" category and 50% into the "small" category. Size selectivity is the relation (usually given graphically) between the particle diameter and percent retained as "large". In an ideal separation, 100% of

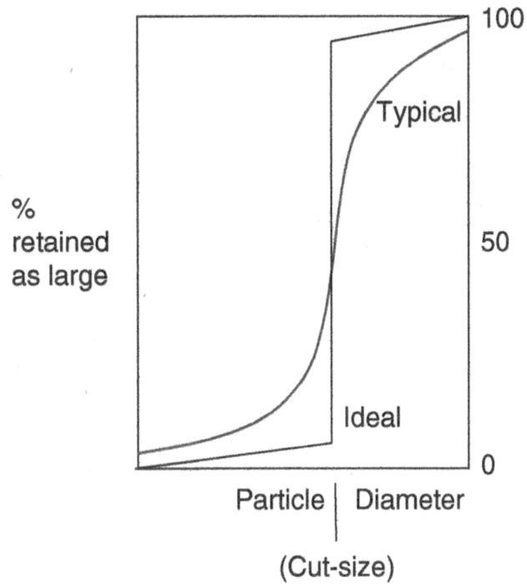

Figure 17.4 Particle size distribution.

particles bigger than the cut size are retained as "large" and zero percent of particles less than the cut size are so retained. In practice, the selectivity curve is more or less "S-shaped" (Figure 17.4).

In designing or analyzing a solid–solid size-separation system, it is necessary to measure the size and size distribution of the particles. Snow & Allen (1992) listed five techniques:

- screening with standard wire-mesh sieves, applicable to particles down to 200 μm and, with special electroformed sieves, down to 5 μm,
- microscopy, applicable down to 1 μm,
- low-angle laser light scattering (diffraction) for particles from 0.05 to 1000 μm,
- electrical sensing and counting, applicable down to 0.7 μm, and
- BET (Brunauer–Emmett–Teller) nitrogen gas adsorption to measure the surface area.

References on solid mixtures and separations are Snow & Allen (1992) and Mayo et al. (2014).

17.2.1 Screening

In continuous processing, a feed stream of particles is continuously deposited onto a horizontal or nearly horizontal screen and conveyed across the surface of the screen by vibratory or gyratory motion of the screen system. The objective may be to "scalp" the oversized particles, in which case everything else passes through the screen, or the objective may be to remove "fines", in which case only the fines pass through the screen, or the objective may be to remove both oversized and undersized particles in which case two screens are used. It may even be desired to divide the pellets into several size "cuts", thus requiring multiple screens.

The effectiveness with which the size divisions are made is dictated by the speed of passage of the pellets over the screen, by the type and vigor of the vibratory or gyratory motion, by the flowability of the particles, by their shape, and by their density. Screening relies on gravity to pull the particles through the screen apertures, so denser particles are easier to screen. Gyratory motion is more effective in spreading particles across the screen surface. The capacity of a screen system is quoted in pounds per hour of feed per unit area of screen, but the actual capacity depends on the type of particles and on what constitutes an adequate degree of separation: how much off-specification material can be left in the product and, conversely, how much wastage of product is acceptable.

Blockage of screen openings reduces system capacity; it can be addressed and minimized if it is a problem for the material at hand.

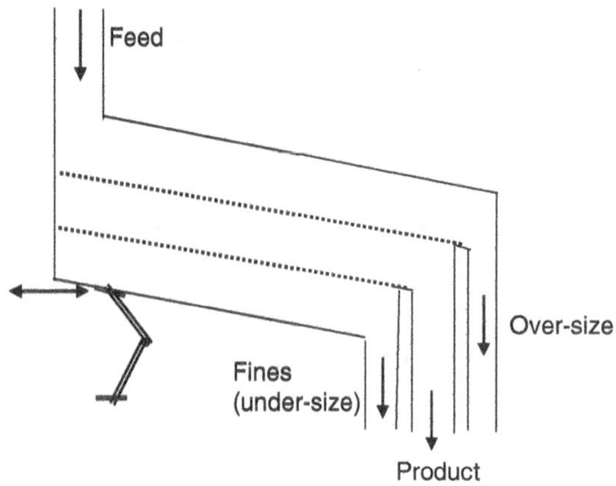

Figure 17.5 Screening of solid particulate matter.

Screening is effective down to about 30–100 μm, which is inadequate for some of today's products that incorporate superfine particles (Figure 17.5).

References on screening are DeCenso (2000), Paulsen (2004), Lower (2006), and Dhodapkar et al. (2007).

17.2.2 Elutriation and Air Classification

A particle immersed in a stationary or moving fluid but with a velocity different from the fluid is acted upon by a drag force which tends to make the particle take the same velocity as the fluid. If an external force (e.g. gravitational or centrifugal) acts on the particle, then the particle adopts a speed and/or direction different from the fluid. The extent to which this happens depends on the relative magnitudes of drag force and the external force. In general, the greater the size (and the greater the density) of the particle, the greater the deviation. This principle is used in centrifugation, cyclone separation, and settling, as well as in elutriation and air classification.

Elutriation itself is a crude separation tool, depending on only gravity as the external force. If particles are introduced into an upward-moving stream of gas, the lighter ones are carried with the gas and the heavier ones fall through the gas. The selectivity of this separation is poor, but elutriation may be a good pre-separation step. Some commercial elutriators are in the form of a zigzag channel.

Air classification uses the rotational motion of air to create centrifugal force. The rotation may come from tangentially introduced air, as in a cyclone, but the more effective and high-throughput classifiers use a mechanical rotor. Cyclones are effective down to particle size around 45 μm; they do not give good "sharpness of cut". Mechanically driven classifiers, incorporating also an elutriation section, operate in the range 5–250 μm. Special high-energy dispersion classifiers operate in the range 1–50 μm.

Classifiers are often used in combination with comminution and sometimes with the online laser measurement of particle average size.

Sharpness of cut is defined on the basis of the *size-selectivity* curve. The particle diameter, D_{75}, is noted at which 75 weight % is classified as "large", and the particle diameter, D_{25}, is noted at which 25% is classified as "large",

$$\text{Sharpness of cut} = D_{25}/D_{75}$$

Values can range from extremes of 0.0 (no separation) to 1.0 (perfect separation). Typical values are 0.3–0.7 but can reach 0.9 for low loading.

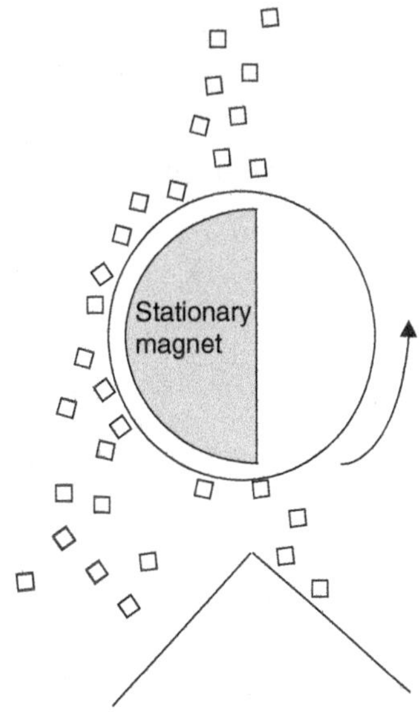

Figure 17.6 Magnetic separation of solid particulate matter.

Klumpar (1992), Hixon (1992), Crawley et al. (2002 April), Muscolino (2010) (Figure 17.6).

17.2.3 Magnetic Separation, Electrostatic Precipitation

If particles of different composition are to be separated, then other properties besides size may be put to use. Differences in density or shape (or size) may be invoked. Magnetic susceptibility is another possibility. Certainly, if one component is a *ferromagnetic*, such as iron, cobalt, or nickel, then because of its very high susceptibility, it is relatively easy to pull free from a mixture of particles. Many other materials – other elements and compounds – are *paramagnetics* and have the ability (susceptibility) to be temporarily magnetized in a magnetic field. If the types of particles in a mixture have different susceptibilities, then they may be separable by magnetic means. Tables of susceptibility are available. Some substances are *diamagnetic* and are repelled by magnetic fields. The equipment whereby separation is carried out is equipped first to attract the susceptible particles and then to deposit them in a different location from the non-attracted (or lesser) particles. A magnetic drum is one such device.

Particles of different composition may sometimes be separated by electrical forces. A charge is induced in the particles by means of contact, conduction or ionic bombardment. Depending on the nature of the material, individual particles pick up more or less charge and are attracted more or less to an electrode of opposite charge. The method becomes ineffective for particles greater than about 4 mm diameter because gravitational forces (proportional to the cube of diameter) become much greater than the electrical forces which are proportional to diameter squared.

17.3 Gas–Liquid Separations

Two general situations are liquid-dispersed-in-gas and gas-dispersed-in-liquid. The first situation arises, for instance, when

- vapor from a boiling or sparged pool of liquid carries (entrains) liquid droplets,
- vapor emerges from a flashing two-phase flow, and
- a gas–vapor mixture is cooled.

The second situation may arise when

- gas passes through, or vapor is generated within, a liquid and is particularly pronounced if the liquid is viscous, and
- a liquid is agitated vigorously.

In general, these gas–liquid mixtures are undesired and require separation.

The liquid-in-gas situation is examined first. The method of separation depends on the size of the liquid droplets and also on the loading in the gas. We examine the methods starting with the largest droplets and the highest loadings.

17.3.1 Gravity Settling

Two-phase mixtures emerging from gas–liquid pipelines often have large liquid loadings and gross dispersion of liquid in the gas. A typical liquid concentration is $50\,\mathrm{l\,m^{-3}}$ of gas. A simple "knock-out" pot can do the initial separation of gas and liquid and should be installed ahead of any more sophisticated device. Gas leaves at the top and liquid leaves at the bottom, both streams probably containing some amount of the other phase.

A reference is Majeed (2019).

17.3.2 Cyclone Flow

For liquid content in the range $1\text{–}50\,\mathrm{l\,m^{-3}}$ a cyclone is effective in separating liquid from gas, especially if the liquid droplets are larger than 10–50 μm. However, smaller droplets are carried out with the gas stream.

17.3.3 Inertial Precipitation: Demisting and Scrubbing

For droplets greater than 5–50 μm, passage of the gas–liquid through a wire-mesh pad can remove most of the liquid. The gas passes around the wire elements but the liquid, having more inertia, collides with the wire and collects (precipitates) on it. The liquid runs together (i.e. coalesces) and forms droplets that are large enough to detach and fall back against the main flow. The *cut size* of the droplets removed (i.e. the diameter of droplet such that 50% are removed from the gas) may be estimated from the formula

$$\text{Cut size} = 2.302 \times 10^6 \times \left[\frac{(\text{gas viscosity} \times \text{wire diameter})}{(\text{liquid density} \times \text{gas velocity})}\right]^{1/2}$$

It is common to install a mesh pad at the top of a knockout pot to remove some of the liquid that still remains in the gas.

Another variety of inertial precipitation is the *scrubber* where the droplets which it is desired to remove collide with larger drops of liquid sprayed into the path of the gas. These drops are large enough to settle by gravity. This type of system is used for droplets 20 μm or less. There are several styles of scrubbers, some of them very thorough. Of course, the introduction of the additional liquid must be compatible with the process materials.

A reference is "Anonymous" (2005).

17.3.4 Electrostatic Precipitation

Droplets as small as 2 μm may be removed from a stream of gas by being given a charge through (ionization of the gas) and then being collected on an electrode. The method works better at larger sizes, e.g., above 50 μm.

17.3.5 Coalescence

When the droplet size is very small, say less than a μm, the inertial techniques become ineffective. Mists and fogs comprise droplets in this size range. Small particles follow the motion of the gas so closely that there is little chance for capture by impingement. In this case, the phenomenon of choice is coalescence. The gas–liquid mixture is passed through a bed of fine fibers onto which the liquid droplets adhere in passing and on which they grow to a size that lets them separate after the gas–liquid mixture leaves the coalescer. The bed is typically a hollow cylinder, either vertical or horizontal. Gas and liquid enter at one end and emerge from the outside of the cylinder all along its length. The fibers are sized to enhance the growth of droplets. They may have a non-wetting fluoropolymer coating which preserves the captured droplets as droplets and enhances their ability to unite with one another.

See Section 17.1.2 for references (Figure 17.7).

17.3.6 Foam-Breaking

The other type of gas–liquid mixture sometimes requiring separation is the mixture where the liquid forms the continuous phase, i.e., froths and foams. Sometimes it is essential to separate the liquid components from the gas components. Often, however, the problem is just that the foam takes up space in a reactor or in the stages of a distillation column. In the extreme, foam may start issuing from the vessel where gas was expected. There are three situations:

1. A liquid is boiling or gas is being passed through it. Even though the gas separates from the liquid by leaving the top surface, it still occupies space within the bubbly mixture. The *in situ* liquid volumetric fraction in the mixture is generally described by an expression of form

   $$\text{Vol frac liquid} = 1/(1 + \text{a} \times U_S)$$

   where U_S is the superficial velocity of the gas (volumetric flow divided by cross-sectional area) and a is a constant, the value of which depends on the system and particularly on liquid viscosity.

Figure 17.7 Coalescer.

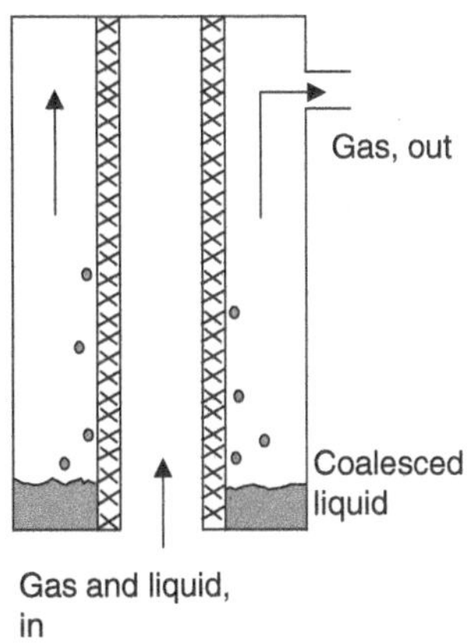

This behavior is simply a fact of life: The expansion of the liquid into a bubbly mixture has to be taken into account when designing or filling a vessel.

2. A gas has been introduced into the liquid during part of a process, but that part is over. While most of the bubbles have risen and disappeared, a population of small slowly rising bubbles remains. This situation occurs most commonly with viscous liquids. The rise velocity is proportional to bubble diameter squared and inversely to liquid viscosity. If the gas must be removed and if it is not practical to wait long enough, then a vacuum may be applied in order to increase the size of the bubbles or the liquid may be heated to reduce its viscosity.
3. A true foam is present, i.e., a layer of stable bubbles at the top of the liquid pool. Generally, a third, trace component is responsible for the stability of the bubbles. Mechanical means, like rotating bars or paddles, are sometimes used to stir and break the bubbles. More commonly a chemical, generally at low concentration (a few parts per million), is added as an *antifoam*. The nature and operative mechanism of these substances vary, but it is thought that they may substitute their activity for that of the component causing the foam. In any case, they cause the liquid membrane between the bubbles to rupture or to thin and drain back into the pool, thus allowing the gas in the bubbles to escape. Silicones are a popular antifoam, as are several different organic compounds and a few inorganics. Antifoams are chosen for their effectiveness in the system at hand and also for the lack of any adverse effects on the process or product. For instance, some antifoams are unacceptable in food or in packaging materials that will come in contact with food.

These topics are discussed in the Chemical Engineers' Handbook: Perry & Green (1997A).

17.4 Gas–Solid Separations

Three possible reasons why gas stream bearing particulates may require separation are as follows:

- the particulates have value,
- the equipment into which the gas is subsequently entering would be damaged by particulates (e.g. a compressor), and
- the gas is being released to atmosphere and must be clean.

A common situation is in pneumatic conveying of pellets, where not only must the pellets be separated from the gas stream but so must any fines that were generated by attrition.

Phillips provided a survey of methods for removing particulates (both solid and liquid) from gases. The following methods are discussed here:

- gravity settling,
- cyclone flow,
- scrubbing,
- filtration, and
- electrostatic precipitation.

General references on gas–solid separations are Phillips (2000), Agarwal (2005), and Jenkins (2020).

17.4.1 Gravity Settling

If particulates are large enough, then they may be removable by providing a zone of low gas velocity and low drag force so that gravity can do its work. The terminal or fall velocity U of small particles of diameter D_p can be calculated by equating the force of gravity to the resistive force of the "stationary" gas:

$$(\pi/6)\, D^3\, (\rho_p - \rho_g)\, g_c = 3\, \pi\, \mu_g\, D_p\, U$$

where ρ_p is the particle density, ρ_g is the gas density, μ_g is the gas viscosity, and g_c is the conversion 9.8 newton per kilogram force. To achieve separation in any reasonable time (a few seconds), the particles have to have diameters in the millimeter range. If the range of particle sizes extends up this high, then settling is a good first stage in the separation sequence.

17.4.2 Cyclone Flow

Cyclones are one of the oldest methods of particulate separation. As mentioned in Section 17.2.2, the size selectivity is mediocre (generally making cyclones inadequate for sorting particles by size) and this fact must be kept in mind in when designing for separation of particles from gases. Typically, for cyclones, the *sharpness of cut* as defined in Section 17.2.2 is 0.2.

Zenz provides the method and equations required for designing a cyclone to remove particles down to a required size. The overall efficiency (percent of entering particles removed) of a given cyclone depends on the entering size distribution. Cyclones are not effective at sizes below about 5 μm. In some devices on the market, the centrifugal action of a cyclone is replaced by mechanically driven rotation and this allows particles as small as 1 μm to be separated.

References are Zenz (2001), Louis (2005), Dhodapkar & Heumann (2011), Muschelknautz (2019), and Sakthivel & Pitchumani (2019).

17.4.3 Scrubbing

Scrubbing with a liquid can knockout particles in the range 0.1–100 μm. The simple spray tower scrubber is not the most efficient design, with poor efficiency especially at particle sizes below 10 μm. More modern designs incorporate cyclone action, packed and fluidized beds, trayed columns, orifice, and venturi flow. In most of these cases, the gas flow is channeled down at selected locations to achieve high velocity so that its energy can be used to atomize the scrubbing liquid and present large liquid surface area for particle capture. These scrubbers have higher pressure drop than simple spray devices (Figure 17.8).

References on scrubbing are Jenkins (2012B), Resetarits (2013), and MacLeod (2018).

17.4.4 Filtration

The most common type of particulate filter used industrially is made of fabric, through which the gas passes and on the surface of which the particles collect. In fact, the cake which forms is the prime filter medium. These filters are sized for gas velocity of 0.3–2.5 m min^{-1} and pressure drop of 0.5–1.5 kPa (2–6 inches of water). Fabric filters are very efficient (greater than 99%,

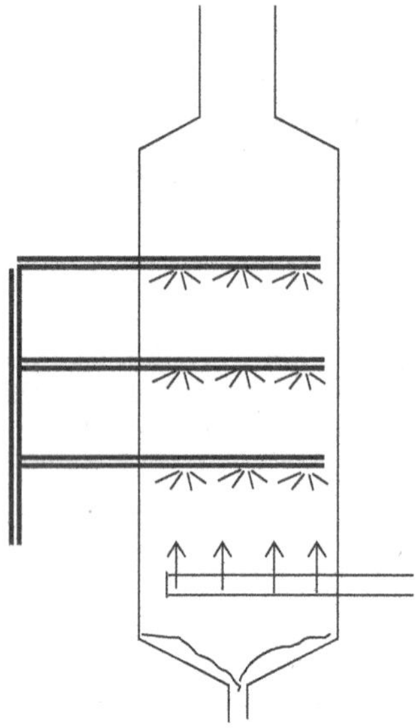

Figure 17.8 Scrubber for solid particles in gas.

approaching 100%) even for particles of submicron size. As the cake of collected particles builds up, the pressure drop increases, so provision is made to periodically remove the cake by shaking or pulsing the fabric or by reversing the flow. This type of filter is limited in the temperature it can tolerate, but new synthetic fibers have raised the limit. A knockout step to remove large hot particles is sometimes provided upstream.

Another approach is to pass the gas-particle mixture through a granular bed. This type of filter can operate at higher temperature. An efficiency of 99.9% can be achieved. As the granules become coated with particles, the bed requires regeneration or replacement. In some case, the bed is moving: Granules are continuously withdrawn, cleaned, and recycled.

For low loading of particulates, *high-efficiency particulate air (HEPA)* filters provide high efficiency down to very small sizes of particles (less than 0.1 μm). The medium is a paper comprising very fine fibers (less than a micron diameter) in a matrix of larger fibers.

A reference on gas filtration is Muller (2017).

17.4.5 Electrostatic Precipitation

Electrical charge can be induced on particles, allowing them to be collected on an oppositely charged surface. The advantages, relative to other separators, are low pressure drop and tolerance of high temperature and pressure. The disadvantage is cost.

References are Grainger & Paulson (2003) and Louis (2006).

17.5 Liquid–Solid Separations

Some of the situations in which a liquid and solid particles are mixed with each other are as follows:

- naturally occurring streams of water bearing silt,
- liquid-phase reaction mixtures containing a catalyst in pellet form,
- mother liquor and crystallized material, and
- waste streams containing sludge or other environmental contaminants.

The most often used separation techniques are centrifugation and filtration. They along with some lesser methods are discussed here.

General references on liquid–solid separations are Butcher (2005), Engel (2015), and Perry & Green (1997C).

17.5.1 Sedimentation Centrifugation

As explained previously, centrifugation is an extension of the technique of *settling*, where the force of gravity is replaced by a much greater *centrifugal* force. The smallest and simplest centrifuge is that found in the chemistry laboratory, where a circular rack of test tubes is whirled at high speed generally to drive a dispersed precipitate to the bottom of the tube.

In Section 17.1.3, the *tubular* (sometimes called *tubular bowl*) centrifuge was mentioned as a device for separating immiscible liquids. This device is also used for liquid–solid mixtures. This is a more common usage of centrifuges in general and, when so designed and used, they are called *sedimentation centrifuges*. There are several styles.

The *solid-bowl basket* (or *solid-bowl batch*) centrifuge is similar to the tubular but is less elongated and is used at larger scales and can tolerate larger solid particles. Like the tubular, it is a batch unit, requiring periodic shutdown to remove the solids that are deposited at the wall. Liquid is continuously discharged. Both centrifuges rotate about a vertical axis.

The *multichamber* (or *chamber bowl*) centrifuge is also a batch vessel rotating about a vertical axis. Internally, it is fitted with concentric vertical partitions such that the slurry has to up and down successive annuli. The effect is to produce more surface area for solids to settle on (Figure 17.9).

A family of *disk* (or *disk stack*) centrifuges are also fitted with internal partitions but inclined at an angle to the vertical. Slurry is initially directed to the outside of the rotating shell and then has to make its way back across the partitions to reach a central exit. Again the objective is to have more area for solids settling. There are three main styles of disk centrifuges:

- The *solid retaining* (or *manual discharge*) disk centrifuge is a batch device in that it must be shut down periodically and cleared of solids.

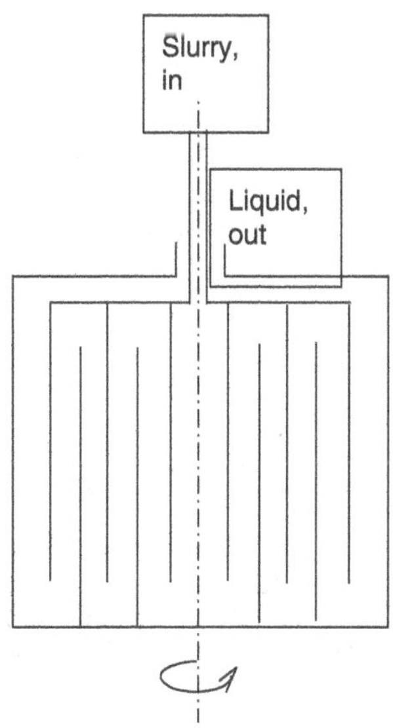

Figure 17.9 Multichamber (or chamber-bowl) centrifuge.

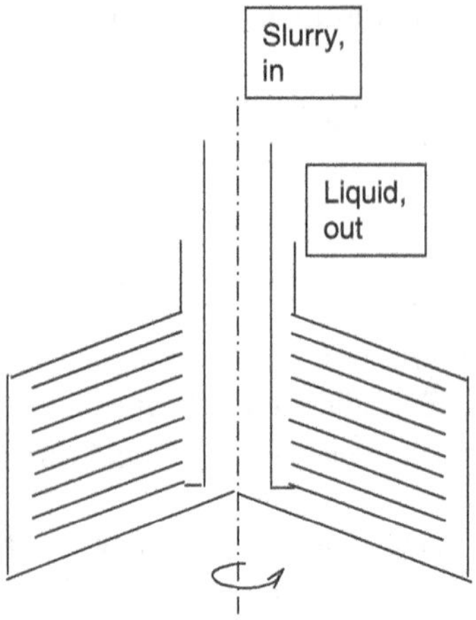

Figure 17.10 Disk centrifuge.

- The *intermittent discharge* (or *solid ejecting*) disk centrifuge allows solids to be expelled automatically from time to time and is thus a continuous operation.
- The *nozzle discharging* disk centrifuge is provided with continuous discharge of solids and is operated continuously. In all of the units, liquid is discharged continuously (Figure 17.10).

The large continuous versatile workhorse of industry is the *scroll decanter* (or *solid-bowl decanter*). Some units rotate about a vertical axis, some about a horizontal axis. The rotating cylinder or bowl is fitted with an internal wall-wiping helical screw which rotates at a slightly different speed. The screw continuously pushes solids to a discharge port as the solids appear on the wall (Figure 17.11).

A hybrid variation on this design is the *screen-bowl* decanter (or *screen-bowl* centrifuge) in which the screw pushes the solids across a cylindrical screen before discharge, allowing the solid to rid itself of free liquid. Along this theme, there are filters which use centrifugal action as the motive force to achieve high rates of filtration.

The references provide comparisons of the operational details and the applicability of the various centrifuge designs.

References are Moir (1988), Christian (1994), Letki (1998), Norton & Wilkie (2004), Eckstein (2004), Torzewski (2008), Schmidt (2010), Patnaik (2012), Peeters (2015), and Gabelman (2016).

Also Perry & Green (1997C) treated this topic in the *Chemical Engineers' Handbook.*

17.5.2 Filtration

Filtration is the separation of solids from a fluid by passage of the fluid through a medium that restrains all or part of the solids. Filtration from gases has already been described in Section 17.4.4. Here, the fluid is liquid.

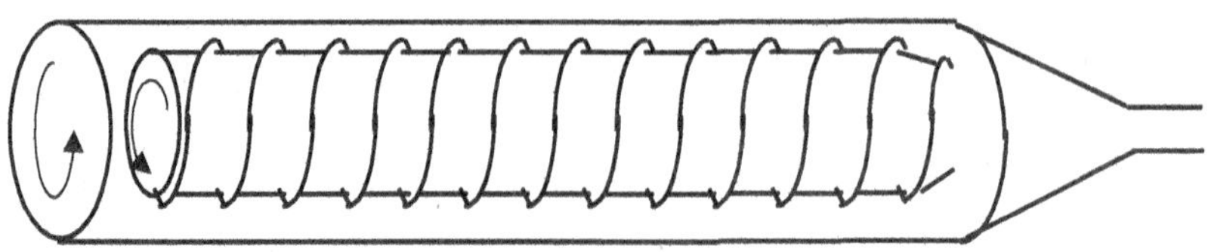

Figure 17.11 Scroll decanter or solid-bowl decanter.

Filtration tends to be used when the solid content in the liquid–solid slurry is relatively low. In most cases, the liquid is the product of value and the removal of solids is done to improve the value and processability of the liquid.

The filtration medium may be any one (or a combination) of woven screens or fabric, or non-woven fabric or paper, or a porous membrane. The configuration may locate the medium in a plate-and-frame apparatus, on horizontal plates, as circular or flat cartridges, as a belt, as a rotary drum, or as the wall of a centrifuge. Granular beds are also sometimes used for filtration.

There are three basic types of filtration:

1. Cake filtration: The solids build up on the surface of the filter medium and form a cake of steadily increasing thickness. This cake actually becomes the filter. A filter aid (e.g. diatomaceous earth) may be added to enhance the filtration. However, it must be removed from time to time as the pressure drop becomes excessive.
2. Depth filtration: The solid particles enter into the filter medium and are trapped between fibers. Gradually, the medium becomes plugged or begins to pass solids out with the *filtrate* (the leaving liquid), so the filter requires periodic cleaning or replacement.
3. Cross-flow membrane filtration: This more recently developed type of filtration is designed to separate out very small particles. Instead of flowing *though* the filter medium (in this case a porous membrane), the liquid–solid mixture moves across it. Some material passes through the membrane, but the surface of the membrane is kept free of solid accumulation. There are four degrees of membrane filtration, as listed by Duffy (2003) in Table 17.2.

The subject of filtration has drawn a great deal of study, and so there are many references to assist the designer. The following list may be helpful:

Dream (2000), Johnston (2000, 2003), Hayes (2001), Hampton (2007), Hampton & Weismantel (2002), Duffy (2003), Hairston (2003), Chen et al. (2004), Cartwright (2004), Repetti & Tseng (2005), Keefe & Dubbin (2005), Cardona & Blosse (2005), Schwartz (2007), Huchler (2008), Jenkins (2010, 2013, 2014), Berg (2011), Sentmanat (2011), Anonymous (2013), Perlmutter (2014), and Gabelman (2015, 2017).

17.5.3 Settling

If the difference in densities of liquid and solid is great enough, then an adequate separation may be achieved simply by letting the particles settle to the bottom of a vessel. The operation may be batch or continuous. If the objective is to produce a clear liquid (e.g. water), then the settler is called a *clarifier*. If the objective is to recover a valuable solid product, then the settler is called a *thickener*. In any case, the behavior of the descending solid phase is complicated by the interaction of particles with one another. One cannot simply extrapolate from the behavior of single particles. In continuous operation, there will always be a gradient of solid concentration from top to bottom of the vessel.

Table 17.2 Membrane filtration degrees.

Type	Pressure drop (bars)	Pore passage size
Micro	0.5–3	0.05–5 μm
Ultra	1.5–10	1000–50 000 mol wt
Nano	6–20	100–300 mol wt
Reverse osmosis	10–60	Allows only water to pass

17.5.4 Flotation

In settling, the more common case and the case usually considered are that of particles heavier than the liquid. Solids settle to the bottom. In the event that the solid material is less dense, then the particles float to the top where they may be skimmed off. However, even denser particles can be made to float if bubbles attach to them. This is the principle of *flotation*. It is used in the metallurgical industry where ore particles are considerably heavier than water, but it is still convenient to remove them from the top rather than the bottom. In waste water treatment, flotation is used to remove fats, greases, and oily material, which would actually float on its own but only very slowly. Bubbles form agglomerates between small globules and help them rise because they get bigger and because they get lighter with the air attached. The process works best when the air bubbles are small, around 2 mm in diameter. Flotation is also used in the recovery of oil from oil sands.

References on flotation are Belhateche (1995), Zinkus et al. (1998), and Hairston (2002).

17.5.5 Flocculation

In settling, where the terminal velocity varies as the square of particle diameter, and in any process where fluid drag forces are involved, it is advantageous to have bigger particles.

A flocculating agent acts to bring small particles together into larger faster-settling entities. Hughes (1977) listed a number of flocculants, all of them organic, most of them polymeric, some nonionic, and some containing acidic or basic functional groups. To be economically practical, a flocculating agent must be effective in low concentrations. They are used in settling and also in decanting centrifuges.

References are Hughes (1977) and Perry & Green (1997B) in *Chemical Engineers' Handbook, Seventh Edition,* Chapter 18, p. 63.

17.5.6 Hydroclone Flow

The hydroclone (or *hydrocyclone* or *hydraulic cyclone*) is the liquid-phase analogue of the gas-phase cyclone. It is used to separate particles continuously from liquid where the particles are more dense than the liquid. Like a cyclone, it has a tangential feed and a swirling motion which drives the particles toward the wall. The initial part of the wall is cylindrical, and the later part narrows down in a conical fashion. Large particles leave along with some of the liquid through a central outlet at the narrow end of the cone. This is the *underflow*. Small particles and most of the liquid leave through a central pipe at the other end, the *overflow*. The design of a hydroclone starts with the specification of the desired *cut size* D_{50}, already defined as that particle diameter for which 50% of those particles leave in the underflow. Gomez (1992) provided an equation for the required diameter of the hydroclone. He provided further equations for liquid capacity and pressure drop. Although hydroclones are usually shown with their axis vertical, this orientation is not mandatory because gravity plays little part in the operation.

Gomez stated the following relationship for the efficiency of separation of particles of diameter D other than the cut size:

$$\text{Percent leaving with the underflow} = E = 100 \times \left[1 - e^{-((D/D50)-0.115)^{\wedge 3}}\right]$$

Based on this expression, the *sharpness of cut* (defined in Section 17.2.2) is calculated as 0.63, which is quite good.

Hydroclones have the advantages of no moving parts and continuous operation. However, the internal motion is vigorous and may damage some materials.

References are Svarovsky (1977), Gomez (1992), and Salcudean et al. (2003).

17.5.7 Expression and Wicking

If the liquid–solid mixture is largely solid and if the liquid is loosely held, then it may be possible to squeeze out or *express* a significant part of the liquid. Wringing a cloth and squeezing an orange are domestic examples, and in fact, expression is used commercially in the making of juices. Expression may be used to get rid of the bulk of the liquid before a more intense step to reach the final desired level.

In papermaking, press-felts are brought into contact with the formed sheet to *wick away* some of the residual water left after the initial formation of the sheet. This step precedes the final drying over steam-heated rolls. Blotting paper is another example of a wicking material.

These unit operations are discussed in Perry & Green (1997D).

17.5.8 Spray Drying

In Chapter 18, Molecular Separations, the technique of spray drying is presented as a way of splitting a dissolved solution into its component parts – the solvent as a liquid and the solute in its solid state. In the present instance, the liquid and solid are in a mechanical mixture and the objective is to split them. The combination of heat and finely divided matter make for high rates.

Some references are Pagcatipunan & Schick (2005), Lipp (2008, 2014), and Jenkins (2012A).

References

Agarwal, A.T. (2005 Feb). Design guide for dust collectors. *Chem. Eng.* 112 (2): 42–49.

Anonymous (2005 Jun). The engineered mist eliminator. *Chem. Eng. Prog.* 101 (6): B1–B13.

Anonymous (2013 Feb). Temporary filtration: more than meets the eye. *Hyd. Proc.* 92 (2): 93.

Buchanan, L.S. (2007 May). Determine the efficiency of air-pollution control devices. *Chem. Eng. Prog.* 103 (5): 37–42.

Belhateche, D.H. (1995 Aug). Choose appropriate wastewater treatment technologies. *Chem. Eng. Prog.* 91 (8): 32–49.

Berg, J.H. (2011 Sep). Reduce costs with 'step-down' filtration. *Chem. Eng.* 118 (9): 50–54.

Bloch, H.P. (2000 Aug). Solve your hydrocracker compressor problems. *Hyd. Proc.* 79 (8): 39–42.

Brown, R.L. and Wines, T.H. (1993 Jan). Improve suspended water removal from fuels. *Hyd. Proc.* 72 (12): 95–100. DOI:10.1007/978-1-4757-1209-4_14.

Buecker, B. (2007 May). Microfiltration for CPI wastewater. *Chem. Eng.* 114 (5): 63–65.

Butcher, C. (2005 Dec). Solids to the left, liquids to the right. *Chem. Eng.* 112 (13): 20–23.

Cardona, M. and Blosse, P. (2005 Mar). Guidelines for selecting sterilizing-grade filters. *Chem. Eng. Prog.* 101 (3): 34–37.

Cartwright, P.S. (2004 Jun). Membranes for process water reuse. *Chem. Eng.* 111 (6): 38–42.

Chen, W., Parma, F., Patkar, A. et al. (2004 Dec). Selecting membrane filtration systems. *Chem. Eng. Prog.* 100 (12): 22–25.

Christian, J.B. (1994 Jul). Improve clarifier and thickener design and operation. *Chem. Eng. Prog.* 90 (4): 50–56.

Cooper, A. and Lyon, M. (2004 Jul). Make the most of MVR. *Chem. Eng.* 111 (7): 44–47.

Crawley, G., Malcolmson, A., Crosley, I., and McLeish, A. (2002 Apr). Particle classification: making the grade. *Chem. Eng.* 109 (4): 54–60.
Cusack, R. (2009 Jun). Rethink your liquid-liquid separations. *Hyd. Proc.* 88 (6): 53–60.
De Censo, A.J. (2000 Apr). Dry screening of granular solids. *Chem. Eng.* 107 (4): 76–83.
Dhodapkar, S., Bates, L., and Klinzing, G. (2007 Sep). Dry screening: sorting out the basic concepts. *Chem. Eng.* 114 (9): 56–61.
Dhodapkar, S. and Heumann, W.L. (2011 May). Harnessing the power of a cyclone. *Chem. Eng.* 118 (5): 34–43.
Dhodapkar, S., Zaltash, A., and Klinzing, G. (2012 Aug). A primer on gas-solids fluidization. *Chem. Eng.* 119 (8): 38–47.
Dream, R.F. (2000 Jan). How to design nano-, ultra- and microfiltration systems. *Chem. Eng.* 107 (1): 84–89.
Duffy, J. (2003 Jun). Putting crossflow filtration to the test. *Chem. Eng.* 110 (6): 35–41.
Eckstein, W.B. (2004 Aug). Operate your filtering centrifuges trouble free. *Chem. Eng.* 111 (8): 48–52.
Engel, D., Burns, H., and Sheilan, M. (2015 Jun). Filtration and separation during chemical process operations: avoid common errors. *Chem. Eng.* 122 (6): 52–57.
Hairston, D. (2002 May). Combing oil from tar sands. *Chem. Eng.* 109 (5): 27–31.
Gabelman, A. (2015 Nov). An overview of filtration. *Chem. Eng.* 122 (12): 50–58.
Gabelman, A. (2016 July). Beyond gravity: centrifugal separations in CPI operations. *Chem. Eng.* 123 (7): 52–59.
Gabelman, A. (2017 Apr). Crossflow membrane filtration essentials. *Chem. Eng.* 124 (4): 49–59.
Gomez, J.V. (1992 Apr). Correlations ease hydroclone selection. *Chem. Eng.* 99 (4): 161–163.
Grainger, C. and Paulson, C. (2003 Jan). Ductwork changes improve ESP performance. *Chem. Eng.* 110 (1): 69–71.
Hairston, D. (2003 Nov). Filtration: choosing cleanable versus disposable systems. *Chem. Eng.* 110 (12): 33–40.
Hampton, J. and Weismantel, G. (2002 Aug). The developments are diverse: filtration. *Chem. Eng.* 109 (8): 115–116.
Hampton, J. (2007 Jan). Cartridge filtration principles for the CPI. *Chem. Eng.* 114 (1): 40–44.
Hayes, K.Q. (2001 Jul). Process filtration: characterizing fluids and medium selection. *Chem. Eng.* 108 (7): 72–78.
Hertle, A. (2012 Feb). Treat oily waste with decanter centrifuge plants. *Hyd. Proc.* 91 (2): 83–85.
Hixon, L. (1992 Jul). Sizing up air classifiers. *Chem. Eng. Prog.* 88 (7): 59–62.
Huchler, L.A. (2008 Aug). Could ultrafiltration or microfiltration replace my gravity clarifier? *Hyd. Proc.* 87 (8): 126.
Hughes, M.A. (1977). Coagulation and flocculation. In: *Solid-Liquid Separation* (ed. L. Svarovsky). London: Buttersworth.
Jenkins, S. (2010 Aug). Heat transfer fluids: system filtration. Facts at your fingertips. *Chem. Eng.* 117 (8): A.
Jenkins, S. (2012A Jun). Spray drying parameters. Facts at your fingertips. *Chem. Eng.* 119 (6): 33.
Jenkins, S. (2012B Dec). Wet scrubbers. Facts at your fingertips. *Chem. Eng.* 119 (13): 27.
Jenkins, S. (2013 Dec). Filtration testing and slurry conditioning. Facts at your fingertips. *Chem. Eng.* 120 (12): 40.
Jenkins, S. (2014 Nov). Wire meshes in filtration. Facts at your fingertips. *Chem. Eng.* 121 (11): 36.
Jenkins, S. (2020 Jan). Removal of particulate matter from industrial processes. Facts at your fingertips. *Chem. Eng.* 127 (1): 21.

Johnston, P.R. (2000 Mar). Understand particle-size distributions when testing filter media. *Chem. Eng. Prog.* 96 (3): 47–50.

Johnston, P.R. (2003 Feb). Filtration. *Chem. Eng. Prog.* 99 (2): 52.

Katrona, A., Darde, T., and Wines, T.H. (2001 Aug). Improve haze removal for FCC gasoline. *Hyd. Proc.* 80 (8): 103–108.

Keefe, R.J. and Dubbin, D.M. (2005 Aug). Specifying microfiltration systems. *Chem. Eng.* 112 (8): 48–51.

Klumpar, I.V. (1992 Apr). Control and scale-up air classifiers. *Chem. Eng. Prog.* 88 (4): 50–55.

Lipp, C.W. (2008 Oct). Spray technology fundamentals ns applications: game winning strategies. *Chem. Eng.* 115 (10): 42–50.

Lipp, C.W. (2014 Aug). Spray nozzles 101: spray nozzle selection. *Chem. Eng. Prog.* 110 (8): 51–58.

Ludwig, E.E. (1999). *Applied Process Design for Chemical and Petrochemical Plants*, Vol. 1 3rd ed. Houston, TX: Gulf Publishing Company.

Letki, A.G. (1998 Sep). Know when to turn to centrifugal separation. *Chem. Eng. Prog.* 94 (9): 29–44.

Louis, T. (2005 Sep). Calculations: cyclone recovery/collection systems. *Chem. Eng. Prog.* 101 (9): 16–17.

Louis, T. (2006 Oct). Factors affecting ESP performance. *Chem. Eng. Prog.* 102 (10): 20–21.

Lower, W.E. (2006 Dec). Factors affecting screening performance. *Chem. Eng.* 113 (13): 53–56.

MacLeod, G.D. (2018 Oct). Consider wet scrubbers for pollution control. *Chem. Eng. Prog.* 114 (10): 47–51.

Majeed, K.T. (2019 Apr). Solve gravity separation challenges. *Hyd. Proc.* 98 (4): 55–57.

Mayo, M., Meadows, C., and Reid, C. (2014 Aug). Dry separation methods. *Chem. Eng.* 121 (8): 61–64.

Moir, D.N. (1988 Mar). Sedimentation centrifuges. *Chem. Eng.* 105 (3): 42–51.

Muller, C. (2017 Apr). Gas-phase filtration media properties: what to know. *Chem. Eng.* 124 (4): 60–63.

Muschelknautz, U. (2019 Mar). Design and calculation methods for uniflow cyclones. *Chem. Eng.* 126 (3): 52–63.

Muscolino, J. (2010 Nov). Mechanical centrifugal air classifiers. *Chem. Eng.* 117 (12): 48–50.

Norton, V. and Wilkie, W. (2004 Aug). Clarifying centrifuge operation and selection. *Chem. Eng. Prog.* 100 (8): 34–39.

Pagcatipunan, C. and Schick, R. (2005 Dec). Maximize the performance of spray nozzle systems. *Chem. Eng. Prog.* 101 (12): 38–44.

Patnaik, T. (2012 Jul). Solid-liquid separation: a guide to centrifuge selection. *Chem. Eng. Prog.* 108 (7): 4.

Paulsen, D. (2004 May). Sorting out screen performance. *Chem. Eng.* 111 (5): 3–36.

Peeters, B. (2015 Apr). Wastewater sludge centrifugation before drying. *Chem. Eng.* 122 (4): 56–60.

Perlmutter, B.A. (2014 Dec). Choosing a fine-particle filtration system. *Chem. Eng. Prog.* 110 (12): 35–39.

Perry, R.H. and Green, D.W. (1997A, Chapter 14). *Perry's Chemical Engineers' Handbook*, Seventhe, 95–98. New York: McGraw Hill Book Company.

Perry R. H., Green D. W., *Perry's Chemical Engineers' Handbook, Seventh Edition*, 1997B, Chapter 18, 63, McGraw Hill Book Company, New York.

Perry, R.H. and Green, D.W. (1997C, Chapter 18). *Perry's Chemical Engineers' Handbook*, Seventhe, 110–125. New York: McGraw Hill Book Company.

Perry, R.H. and Green, D.W. (1997D, Chapter 18). *Perry's Chemical Engineers' Handbook*, Seventhe, 125–130. New York: McGraw Hill Book Company.

Phillips, H.W. (2000 Sep). Select the proper gas cleaning equipment. *Chem. Eng. Prog.* 96 (9): 19–38.

Repetti, R.V. and Tseng, M.H.-S. (2005 Aug). Predict filter performance with beta ratios. *Chem. Eng.* 112 (8): 44–46.

Resetarits, M. (2013 Sep). The science of droplets. *Chem. Eng.* 120 (9): 73.

Sakthivel, S. and Pitchumani, B. (2019 May). Separate fine solid particles from dusty air with a uniflow cyclone. *Hyd. Proc.* 98 (5): 51–54.

Salcudean, M., Gartshore, I., and Statie, E.C. (2003 Apr). Test hydroclones before they are built. *Chem. Eng.* 110 (4): 66–71.

Schmidt, P. (2010 Dec). Filtration centrifuges: an overview. *Chem. Eng.* 117 (13): 34–38.

Schultz, T.E. (2005 Jul). Get the most out of API separators. *Chem. Eng.* 112 (7): 38–42.

Schwartz, L. (2007 Sep). Clarifying filtration terminology. *Chem. Eng. Prog.* 103 (9): 18–20.

Sentmanat, J.M. (2011 Oct). Clarifying liquid filtration. *Chem. Eng.* 118 (10): 38–47.

Sigales, B. (1975 June). How to design settling drums. *Chem. Eng.* 82 (6): 141–144.

Snow, R.H. and Allen, T. (1992 May). Effectively measure particle-size-classifier performance. *Chem. Eng. Prog.* 88 (5): 29–33.

Svarovsky, L. (1977). Hydroclones. In: *Solid-Liquid Separation* (ed. L. Svarovsky). London: Buttersworth.

Torzewski, K. (2008 Jan). Sedimentation centrifuging. *Chem. Eng.* 115 (1): 27.

Wines, T.H. and Brown, R.L. (1993 Dec). Difficult liquid-liquid separations. *Chem. Eng.* 104 (13): 104–109.

Wines, T.H. (2000 Jan). Improve liquid/gas coalescer performance. *Hyd. Proc.* 79 (1): 89–96.

Wines, T.H., Whitney, S., and Arshad, A. (2011 Jul). Liquid-gas coalescers: demystifying performance ratings. *Chem. Eng.* 118 (7): 38–43.

Van Ommen, J.R. (2009 Jul). Reshaping the structure of fluidized beds. *Chem. Eng. Prog.* 105 (7): 49–62.

Zenz, F.A. (2001 Jan). Cyclone-design tips. *Chem. Eng.* 108 (1): 60–64.

Zinkus, G.A., Byers, W.D., and Doerr, W.W. (1998 May). Identify appropriate water reclamation technologies. *Chem. Eng. Prog.* 94 (5): 19–31.

18

Molecular Separations

Molecular separations consist of the transfer of components, molecule by molecule, from one phase to another. They are sometimes used in sequence with one another or in sequence with mechanical separations. An example is the manufacture of paper, where the stock slurry is first filtered through a moving wire, then compressed and dewatered between press felts – both operations being mechanical separations, and finally dried by passage over heated rolls – a molecular separation. Molecular separations are part of the chemical engineering domain of *mass transfer*, with mechanical separations being part of *momentum transfer* (and, for completeness, most molecular and mechanical separations also involve *energy transfer*).

The discussion is organized around the process-design situations that call for treatment by molecular separation. The above techniques are discussed as they crop up as one of the appropriate approaches to particular situations. The separation situations to be addressed are as follows:

1. mixtures of permanent gases,
2. mixtures of vapors and permanent gases,
3. mixtures of vapors,
4. mixtures of liquids,
5. liquid solutions containing dissolved solids, and
6. solid solutions and solids containing liquids or vapor.

There are many techniques for molecular separations. An all-inclusive list cannot be given because new techniques are developed from time to time. The ones that are considered here, to a greater or lesser extent, are listed in Table 18.1.

The following two-volume reference, although somewhat out of date, describes several of the more modern molecular-separation techniques that have been developed Li (1972).

18.1 Separation of Permanent Gases

Permanent gases are substances of very low critical (liquefaction) temperature and relatively low reactivity, such as hydrogen, helium, nitrogen, oxygen as well as argon, and the other inert gases. Sometimes a gas stream just needs to be enriched, and at other times, high purity is needed. There are three methods generally used to separate mixtures.

18.1.1 Cryogenic Distillation

Although the gases are "permanent", they can still be liquefied at sufficiently low temperature accompanied by high pressure. The mixture is then distilled. For the separation of air into its

Practical Process Design for Chemical Engineers, First Edition. Keith Marchildon and David Mody.

Table 18.1 Molecular separation operations.

	Gas-gas	Gas-vapor	Vapor-vapor	Liquid-liquid	Liquid-solid	Solid-liquid
Case: Operation						
Distillation	1		3	4		
Adsorption	1	2	3	4		
Membrane	1	2	3	4		
Condensation		2				
Absorption		2				
Stripping				4		
Extraction				4		
Melt crystallization				4		6
Vaporization					5	
Solution crystallization					5	
Ion exchange					5	
Reverse osmosis					5	
Dialysis					5	
Drying						6
Leaching						6

components, this is the oldest and most common method. Great purity can be attained: for instance, nitrogen containing one part per billion of contaminant. Such purity is needed in certain applications. Propane and propylene are commonly separated cryogenically because of the need for high-purity propylene. The method is energy intensive, but the energy can be reduced by management with heat exchangers. Alcantara-Avila et al. (2014). calculated a potential reduction of 75%.

18.1.2 Adsorption

The phenomenon of surface adsorption can be used to condense permanent gases and to do it preferentially. Treybal (1980) showed a case where a 1 : 2 mixture of oxygen and nitrogen was adsorbed in the ratio 1 : 1 on activated carbon because of the preference of oxygen for adsorption.

Figure 18.1 shows a continuous process for carrying out the adsorption. Adsorber #1 is currently removing an unwanted component from a gas stream, the rate of adsorption being increased by high pressure. Adsorber #2, which had been adsorbing until becoming saturated, is currently desorbing at low pressure aided by a small stream from adsorber #1.

The subject of gas fractionation by adsorption is treated in textbooks by Yang (1997, 2003). An introduction to the technology is given by Broom (2018) and Gabelman (2017). Some calculational details are provided by Torzewski (2009B) and Louis (2005A). Sircar et al. (1995) discussed the most widely used adsorbent, activated carbon, as a tool for separating gas mixtures. Vermesse et al. (1996) measured the high-pressure adsorption on zeolites of nitrogen, methane, and some noble gases. Siperstein & Myers (2001) explored the prediction of the equilibrium of individual components when part of a multicomponent mixture.

Figure 18.1 Pressure swing adsorption of gases.

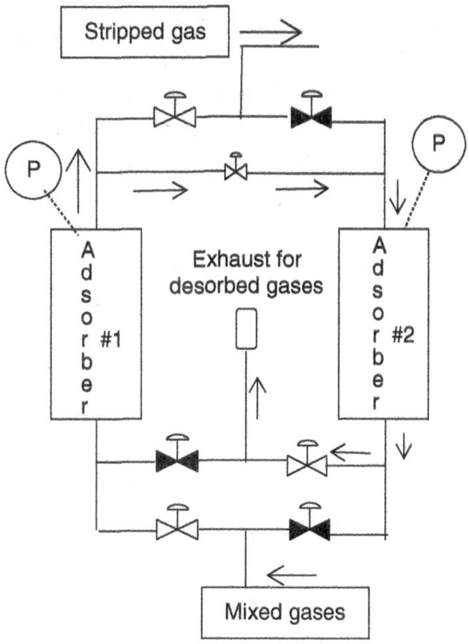

18.1.3 Membrane Permeation

Preferential diffusion through a membrane of one component over another is the principle of this method. Hollow-fiber polymeric membranes are commonly used. Ceramic membranes are being developed that will be more selective and that will allow higher temperature. The purity of the separated components is not as high as that obtainable in cryogenic separation but is adequate for many purposes. The equipment is less expensive and more portable, and the process uses less energy than distillation. It is coming into common use.

Figure 18.2 illustrates the principle of separation. If the initial gas is air, then "A" would be oxygen and "B" would be nitrogen.

Loprete (2019) provided an introduction to the technology of separating gases with membranes. Shelley (2009) described the use of membranes in capturing carbon dioxide from coal-burning plants. Hairston (2000) discussed the merits of membrane separation relative to cryogenics.

A commercial product based on membrane separation is the PRISM® nitrogen membrane separator. It can produce nitrogen at purities to 99.5%, which is needed for nitrogen to provide inert atmospheres. Enriched air (in the permeate) is a by-product. Pressure is typically in the 10–15 bar gauge, temperature at or above ambient. The high pressure makes for easy handling and loading.

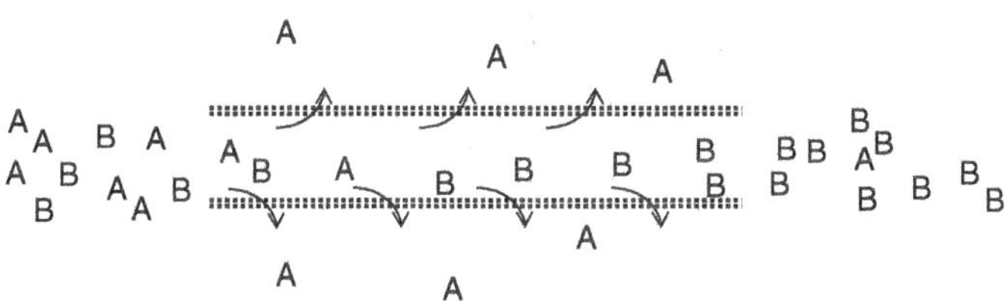

Figure 18.2 Principle of separation via membranes.

18.2 Separation of Gas–Vapor Mixtures

The most common and most widely studied of the gas–vapor systems is air–water. A large amount of graphical data exist. It should be noted that the coincidence of wet-bulb temperature and adiabatic-saturation temperature for the air–water system is not generally true for gas–vapor systems.

In general, the separation of vapor from a gas is easier than a gas from a gas. The following are methods that are used alone or in sequence.

18.2.1 Condensation

If the vapor content is significant and if very complete removal of vapor is not required, then simple condensation may be adequate. This operation may be carried out in the type of condenser shown in Figure 18.3, or it may be done using a cool spray of already condensed material. Condensation may be used to remove the bulk of the vapor prior to a more thorough method.

18.2.2 Absorption

Absorption of organic components out of a gas stream is a traditional chemical engineering unit operation. It is commonly carried out in a packed or a trayed countercurrent column. Figure 18.4 shows the gas–vapor mixture entering the bottom and emerging from the top free, or substantially free, of vapor. The absorbing liquid (or "oil") enters at the top and leaves at the bottom as a solution containing the absorbed vapor component. This stream must subsequently be stripped of the condensed vapor component so that the oil can be reused in the absorber.

Absorption is a "senior member" of the chemical engineer's trusted company of operations. However, it still enjoys efforts to upgrade the technology and to improve the accuracy of designs. LePree (2015), Torzewski (2008B), Louis (2005), and Bennett (2000) suggested areas to examine, including the shape of packing in packed columns. Graf (2011), Kister et al. (2007), and Bennett provided design procedures and equations.

18.2.3 Adsorption

Adsorption on a solid is commonly used for vapor removal. For instance, in the lab, air is dried by contact with silica gel. Common adsorbents are activated carbon (a highly porous material),

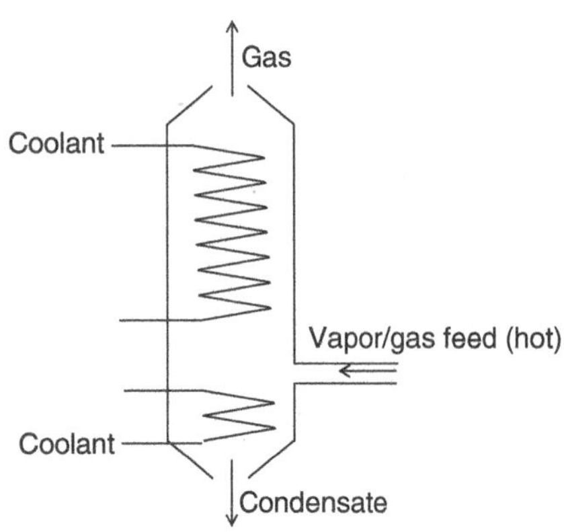

Figure 18.3 Separation via condensation.

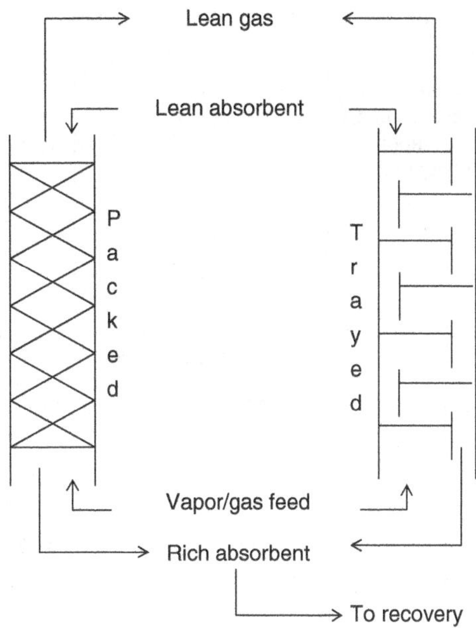

Figure 18.4 Absorption.

activated alumina, silica gel, and crystalline zeolites (commonly known as *molecular sieves*). The last named adsorbents are distinguished by the following qualities:

- they can separate materials on the basis of molecular size and configuration;
- they are very tenacious, especially for polar and polarizable molecules;
- they have relatively high adsorption capacity even at low adsorbate concentration and elevated temperature.

In some cases, activated carbon is preferred precisely because of its weaker bonding with adsorbents, which allows easier regeneration. A reference to adsorption on zeolites is Breck (1974).

18.2.4 Membrane Separation

The composition and construction of a membrane can be tailored to allow preferred passage of a vapor component, similar to what is shown in Figure 18.2. This operation is called *vapor permeation*.

18.3 Separation of Vapor Mixtures

Vapors can be separated from one another by three methods.

18.3.1 Distillation

Because vapors can be condensed, the traditional unit operation of distillation is a method of separation. Two extreme cases are as follows:

1. the higher-condensing (i.e. less volatile) component is the major component;
2. the lower-condensing component is the major component.

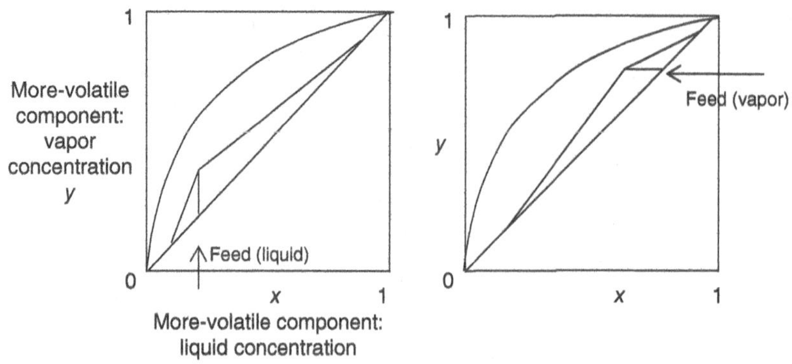

Figure 18.5 Separation of vapor mixture by distillation.

In the former case, the total vapor should be condensed before feeding to the distillation column, as shown on the left of Figure 18.5. In the latter case, the vapor should be fed as such to the column, as shown on the right. These practices avoid excessive and unnecessary vapor or liquid flows respectively within the column.

18.3.2 Adsorption

Adsorption is frequently used to remove contaminants from vapor streams. Solid adsorbers can be tailored to remove specific components from vapor (and liquid). In the case of activated carbon, two aspects can be varied: the *functionality* and the *energy*. For instance, a catalytic functionality can be imparted to the surface which allows oxygen to react with contaminants that would otherwise adsorb only weakly and to convert them into products that adsorb well. The energy of an adsorber is a measure of how strongly it adsorbs. The smaller the pore size, the greater the energy. High energy is required for trace components and components that adsorb weakly Carr & Vaughn (2003).

18.3.3 Membrane Permeation

The term *pervaporation* takes its name from the fact that liquid components permeating through the membrane exit on the other side in the form of vapor. If the component is already a vapor, then the process is called *vapor permeation*. In either case, the process is distinguished by the fact that the minor component (rather than the major component as in *reverse osmosis*) diffuses through the membrane. One application is in the separation of water and ethanol where vapor permeation is used to remove water from the (mainly ethanol) azeotropic vapor mixture leaving the top of a distillation column.

A description of pervaporation is given by Wynn (2001).

18.4 Separation of Liquid Mixtures

The separation of a liquid mixture into its components is one of the most widely studied operations in chemical engineering. While distillation is the first method that comes to mind, it is not always suitable and it is always a considerable consumer of energy. Other methods have been developed that are either more selective, more gentle, or less energy demanding.

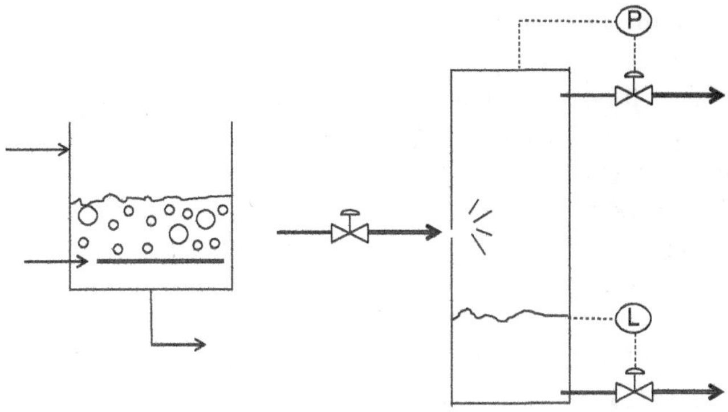

Figure 18.6 Liquid separation by sparging and flashing.

18.4.1 Distillation and Stripping

The effectiveness of distillation depends on volatility differences between components. As a rule of thumb, if the ratio of volatilities is less than 2, then the separation begins to require an excessive number of stages and/or to require an excessive amount of reflux (which needs energy to produce). On the other hand, if the volatility ratio is high and if the starting concentration of the more volatile component is low, then it may be practical to remove it by a single stage of flashing or by stripping with a permanent gas, as shown in Figure 18.6.

Two techniques are sometimes applied to extend the effectiveness of distillation down to volatility ratios less than 2 and also to extend it to cases where the components form an azeotrope. One is *azeotropic distillation*, where an extraneous liquid, an *entrainer*, is added part way up the column to form a low-boiling azeotrope with one of the components, thus allowing that component to free itself from the other and to leave the column as a vapor. The other is *extractive distillation* in which, again, an extraneous liquid is added but, in this case, to form a low-boiling solution with one component, thus allowing that component to separate from the other and leave the bottom of the column.

If the starting liquid comprises more than two components and if all components are to be recovered individually, then the required number of columns (or separation devices of some sort) equals the number of components minus one. There is a choice in the order in which the separations are done: For instance, with a mixture of A, B, and C, we could separate A from B and C in the first column and then separate B from C in the second column, or we could separate B first, or we could separate C first; i.e., there are three choices. For 4 components, there are 12 choices. The number of choices grows rapidly with the number of components. Some *heuristic* rules have been developed to assist in setting up the sequence that has lowest cost of capital and of energy:

I. Do the easy splits first, i.e., where the volatility ratio is high. Easy splits generally consume less energy. Subsequently, the more difficult splits, which require more energy, will be done with lower flow rates, reducing the overall energy needed by the system.
II. Separate out the large-fraction components early.
III. Run the columns with vapor–liquid splits as close as possible to 1 : 1.
IV. If possible, use the “direct” sequence, i.e., remove the isolated component from the top of the column at each step.
V. Avoid refrigeration and vacuum, possibly by choosing the operating pressure of the column. Avoid difficult splits. Seek alternative separation techniques.

Obviously, these rules will sometimes conflict with one another, but they provide a helpful starting point.

In dealing with a provider of distillation columns, it should be noted that even the ancient art of distillation has seen some significant improvements in the equipment of which the vendor should be aware. There is a wealth of publications of which a few recent articles are listed here:

O'Connell et al. (2019) using a membrane to separate heavy components out of the reflux stream, Duss & Taylor (2018) predicting tray efficiency, Resetarits et al. (2015) column design, especially trays, Rumyantseva & Beck (2014) using software to achieve thermal optimization, Bouck (2014) errors to avoid in column design, Choudhari et al. (2012) energy integration by vapor compression, Pilling & Summers (2012) column design, White (2012) optimization of energy use, Julka (2009) azeotropic distillation, Taylor et al. (2003) rate-vs-equilibrium tray calculations, and Lee & Gentry (1997) extractive distillation.

18.4.2 Extraction

Extraction is another of the traditional chemical engineering operations, relying on the difference of solubility, in an introduced solvent, of one component over another. The precise arrangement of the system may depend on whether the component being extracted is a valuable product or whether it is a contaminant in the non-extracted (or lesser-extracted) component. The simplest extraction device is the laboratory separatory funnel. Generally, more than one stage of extraction is needed in order to get the required degree of separation, each stage consisting of intimate dispersion of the two liquid phases with each other in order to allow the transfer of extract, followed by the separation of the two phases. The operation may be carried out in stirred tanks followed by decanters, or it may be done in a multistage agitated column. The example in Figure 18.7 shows a three-stage system with 70% separation efficiency for each stage resulting in 90% overall efficiency.

Gupta and Jain (2018) presented an optimized approach to choosing solvent. Koch & Shiveler (2015) reviewed design principles for extraction. Glatz & Parker (2004) discussed improvements that could improve extraction systems.

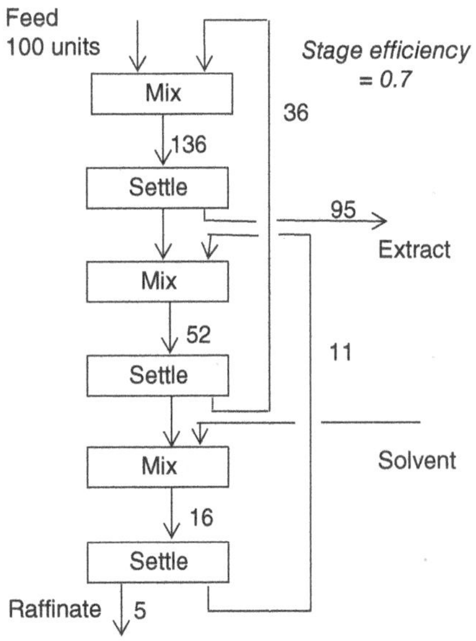

Figure 18.7 Multistage liquid–liquid extraction.

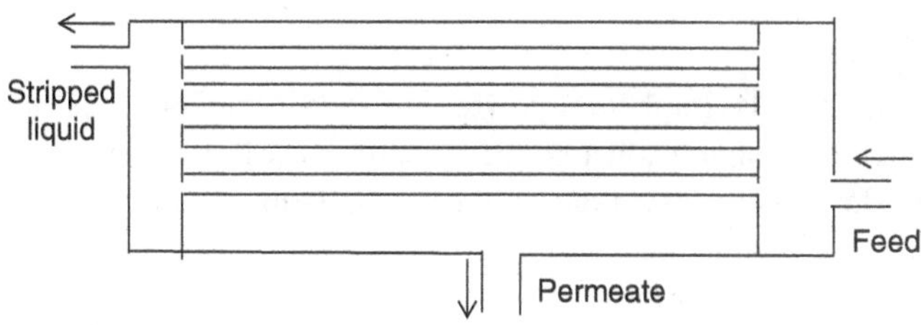

Figure 18.8 Pervaporation.

18.4.3 Adsorption

Adsorption on solids is frequently used to remove impurities (often color-producing impurities) from products. Carr & Vaughn (2003 April) presented examples of the use of activated carbon in this service.

18.4.4 Membrane Permeation

Pervaporation, as explained in Section 18.3.3, is a technique whereby a lesser component diffuses, out of a main stream, through a membrane (Figure 18.8). Wynn (2001) presented examples. Prime applications have been in the dehydration of organic liquids. Lower alcohols have also been removed this way. For instance, in a biological process for producing ethanol, where the ethanol is subsequently a poison for the organism, the ethanol is removed continuously by pervaporation.

18.4.5 Melt Crystallization

Crystallization from aqueous solutions is a well-practiced art. In that case, an innocuous solvent (i.e. water) is being used. When the mixture consists of organic components, then any solvent that is introduced will be another organic, which must subsequently be disposed of. To avoid solvents is the purpose of *melt crystallization*: The organic mixture itself is the solvent. The technique is effective when one of the components begins to solidify before the others and where there are no solid solutions formed among components. The component must solidify into pure crystals, which is generally the case because as crystals form they reject or expel extraneous substances. There remains the need to separate the crystals from the residual organic melt.

One embodiment of a melt crystallizer is a batch system of vertical tubes inside of which the starting organic mixture is loaded. Cooling is applied to the outside of the tubes and the desired component freezes onto the inside walls. After a time, the residual melt is drained off. The tube walls are heated somewhat to induce "sweating" of some liquid from the crystals: This liquid contains most of the impurities that were trapped in the crystal film. It is drained off. Then the heat is increased to melt the crystals, and they are drained off as a product.

A melt-crystallization system may be operated in a multistage manner, where the product (i.e. the crystal-forming component) is recrystallized to further purify it and where the residual liquid is put through another stage of crystallization in order to recover more useful product.

References to this process are Ondrey & Hairston (2001), Ondrey (2000), and Wynn (1992).

18.5 Separation of Liquid Solutions from Dissolved Solids

This section examines the class of homogeneous liquid mixtures where one of the components would be a solid if in its pure state. When dissolved in the liquid, the solid may retain its molecular form or it may dissociate into ions.

18.5.1 Vaporization

One way to separate the dissolved component is simply to vaporize the liquid component(s) from the mixture. However, this process consumes a significant amount of energy if the liquid is predominant. We reserve vaporization where the object is to rid the solid of a small amount of liquid for section 18.6.1. The process may consist of a boiling operation followed by a diffusional removal of the last traces of liquid. An alternative is to *spray dry* by atomizing the solution and then exposing it to elevated temperature. The temperature may be provided by a hot gas, or it may come by radiation from hot walls of a vessel. The very high surface-to-volume ratio of the finely divided droplets provides excellent heat transfer and vaporization. Usually, the desired product is the solid, which collects at the bottom of the device.

18.5.2 Solution Crystallization

Crystallization from solution is a well-known operation, capable of producing a solid product of great purity. Of course, if the remaining liquid is the product of interest and if the crystallization is carried out just to eliminate an unwanted component, then the crystal purity is of no interest.

The operation requires creating a *metastable* state in the liquid, where the concentration of dissolved solute exceeds the solubility limit. The difference between the actual concentration and the soluble limit is the driving force for the dissolved solid to come out of solution. The objective is for it to come out of solution in the form of large uniform crystals (Figure 18.9). The other objective is for it to do so fairly rapidly.

The metastable state is created in one (or both, simultaneously) of two ways:

- cooling the solution or
- evaporating some of the solution.

It may be necessary to "seed" the solution with some initial crystals, with the goal that newly solidifying material will adhere to and grow these initial crystals rather than form a multiplicity of small new crystals.

Crystallization may be carried out batch-wise or carried out continuously (often in two or more stages). It is important that the vessel be uniformly mixed but without damaging the crystals through overly vigorous agitation: This condition is generally easier to achieve at the lab or pilot-plant scale than at the commercial scale.

Recent references are Malfand (2017), Albin & Moyers (2017), Wibowo (2011, 2014) Wibowo et al. (2004), and Genck (2011, 2000).

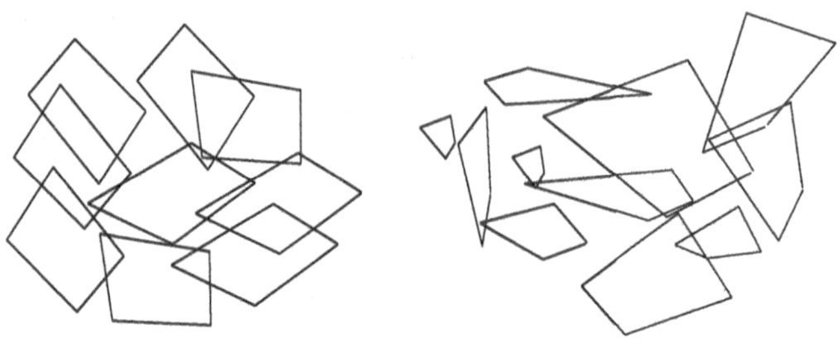

Figure 18.9 Regular and irregular crystals.

18.5.3 Ion Exchange

Ion exchange may be considered as a form of adsorption, an adsorption in which ionic force binds the adsorbate and adsorbent. It is used to remove ionic species from solution. The original ion exchangers were zeolites in which relatively loosely held sodium ions were exchanged for heavier ions in solution. The application was primarily the softening of water, and the heavier ions are primarily calcium and magnesium. This is still a major application of ion exchange.

Modern day ion-exchange substances comprise a much wider variety and are tailored to specific uses. Polymeric gels are used in which functionality is included to trap either positive or negative ions. For positive ions, the functional groups are sulfonic, carboxylic, or phenolic: These are *cationic* exchangers. For negative ions, the functional groups may be amines: These are *anionic* exchangers. In some applications, a cationic exchanger and an anionic exchanger are used in series. In some other exchangers, a mixed bed is used.

Ion exchangers have to be regenerated periodically to remove the trapped ions and replace them with the original, exchangeable ions. This is done be exposing the gel beads to a concentrated solution of the original ions. Those ions may be typically sodium, or they may be hydrogen or hydroxyl. In the latter two cases, in use the exchanger releases into the treated stream only hydrogen or hydroxyl rather than an extraneous material like sodium.

McNulty (1997) and Shanley (2000) are references for ion exchange.

18.5.4 Reverse Osmosis and Dialysis

Dialysis is a general term for separating molecules on the basis of size using semipermeable nonporous membranes. Pervaporation and vapor permeation are examples. Reverse osmosis (RO) is another variant.

In the natural process of osmosis, if a solution and a solvent are on opposite sides of a membrane, the solvent molecules will diffuse into the solution. In reverse osmosis, a pressure is applied to the solution side, high enough to overcome the natural osmotic force and to force the solvent out of the solution and across the membrane. Thus, in an application like the purification of saline or brackish water, the major component, water, passes through the membrane. This contrasts with pervaporation where the minor component passes through.

In modern practice, the membrane is the wall of hollow fibers as in DuPont's PERMASEP® permeator. The substance may be cellulose acetate or, in the PERMASEP, an aromatic polyamide.

Considerations are as follows:

- the rate of permeation,
- the degree ("rejection") to which the unwanted solute is left behind, and
- the extent of recovery of the desired solvent.

These factors depend on the nature of the solution and of the membrane but also on the differential pressure imposed across the membrane. This pressure is typically hundreds or thousands of pounds per square inch, requiring the membrane to be adequately supported physically.

Reverting to processes where only a minor component crosses the membrane, a dialysis process known as *electrodialysis* applies an electric potential across the membrane to facilitate the transfer of specific charged ions.

Introduction to the technology of reverse osmosis is provided by Stover (2014), Huchler (2002, 2010, 2014), and Kucera (2008, 2019). Moftah (2003) discussed water purification by RO.

18.6 Separations: Solid–Solid, Dissolved Fluids from Solid

By contrast with liquid solutions, which may contain normally solid solutes, the solid solution is primarily solid but may contain normally liquid components. The prime example is a solid containing dissolved water. On the other hand, the solid may be composed entirely of components that would be solid in their pure state. Where the word "moisture" is used, in this case, it denotes any liquid.

18.6.1 Drying

The term *drying* is used here for the removal of any component (not just water) in the form of vapor. Drying is a major operation in the processing industries, and there are many types of dryers. The choice depends on several factors.

Parikh (2014) reviewed drying in general. For guidance in choosing a dryer, see Moyers (2003).

The solid may be in different forms. A finely divided solid that is wetted by the liquid may be in the form of a slurry or paste. If only surface moisture is present, then a heated carrier gas may be sufficient to flash away the moisture, Christiansen & Sardo (2001). Spreading on a tray and sweeping with an inert gas, possibly with heating, can be satisfactory. A more effective device is the wiped-film evaporator, which continually refreshes the surface of the mixture.

Glover (2004), Raouzeos (2012), and Bethge (2016) discussed wiped-film evaporators.

If a slurry or paste can be atomized, then spraying the mixture through a hot carrier gas, possibly with wall heating as well, is an effective way of reducing moisture. The small size of the solid particles effectively eliminates internal mass transfer resistance. Spray drying is discussed by Langrish (2009) and Moller & Fredsted (2009).

Alternatively, the solid may be in fixed form as a thread, as a sheet, or as pellets. One key consideration is whether the prime resistance to mass transfer is on the inside or the outside of the solid. If on the outside, the governing process may be heat transfer. If the prime resistance is on the inside, then time and temperature are the main requirement for drying. If on the outside, then agitation will assist as in a fluidized bed or as in a tumbling or rotary drier. In many cases, the ratio of internal-to-external resistance changes during the operation: At first, during the *constant-rate* period, the vaporizing component is at or near the surface of the solid and requires only heat to vaporize; later, in the *falling-rate* period, the remainder of the vaporizing component is in pores or is truly dissolved and requires time to diffuse to the surface. If a high degree of drying is required, encompassing both of these regimes, then a two stage system may be needed. For instance, in drying of pellets, a fluidized bed may do the initial drying and a slowly moving bed may do the final drying.

The characteristics of the solid dictate the available forms of dryers. Fragility may rule out the more vigorous styles of the dryer. A gravity dryer can be more suitable, with pellets descending and a drying gas ascending in a tubular vessel (Mehos 2009). Vacuum-assisted drying is discussed by Parikh (2015) and Tritschler (2017).

If the vapor is innocuous or of no value, like water, then *direct drying* using a stream of gas (e.g. air, nitrogen, and superheated steam) is acceptable and effective. The main purpose of the gas is to provide the heat and also to convey away the vapor.

If the vapor must be recovered or disposed of, then it may be preferable to use *indirect drying* in which the heat is supplied through the dryer surfaces. A small amount of gas may be used to convey away the vapor. In this type of dryer, it is necessary, in order to order to achieve good heat transfer, to provide agitation. For slurries and pastes, a wiped-film evaporator is appropriate. Pellets can be agitated in a *thermal screw*. Kimball (2001), Raouzeos (2003), and Lattman & Laible (2005) provided guidance on the subject of direct versus indirect drying.

18.6.2 Leaching

In the operation of leaching, the component to be removed from the solid is taken out in the form of liquid, using a solvent that penetrates the solid, dissolves the component, and carries it out of the solid. This operation is akin to that of liquid–liquid extraction. Leaching is very prominent in metallurgical practice and also in the food industry. Its rate is governed primarily by the diffusion of solvent in and out of the solid Karnofsky (2005).

18.6.3 Melt Crystallization: Zone Melting

If the solid is truly a mixture of solids, then it may be advantageous to fully or partially melt it and to use one of the liquid separation methods to separate the components. A process called *zone melting* is described by Atwood (1972) in which a solid solution is successively melted and frozen. During each molten period, there is countercurrent diffusion of components.

Atwood (1969, 1972) described this process.

References

Adler, S.F. (2001 Apr). Biofiltration – a primer. *Chem. Eng. Prog.* 97 (4): 33–41.

Albin, B. and Moyers, C. (2017 Nov). Confronting crystallization issues. *Chem. Eng.* 124: 30–37.

AIChE's Equipment Testing Procedures Committee (2013 Jun). Evaluating distillation column performance. *Chem. Eng. Prog.* 109 (6): 27–35.

Alcantara-Avila J. R., Gomez-Castro F. I., Segovia-Hernandez G., Sotawa K. I., Horikawa T., Energy minimization in cryogenic distillation columns through intermediate side heat exchangers, in *Computer Aided Chemical Engineering*, 2014, 33, 1501–1506 doi: https://doi.org/10.1016/B978-0-444-63455-9.50085-4.

Anonymous (2002 Jun). Understanding flooding. *Chem. Eng.* 109 (6): 61.

Atwood G. R., Crystallization from solutions and melts, Chem. Eng. Prog. Symposium series No. 95, 1969, vol. *65*.

Atwood, G.R. (1972). Developments in melt crystallization. In: *Recent Developments in Separation Science* (ed. N.N. Li). CRC Press.

Bennett, D.L. (2000 May). Packed columns. *Chem. Eng. Prog.* 96 (5): 27–34.

Bennett, D.L. and Kovak, K.W. (2000 May). Optimize distillation columns. *Chem. Eng. Prog.* 96 (5): 19–34.

Bernard, A. (2012 Jun). Distillation: avoid problems during tower startup. *Chem. Eng.* 119 (6): 48–59.

Bethge, D. (2016 Nov). Heat transfer in wiped film evaporators. *Chem. Eng.* 123 (11): 44–47.

Bouck, D. (2014 Feb). Distillation revamp pitfalls to avoid. *Chem. Eng. Prog.* 110 (2): 31–38.

Bravo, J.L. and Kusters, K.A. (2000 Dec). Tray technology for the new millennium. *Chem. Eng. Prog.* 96 (12): 33–37.

Breck, D.W. (1974). *Zeolite Molecular Sieves*. New York: John Wiley & Sons.

Broom, D. (2018 Mar). Characterizing adsorbents for gas separations. *Chem. Eng. Prog.* 114 (3): 30–37.

Carr, S. and Vaughn, R. (2003 Apr). Taking stock of activated carbon's many talents. *Chem. Eng.* 110 (4): 76–80.

Cartwright, P.S. (2004 Jun). Membranes for process water reuse. *Chem. Eng.* 111 (6): 38–42.
Caruana, C.M. (2000 Jan). Melt crystallization heats up. *Chem. Eng. Prog.* 96 (1): 11–12.
Chen, W., Parme, F., Patkar, A. et al. (2004 Dec). Selecting membrane filtration systems. *Chem. Eng. Prog.* 100 (12): 22–25.
Chin, K. (2003 Jul). Today's distillation modeling. *Chem. Eng. Prog.* 99 (7): 5.
Choudhari, A., Gune, P., and Divey, J. (2012 Mar). Distillation optimization by vapor recompression. *Chem. Eng.* 119 (3): 43–47.
Christiansen, O.B. and Sardo, M.S. (2001 Aug). Find the optimum flash dryer to remove surface moisture. *Chem. Eng. Prog.* 97 (8): 54–58.
D'Aquino, R.L. (2007 Jun). Controlling crystal growth. *Chem. Eng.* 114 (6): 71.
Duss, M. and Taylor, R. (2018 Jul). Predict distillation tray efficiency. *Chem. Eng. Prog.* 114 (7): 24–30.
Fair, J.B. (2002 Aug). The workhorse today & tomorrow – distillation. *Chem. Eng.* 109 (8): 108–112.
Fewel, K.J. (2015 Apr). Use vapor distributors to improve fractiontion column. *Hyd. Proc.* 94 (4): 49–52.
Fox, B. (2001 Oct). Vacuum and microwaves dry granulated products. *Chem. Eng.* 108 (10): 135–141.
Fraud, N. and Gottschalk, U. (2010 Dec). Using chromatography to separate complex mixtures. *Chem. Eng. Prog.* 106 (12): 27–31.
Friedman, Y.Z. (2010 Jun). Total- or partial-draw configuration, Hyd. *Proc.* 89 (6): 1.
Gabelman, A. (2017 Jul). Adsorption basics: part 1. *Chem. Eng. Prog.* 113 (7): 48–53.
Gabelman, A. (2017 Aug). Adsorption basics: part 2. *Chem. Eng. Prog.* 113 (8): 38–45.
Gallot, J.C. and Schur, A. (2006 Dec). Crystallizers and agitators. *Chem. Eng.* 113 (13): 38–41.
Genck, W.J. (2000 Aug). *Better Growth In Batch Crystallizers. Chem. Eng.* 107 (8): 90–95
Genck, W.J. (2003 Jun). Optimizing crystallizer scaleup. *Chem. Eng. Prog.* 99 (6): 36–44.
Genck, W.J. (2004 Oct). Guidelines for crystallizer selection and operation. *Chem. Eng. Prog.* 100 (10): 26–32.
Genck, W. (2011 Jul). A clearer view of crystallizers. *Chem. Eng.* 118 (7): 28–32.
Glatz, D. and Parker, W. (2004 Nov). Enriching liquid-liquid extraction. *Chem. Eng.* 111 (12): 44–51.
Glover, W.B. (2004 Apr). Scaleup of agitated thin-film evaporators. *Chem. Eng.* 111 (4): 55–58.
Graf, K. (2011 Oct). Determining packing height with accuracy. *Chem. Eng.* 118 (10): 55–63.
Grossel, S.S. (2006 Jul). Distillation troubleshooting. *Chem. Eng. Prog.* 102 (7): 53.
Gupta, A. and Jain, S. (2018 Apr). An optimized approach to choosing solvent for the liquid-liquid extraction process. *Hyd. Proc.* 97 (4): 69–71.
Hairston, D. (2000 Mar). Membranes put the squeeze on cryogenics. *Chem. Eng.* 107 (3): 33–39.
Hairston, D. (2002 Sep). The bead on ion exchange resins. *Chem. Eng.* 109 (10): 37–40.
Huchler, L.A. (2002 Apr). What about reverse osmosis. *Hyd. Proc.* 81 (4): 107–108.
Huchler, L.A. (2010 Sep). Improving reverse osmosis performance. *Cem. Eng. Prog.* 106 (9): 40–44.
Huchler, L.A. (2014 Jan). Best practices for RO operations. *Hyd. Proc.* 93 (1): 98.
Herbert, S. and Sandford, N. (2016 May). Consider moving to fixed valves. *Chem. Eng. Prog.* 112 (5): 34–41.
Jenkins, S. (2010A Jun). Distillation tray design. *Chem. Eng.* 117 (6): 27.
Jenkins, S. (2010B Oct). MSMPR crystallization equipment. *Chem. Eng.* 117 (10): 3.
Jenkins, S. (2015 Jun). Liquid-liquid extraction. *Chem. Eng.* 122 (6): 33.
Jenkins, S. (2016 Aug). Distillation column design factors. *Chem. Eng.* 123 (8): 38.
Jevric, J. and Fayed, M.E. (2002 Dec). Shortcut distillation calculations via spreadsheets. *Chem. Eng. Prog.* 98 (12): 60–67.
Julka, V., Chiplunkar, M., and O'Young, L. (2009 Mar). Selecting entrainers for azeotropic distillation. *Chem. Eng. Prog.* 105 (3): 47–53.

Ivakpour, J. and Kasiri, N. (2008 Oct). Improve sped and convergence of distillation column simulation. *Hyd. Proc.* 87 (10): 75–79.

Karnofsky, G. (2005 Dec). Some time-tested tips for washing and leaching. *Chem. Eng.* 112 (13): 52–55.

Kasiri, N., Jouybanpour, P., and Ehsani, R. (2016 Aug). Utilize genetic programming to develop new point efficiency correlation. *Hyd. Proc.* 95 (8): 59–62.

Kimball, G. (2001 May). Direct vs indirect drying: optimizing the process. *Chem. Eng.* 108 (5): 74–81.

Kister, H.Z. (2002 Oct). Can we believe the simulation results? *Chem. Eng. Prog.* 98 (10): 52–58.

Kister, H.Z. and Litchfield, J.F. (2004 Sep). Distillation: diagnosing instabilities in the column overhead. *Chem. Eng.* 111 (9): 55–59.

Kister, H.Z. (2006 Mar). Distillation: introducing reboiler return. *Chem. Eng. Prog.* 102 (3): 16–17.

Kister, H.Z. (2006 Jun). Acid-gas absorption. *Chem. Eng. Prog.* 102 (6): 16–17.

Kister, H.Z., Scherffius, J., Afshar, K., and Abkar, E. (2007 Jul). Realistically predict capacity and pressure drop for packed columns. *Chem. Eng. Prog.* 103 (7): 28–38.

Kister, H.Z. (2008 Jun). Effects of design on tray efficiency in commercial towers. *Chem. Eng. Prog.* 104 (6): 39–47.

Kister, H.Z., Dionne, R., Stupin, W.J., and Olsson, M.R. (2010 Apr). Preventing maldistribution in multi-pass trays. *Chem. Eng. Prog.* 106 (4): 32–41.

Kister, H.Z. (2013A Feb). Use quantitative gamma scans to troubleshoot maldistribution on trays. *Chem. Eng. Prog.* 109 (2): 33–42.

Kister, H.Z. (2013B Apr). Apply quantitative gamma scanning to high-capacity trays. *Chem. Eng. Prog.* 109 (4): 45–49.

Kister, H.Z. (2016 Jan). Flooded condenser controls: principles and troubleshooting. *Chem. Eng.* 123 (1): 37–49.

Koch, J. and Shiveler, G. (2015 Nov). Design principles for liquid-liquid extraction. *Chem. Eng. Prog.* 111 (11): 22–30.

Kohler, S.D., Padmanabhan, L., Murphy, E.C. et al. (2011 Apr). Distillation back mixing: impact on batch yields. *Chem. Eng.* 118 (4): 40–43.

Kucera, J. (2008 May). Understanding RO membrane performance. *Chem. Eng. Prog.* 104 (5): 30–34.

Kucera, J. (2019 Feb). The essentials of reverse osmosis. *Chem. Eng. Prog.* 115 (2): 30–37.

Langrish, T. (2009 Dec). Applying mass and energy balances to spray drying. *Chem. Eng. Prog.* 105 (12): 30–34.

Lattman, M. and Laible, R. (2005 Nov). Batch drying: the "indirect" solution to sensitive drying problems. *Chem. Eng.* 112 (11): 34–39.

Lee, S.H. and Binkley, M.J. (2011 Jun). Optimize design for distillation feed. *Hyd. Proc.* 90 (6): 101–105.

Lee, F.-M. and Gentry, J.C. (1997 Oct). Don't overlook extractive distillation. *Chem. Eng. Prog.* 93 (10): 56–64.

Lek, C.M., Rangaiah, G.P., and Hidajat, K. (2004 Sep). Distillation: revisiting some rules of thumb. *Chem. Eng.* 111 (9): 50–55.

LePree, J. (2015 Sep). Improve efficiency with better evaporation and drying. *Chem. Eng.* 122 (9): 26–28, 30–31.

LePree, J. (2015A Jun). High-performance internals boost tower capacity. *Chem. Eng.* 122 (6): 24–28.

LePree, J. (2015B Sep). Improve efficiency with better evaporation and drying. *Chem. Eng.* 122 (9): 26–31.

Li, N.N. (1972). *Recent Developments in Separation Science*. Cleveland, OH: CRC Press, The Chemical Rubber Company.

Lipnizki, J. (2007 Sep). Strategies for controlling membrane fouling. *Chem. Eng.* 114 (9): 62–64.

Loprete, K. (2019 Sep). Introduction to gas separation membrane technologies. *Chem. Eng. Prog.* 115 (9): 31–37. *86(11)*, 81–86.

Louis, T. (2005A). Adsorber sizing made easy. *Chem. Eng. Prog.* 101 (3): 16.

Louis, T. (2005B May). Sizing packed towers – absorbers without data. *Chem. Eng. Prog.* 101 (5): 18.

Malfand, M. (2017 Nov). Industrial crystallization for the CPI. *Chem. Eng.* 124 (12): 38–41.

Marshall, R. (2006 Aug). Distillation. *Chem. Eng.* 113 (8): 59.

Mathias, P.M. (2009 Dec). Visualizing the McCabe-Thiele diagram. *Chem. Eng. Prog.* 105 (12): 36–44.

McCausland, L.J., Cains, P.W., and Martin, P.D. (2001 Jul). Use the power of sonocrystallization for improved properties. *Chem. Eng. Prog.* 97 (7): 56–61.

McKeithan, P.D. (2000 Nov). Troubleshooting the convection-conveyor dryer. *Chem. Eng.* 107 (11): 125–128.

McNulty, J.T. (1997 Jun). The many faces of ion-exchange resins. *Chem. Eng.* 104 (6): 94–100.

Mehos, G.J. (2009 May). Designing and operating gravity dryers. *Chem. Eng.* 116 (5): 34–39.

Meszaros, I. and Meili, A. (2002 Mar). Optimize distillation operations with ejectors. *Hyd. Proc.* 81 (3): 51–55.

Moftah, K. (2003 Sep). For water treatment, consider high-pH reverse osmosis. *Chem. Eng.* 110 (9): 62–69.

Moller, J.T. and Fredsted, S. (2009 Nov). A primer on spray drying. *Chem. Eng.* 116 (12): 34–40.

Moyers, C.G. (2003 Dec). Evaluating dryers for new services. *Chem. Eng. Prog.* 99 (12): 51–56.

Mukherjee, S. (2005 Sep). Tray column design: keep control of the details. *Chem. Eng.* 112 (9): 52–58.

Nino, S. (2011 Nov). Improve material balance in high-purity distillation control. *Hyd. Proc.* 90 (11): 83–88.

Nye, J.O., Herzog, K., and Cheaney, S. (1999 Sep). Use a side reboiler to increase tower capacity. *Hyd. Proc.* 78 (9): 51–56.

O'Brien, D. and Schultz, M.A. (2004 Feb). Distillation. *Chem. Eng. Prog.* 100 (2): 14.

O'Connell, J.P., Eden, M.R., Tula, A.K., and Gani, R. (2019 Dec). Retrofitting distillation columns with membranes. *Chem. Eng. Prog.* 115 (12): 41–49.

Ondrey, G.S. (2000 Jan). Crystallization *sans* solvents. *Chem. Eng.* 107 (1): 30–33.

Ondrey, G.S. and Hairston, D. (2001 Jul). Crystallization: a melting pot. *Chem. Eng.* 108 (9): 43–46.

Ondrey, G. (2005 Sep). Drying with superheated steam. *Chem. Eng.* 112 (9): 15.

Ortiz-Vega, D., Dowdle, J., Cristancho, D., and Badhwar, A. (2015 Jun). Accurate rate-based modeling of acid gas and mercaptan removal using hybrid solvents. *Hyd. Proc.* 94 (6): 53–56.

O'Young, L. (2005 Jun). Crystallization. *Chem. Eng. Prog.* 101 (6): 18.

Panagiotou, T. and Fisher, R.J. (2008 Oct). Form nanoparticles via controlled crystallization. *Chem. Eng. Prog.* 104 (10): 33–39.

Parikh, D.M. (2014 Apr). Solids drying: basics and applications. *Chem. Eng.* 121 (4): 42–45.

Parikh, D.M. (2015 Apr). Vacuum drying: basics and application. *Chem. Eng.* 122 (4): 48–54.

Peeters, J.G. and Theodoulou, S.L. (2007 Nov). Consider membrane technologies to treat oily wastewater. *Hyd. Proc.* 86 (11): 81–86.

Perkins, E.A. and Schad, R.C. (2002 Feb). Get more out of single-stage distillation. *Chem. Eng. Prog.* 98 (2): 48–52.

Pilling, M. (2005 Jun). Ensure proper design and operation of multi-pass trays. *Chem. Eng. Prog.* 101 (6): 22–27.

Pilling, M. and Holden, B.S. (2009 Sep). Choosing trays and packing for distillation. *Chem. Eng. Prog.* 105 (9): 44–50.

Pilling, M. and Summers, D.R. (2012 Nov). Be smart about column design. *Chem. Eng. Prog.* 108 (11): 32–38.

Ramkrishna, D. (2002). *Population Balances Theory and Applications to Particulate Systems Engineering*. San Diego, CA: Academic Press.
Rangaiah, G.P. and Yue, C.T. (2006 Sep). Optimization of complex distillation columns. *Chem. Eng.* 113 (9): 60–64.
Raouzeos, G. (2003 Dec). The ins and outs of indirect drying. *Chem. Eng.* 110 (13): 30–37.
Raouzeos, G. (2012 Sep). Thin-film drying offers deep benefits. *Chem. Eng.* 119 (9): 40–45.
Resetarits, M., ReBeau, T., and Thurber, T. (2015 Apr). Column revamps: from outside to inside. *Chem. Eng.* 122 (4): 38–47.
Rumyantseva, I. and Beck, R. (2014 Aug). Distillation column thermal optimization: employing simulation software. *Chem. Eng.* 121 (8): 49–53.
Rusche, F.A. (1999 Feb). Gilliland plot revisited. *Hyd. Proc.* 78 (2): 79 ff.
Ryan, J. (2001). Replotting the McCabe-Thiele diagram. *Chem. Eng.* 108 (5): 109–113.
Samant, K.D. and O'Young, L. (2006 Oct). Understanding crystallization and crystallizers. *Chem. Eng. Prog.* 102 (10): 28–37.
Sands, R.R. (2006 Apr). Distillation: how to specify and install cartridge trays. *Chem. Eng.* 113 (4): 86–92.
Shah, P.B. (2002 Jul). Squeeze more out of complex columns. *Chem. Eng. Prog.* 98 (7): 46–55.
Shanley, A. (2000 Jan). New frontiers for ion exchange. *Chem. Eng.* 107 (1): 61–64.
Shelley, S. (2009 Apr). Capturing CO_2 : Membrane systems move forward. *Chem. Eng. Prog.* 105 (4): 42–47.
Siperstein, F.R. and Myers, A.L. (2001 May). Mixed-gas adsorption. *AICHE J.* 47 (5): 1141–1159.
Sircar, S., Golden, T.C., and Rao, M.B. (1995). Activated carbon for gas separation and storage. *Carbon* 34 (1): 1–12.
Sloley, A.W. (2001 Jan). Effectively control column pressure. *Chem. Eng. Prog.* 97 (1): 38–48.
Stover, R.L. (2014 July). A primer on reverse osmosis technology. *Chem. Eng.* 121 (7): 38–44.
Summers, D.R. and Sloley, A. (2007 Jan). How to handle low-liquid loadings. *Hyd. Proc.* 86 (1): 67–78.
Summers, D.R. (2009 Jun). Dry tray pressure drop of sieve trays. *Chem. Eng.* 116 (6): 36–39.
Summers, D.R. (2010A Feb). Evaluating and documenting tower performance. *Chem. Eng. Prog.* 106 (2): 38–45.
Summers, D.R. (2010B Apr). Designing four-pass trays. *Chem. Eng. Prog.* 106 (4): 26–31.
Summers, D.R. and Cai, T.J. (2017 Aug). Dry-tray pressure drop of sieve trays revisited. *Chem. Eng.* 124 (8): 38–41.
Summers, D.R. (2018 Feb). Bubble-cap tray vapor turndown. *Chem. Eng.* 125 (2): 38–41.
Sutradhar, B.C. (2004 Mar). Coping with crystallization problems. *Chem. Eng.* 111 (3): 46–52.
Taylor, R., Krishna, R., and Kooijman, H. (2003 Jul). Modeling of distillation. *Chem. Eng. Prog.* 99 (7): 28–39.
Torzewski, K. (2008A Mar). Membranes. *Chem. Eng.* 115 (3): 33.
Torzewski, K. (2008B Apr). Random tower packing. *Chem. Eng.* 115 (4): 33.
Torzewski, K. (2008C Jun). Humidity control. *Chem. Eng.* 115 (6): 55.
Torzewski, K. (2008D Sep). Crystallization. *Chem. Eng.* 115 (9): 51.
Torzewski, K. (2009A Jan). Tray column design. *Chem. Eng.* 116 (1): 29.
Torzewski, K. (2009B Mar). Membrane configurations. *Chem. Eng.* 116 (3): 27.
Torzewski, K. (2009C Aug). Adsorption. *Chem. Eng.* 116 (8): 23.
Treybal, R.E. (1980). *Mass Transfer Operations*, 3e. New York: McGraw-Hill Book Company.
Tritschler, E. (2017 Apr). Batch drying with vacuum contact dryers. *Chem. Eng.* 124 (4): 36–42.
Vermesse, J., Vidal, D., and Malbrunot, P. (1996). Gas adsorption on zeolites at high pressure. *Langmuir* 12: 4190–4196.

Wang, H. and Zhou, H. (2013 Apr). Understand the basics of membrane filtration. *Chem. Eng. Prog.* 109 (4): 33–40.

White, D.C. (2012 Mar). Optimize energy use in distillation. *Chem. Eng. Prog.* 108 (3): 35–41.

Wibowo, C., O'Young, L., and Ng, K.M. (2004 Jan). Streamlining crystallization process design. *Chem. Eng. Prog.* 100 (1): 30–39.

Wibowo, C. (2011 Mar). Developing crystallization processes. *Chem. Eng. Prog.* 107 (3): 21–31.

Wibowo, C. (2014 Mar). Solid-liquid equilibrium: the foundation of crystallization process design. *Chem. Eng. Prog.* 110 (3): 37–45.

Wynn, N.P. (1992 Mar). Separate organics by melt crystallization. *Chem. Eng. Prog.* 88 (3): 52–60.

Wynn, N. (2001 Oct). Pervaporation comes of age. *Chem. Eng. Prog.* 97 (10): 66–72.

Xu, S.X. and Mixon, W. (2007 May). Diagnosing maldistribution in towers. *Chem. Eng. Prog.* 103 (5): 28–35.

Xy, S. and Pless, L. (2002 Jun). Distillation tower flooding – more complex than you think. *Chem. Eng.* 109 (6): 60–67.

Yahyaabadi, R. (2007 Mar). Increase distillation capacity with side heaters and coolers. *Hyd. Proc.* 86 (3): 99–105.

Yang, R.T. (1997 June). *Gas Separation by Adsorption Processes*. London: Imperial College Press.

Yang, R.T. (2003). *Adsorbents: Fundamentals and Applications*. Hoboken, NJ: John Wiley & Sons Inc.

Yohe, T. (2002 Aug). Crystallization. *Chem. Eng. Prog.* 98 (8): 76.

19

Chemical Reactions

A chemical engineer may expect, at some point in her or his career, to have to arrange for the conduct of a chemical reaction. Reactors can take many forms depending on several factors but primarily on the phase or phases of the reactants and products. The following seven situations are examined here

1. gas phase,
2. liquid phase,
3. gas–liquid,
4. immiscible liquids,
5. fluid and solid, non-catalytic,
6. solid-catalyzed, and
7. bio-reactions.

However, before any thought of choosing and designing a reactor, the engineer must ascertain two general facts about the proposed reaction:

A. The *chemical* characteristics of the reaction.

- Is it reversible?
- What is the order of the reaction with respect to each of the reactants?
- What is the rate constant? What is the equilibrium constant?
- Is it a simple reaction, following some simple path such as

Reactant → product

or are there intermediate species between the feed material and the desired product?

- Is the desired product subject to further, undesired reaction?
- Are there other (desirable or undesirable) reactions, parallel to the reaction that produces the desired product?
- Are there inhibitors to the reaction or poisons to the catalyst that degrade or inhibit the reaction, either products, intermediates, or temperature?

Ascertaining all of these facts may be very difficult. For a new reaction, extensive experiments may be required, which, hopefully and ideally, can be carried out at the laboratory bench, i.e., at a small scale. Understanding the chemistry is absolutely vital to successful reactor design. The scale of a whole plant, for instance, can depend on the value of a single kinetic constant.

Practical Process Design for Chemical Engineers, First Edition. Keith Marchildon and David Mody.

B. The *physical* phenomena that accompany the chemical reaction.

- Is the reaction significantly endothermic so that heat must be continuously supplied in order to sustain its progress? Is there a volatile product or by-product that must be continuously removed by vaporization?
- Is the reaction significantly exothermic so that heat must be continuously removed in order to maintain temperature and to prevent runaway?
- Is there a significant resistance to mass transfer of a reactant or product from one phase to another?
- If a porous catalyst is involved, is the diffusion of components to and from the reactive sites a limiting rate?

It is not unusual for the design of a particular reactor design to be focused more on heat, mass, and momentum transfer rather than on chemistry. There may be more facts to learn, such as the toxicity of reactants, intermediates, or products, and the tendency for runaway reaction.

The subject of reaction and reactors is important enough to attract a number of publications over the years. Cusack (1999A,B, 2000) and Bartholomew & Farrauto (2006) provided a good overview. Worstell (2001) and Ranade (2002) stated the questions which should be initially asked by anyone before starting to design a reactor. The book by Conesa (2019) explores the many aspects of the mathematical modeling of chemical reactors. Dutta & Gualy (2000) proposed a 14-step modeling approach. Bakker et al. (2001) look at the physical side of reactors and their design by computational fluid dynamics.

19.1 Gas-Phase Reactions

We are asked to design a reactor to react a gas or a mixture of gases. Likely, the choice will be to operate at

- high temperature because reaction occurs faster and because there is no need or desire as there would probably be in liquid-phase processing to keep materials below their boiling point, and
- high pressure because reaction occurs faster when the volumetric concentrations of the components are higher.

Reactions within gasses are very frequently carried out in the presence of a solid catalyst, but that aspect is considered later. For non-catalyzed reactions, the most efficient reactor is continuous plug flow rather than back-mixed.

Some possibilities are as follows:

1. a simple long tube, operated in turbulent flow in order to approximate plug flow and to achieve good mixing and heat transfer,
2. a simple tube fully or partially filled with static mixers to achieve even better plug flow, mixing, and heat transfer (see Figure 19.1),
3. a vertical vessel (Figure 19.2) filled with a packing (but the packing uses up reaction volume),
4. a vertical vessel fitted with periodic sieve-plate baffles to achieve a staged, close-to-plug-flow state, and
5. a multi-tube vertical vessel to achieve more surface area if heat rejection is a major concern, which it often is with gaseous reactions.

For a reaction in any phase, plug flow is attractive because it makes the most efficient use of reactor volume compared with any other continuous-flow configuration. Comparing, for instance,

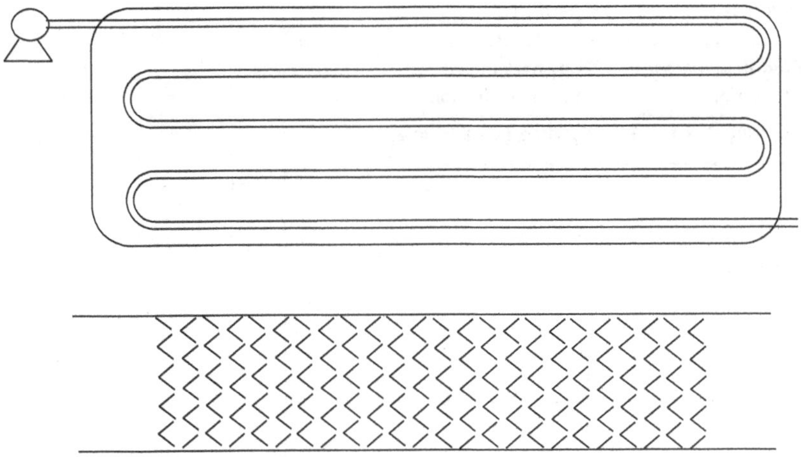

Figure 19.1 Continuous plug-flow reactor including static mixers.

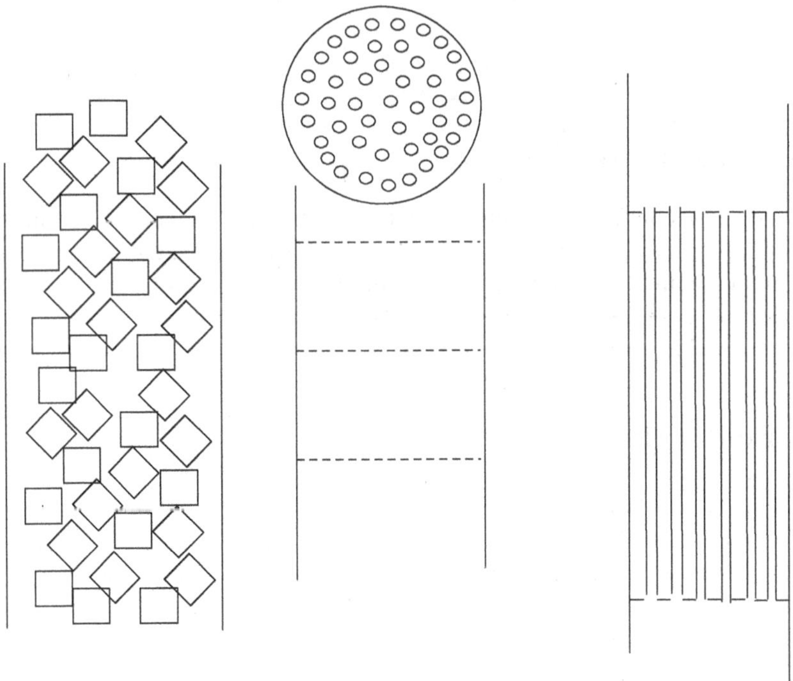

Figure 19.2 Means for radial dispersion of continuous fluid flow.

with a back-mixed reactor, for a simple reaction of

$$\text{Rate} = -\text{k} \times [X]$$

where $d[X]/d(\text{time}) = \text{rate} = -\text{k} \times [X]$,

i.e., the consumption of component "X" is first order in "X" and irreversible, then the following conversions are achieved, as shown in Table 19.1.

Unfortunately, perfect plug flow is an ideal that is never fully attained in practice. In the flow in a tube, there is always some degree of axial dispersion. This is obvious for laminar flow (see Figure 19.3) because, although the axial dispersion due to molecular motion is very small, the

Table 19.1 Conversion versus different flow patterns.

Kinetic constant × residence time	Plug-flow reactor conversion	Fully back mixed reactor conversion
k × time	1 − Exp(−k × time)	k × time/(1 + k × time)
0	0	0
0.2	0.181	0.167
0.5	0.393	0.333
1	0.632	0.5
2	0.865	0.667
Infinity	1	1

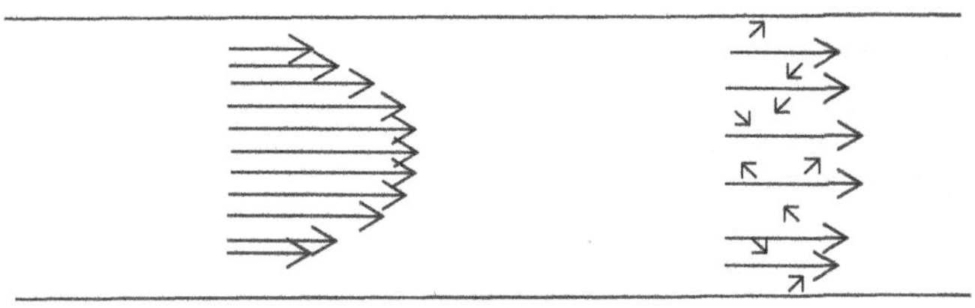

Figure 19.3 Axial dispersion in laminar and turbulent flow.

parabolic velocity profile creates the effect of a huge axial dispersion. For turbulent flow, which has a much flatter velocity profile, axial dispersion is due to the turbulent fluctuations themselves and may significantly detract from the efficiency of a nominally plug-flow tubular reactor. Cusack (2000 February) provided the graphical relationships of Levenspiel (1999) that show how a tracer test can be used to determine this loss of reaction efficiency.

19.2 Liquid-Phase Reactions

One or more miscible liquids are to be reacted. The product or a by-product may be a vapor, but for purposes of this discussion, the system is still considered liquid phase. Also, the case of immiscible liquids is considered later. Broadly, the reactor choices are *batch, continuous plug-flow,* and *continuous back-mixed.* These options are discussed here. Milne et al. (2006B) offered some thoughts on making this choice. Fontes (2015) compared the mathematical modeling of all three types.

19.2.1 Batch Reactor and Fed-Batch Reactor

A batch reactor operates in a time-sequence of events: an amount of reactant is supplied, it is converted chemically to product, and the product is discharged. Most new chemical technologies start out as batch processes before ever being converted to continuous processing. Batch processes have advantages, real or potential:

- relatively low capital cost,
- ease of incremental expansion (build more reactors),
- flexibility for different products, and
- transparency and ease of control,

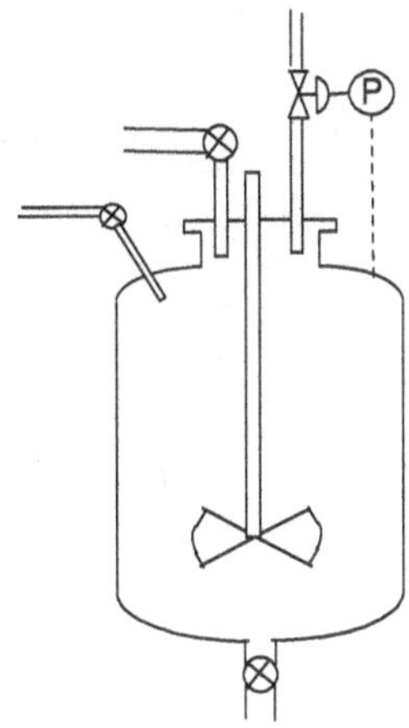

Figure 19.4 Typical batch reactor with agitator, pressure control, and additive port.

and they have real or potential disadvantages:

- generally higher labor cost (which may be relieved by modern sequencing control),
- batch-to-batch nonuniformity of product, and
- internal nonuniformity if sufficient agitation is not supplied for adequate mixing of reactants or for adequate suspension of solid reactants or catalyst particles.

Batch reactors are sometimes operated in *fed-batch* mode, where additions are made to the batch during the course of the process (see Figure 19.4). The designation particularly applies to the case where a significant amount of a reacting component is added. Some situations where this approach is applicable are

- a reacting component is susceptible to an undesirable side-reaction or to volatilization,
- the reaction is a large producer or consumer of heat, so it is desired to moderate its rate, and
- a reaction by-product is continuously volatilized from the vessel, making room for more liquid.

McConville (2007) presented the Do's and Don'ts of scaling up batch reactors. One of the possibilities with batch processing is for an exothermic reaction to get out of hand. Venugopal (2002), Shinskey (2004), Marshall (2007), and Kwasny (2008) dealt with the prevention of runaway reactions.

19.2.2 Continuous Plug-Flow Reactor

For the advantage and with the considerations explained in Section 19.1, plug-flow or staged plug-flow is a candidate for liquid-phase as well as gas-phase reactions. If a volatile product or by-product results from the reaction, it can be carried along with the liquid or move ahead of it: Prediction of the relative velocities of the two phases and prediction of the *in situ* volume fraction occupied by liquid (and therefore available for reaction) is the province of two-phase flow relationships (Figure 19.5).

19.2.3 Continuous Back-Mixed Reactor

The back-mixed reactor or *Continuous Stirred Tank Reactor* (CSTR) is very common in chemical processing. It has the following advantages:

- simple fluid dynamics, dependent only on the ability to be agitated,
- easy to scale up, and
- uniformity, absence of hot spots.

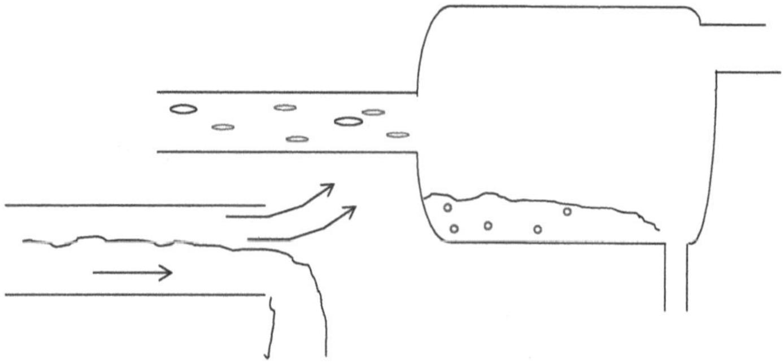

Figure 19.5 Two-phase flow – homogeneous and stratified.

The disadvantages are as follows:

- inefficient use of reactor volume because reactant concentration is all at the final values and therefore low, minimizing the driving force, and
- possibility of inadequate mixing and therefore channeling and zones of stagnation.

The "Stirred" in CSTR tends to imply a mechanical agitator or stirrer, which is frequently the case (Figure 19.6). Such an arrangement, with its requirement for a rotating seal, may be inadvisable when the pressure differs significantly from atmospheric and/or when it is imperative

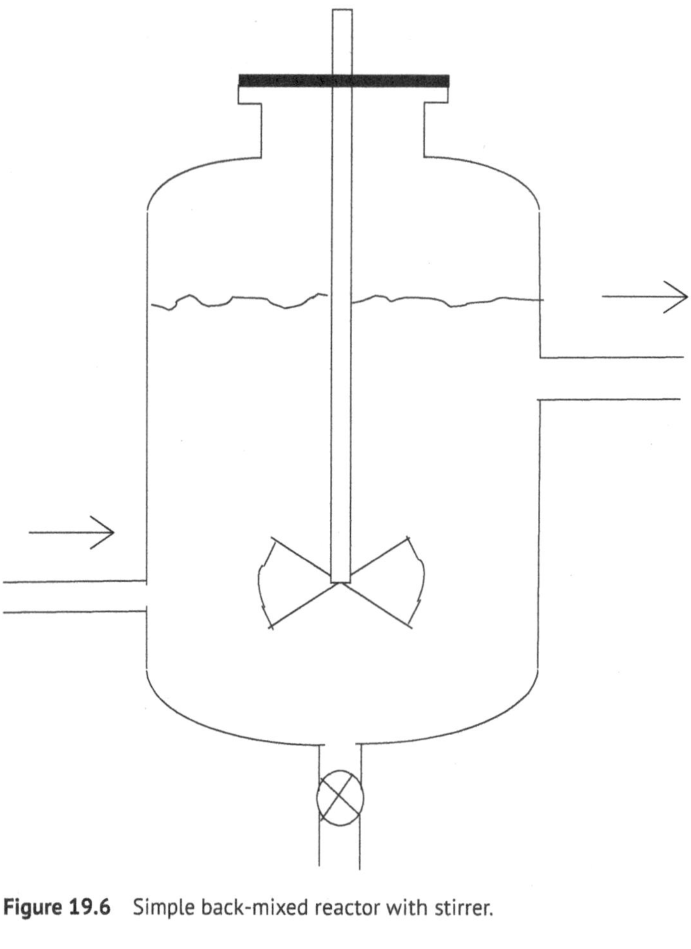

Figure 19.6 Simple back-mixed reactor with stirrer.

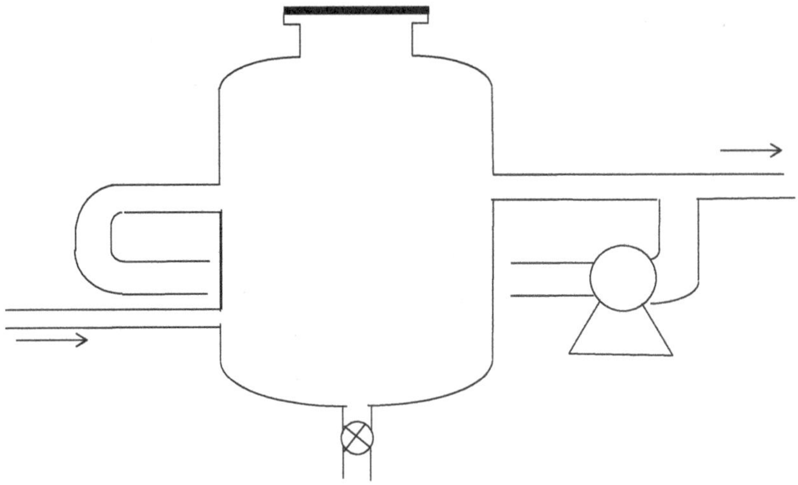

Figure 19.7 Back-mixed reactor with circulating pump-around.

Figure 19.8 Reaction with recycle.

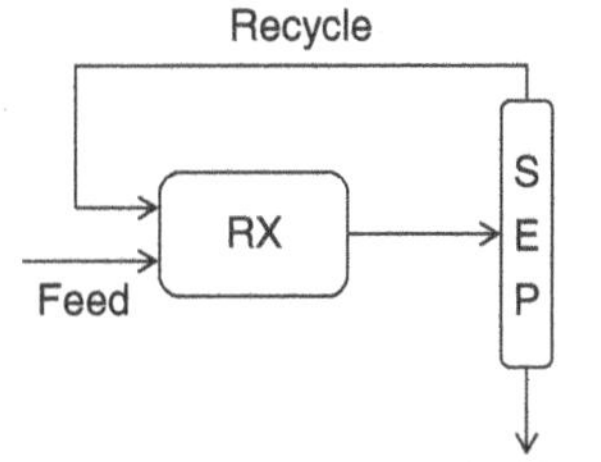

that the contents of the vessel be isolated from the atmosphere. In these cases, an external pump-around loop (Figure 19.7), flowing at a rate of 10 (or more) times the net flow through the vessel, or the introduction of a sparging gas may produce adequate mixing. The pump-around gives the opportunity for including heat transfer along its length.

A very common configuration (see Figure 19.8) in chemical processing is the back-mixed reactor followed by a separator which removes product and recycles unreacted material to the reactor.

There are four circumstances where this configuration should be considered:

1. The reaction must be taken to a very high degree of conversion. This is difficult in a CSTR, where the driving force for reaction becomes very low when the starting reactants have been almost all consumed. With separation and recycle, the CSTR can be operated at a modest conversion and still achieve high conversion by the system as a whole.
2. The reaction is reversible. If product is not removed, then the reaction reaches a standstill.
3. The desired product is susceptible to further, unwanted reaction.
4. The product causes poisoning of a catalyst.

One of the problems in the scale-up of reactors is the provision (or removal) of heat. At the bench scale or in a small pilot plant, the wall of the vessel may provide sufficient surface area for heat transfer. As the scale gets larger, the ratio of wall surface area to reactor volume gets less. In this case, two options for increased heat transfer are as follows:

- provide an internal heating coil or a bank of heating coils (Figure 19.9);
- provide an external heat exchanger with liquid pumped through it or with a boiling mixture being circulated by gravitational force with no need for a pump (Figure 19.10).

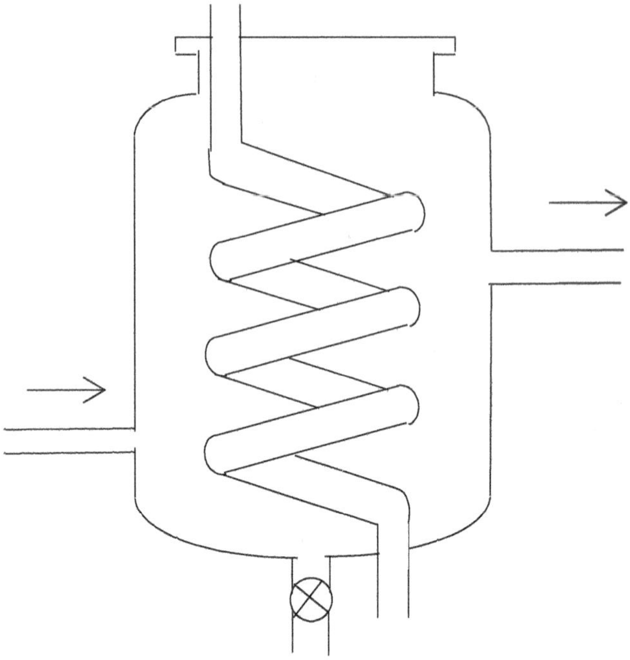

Figure 19.9 Internally heated reactor.

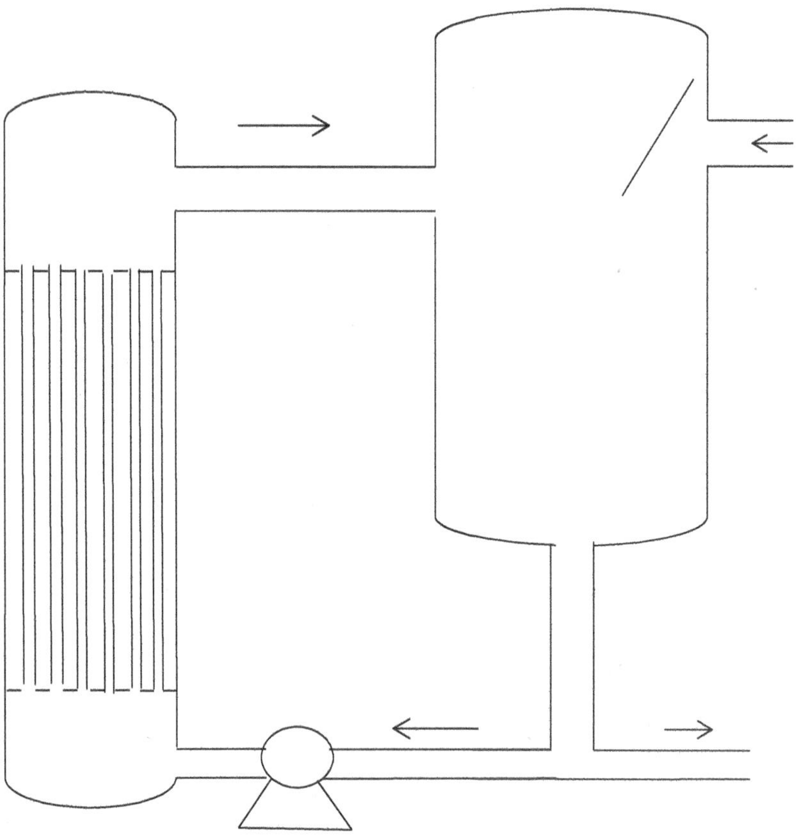

Figure 19.10 Externally heated reactor.

Simply cranking up the temperature of the heating system may not be possible, and it may be undesirable because the hotter surfaces may degrade the product

19.3 Gas–Liquid Reactions

Not uncommonly, one reactant is a liquid and the other is a gas. It may be impractical to raise or lower the temperature (and pressure) to bring all reactants into the same phase. Assuming that reaction takes place in only the liquid phase, it is necessary to transfer the gaseous component into this phase: that is, mass transfer becomes a consideration. The relative roles of mass transfer and of chemical reaction in dictating the overall speed of the process are illustrated in Figure 19.11.

In this simple example, the component **C** is being transferred from the gas to the liquid and the rate of reaction depends on the liquid-phase concentration **[C]** in some simple manner, as shown by line "R1". The rate of transfer of **C** into the liquid is proportional to the difference between the concentration **[C]vle** that would be in equilibrium with the vapor and the actual liquid-phase concentration **[C]**, as shown by the line "M1". The rate of transfer and the rate of reaction have to be equal at steady state, so the concentration **[C]** and the overall rate of the process are determined by the point at which the two lines cross. In the case shown, both the reaction and mass transfer are exerting a significant influence and neither phenomenon can be ignored.

By contrast, the combination of lines "R2" and "M2" show a case where mass transfer has a negligible effect and the combination of lines "R3" and "M3" show a case where mass transfer has the dominating effect.

Some considerations are as follows:

- For the particular system, it is necessary to establish what is the actual relation between the chemical rate and the transfer rate. This requires reference to the literature and also careful pilot-plant study.
- On scale-up, the chemical rate stays the same, but the mass transfer coefficient or rate factor may change.
- Very fast reaction rates may actually enhance mass transfer, by continually lowering **[C]** at the interface to zero (or by continually raising it if [C] is being transferred out of the liquid phase).
- In cases where mass transfer effects are significant, there is an opportunity to use agitation intensity to speed up the process.
- In cases where the chemical reaction rate is a significant limitation, it may be profitable to look for a better catalyst.

The definitive discussion of this type of reactor is given by Fair (1967).

Figure 19.11 Rates of chemical reaction and of mass transfer.

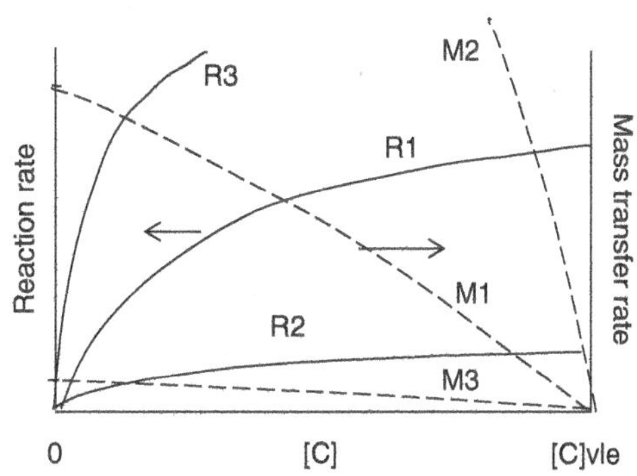

19.4 Reaction of Immiscible Liquids

Two common situations are as follows:

1. The reactants themselves are immiscible and react only at their interface.
2. A non-reacting immiscible liquid is added to the reaction mixture in order to extract the product as it is made. This removal of product may be desirable to protect it from further (or reverse) reaction.

If the reaction is being carried out in a batch vessel or in a CSTR, the usual procedure is to provide high-intensity agitation to break up the dispersed phase and increase the interfacial surface area. If the reaction is being carried out in a pipeline, then high-intensity, high-pressure-drop static mixers will maximize the dispersion. In either case, energy must be provided either to rotate an agitator or to pump the liquid through the static mixers. In-line high-shear mixers that resemble centrifugal pumps are also commercially available.

The interplay between the reaction rate and mass transfer is similar to that in a gas–liquid reactor.

Figure 19.12 shows a pipeline reactor with two streams of entering reactants. A third component is added part way along the reaction path. The in-line static mixer which is shown is of only mild intensity. It would be suitable for making the distribution of the two phases radially uniform, but not to finely disperse the phases. For that purpose, a static mixer of the type shown in Figure 19.1 would be necessary.

Streiff & Rogers (1994) discussed the benefits and the technology for all pipeline reactors in which static mixers fill part or all of the pipe along its way.

19.5 Fluid–Solid Reactions, Non-Catalytic

This type of system occurs mainly in combustors and in ore smelting. Both the solids and the fluids are reactants. The overall rate of reaction depends, again, on a combination of mass transfer and chemical reaction rate, the individual actions being

1. diffusion of fluid (gas or liquid) from the bulk of the fluid to the surface of the particles,
2. diffusion through any crust that exists at the surface,

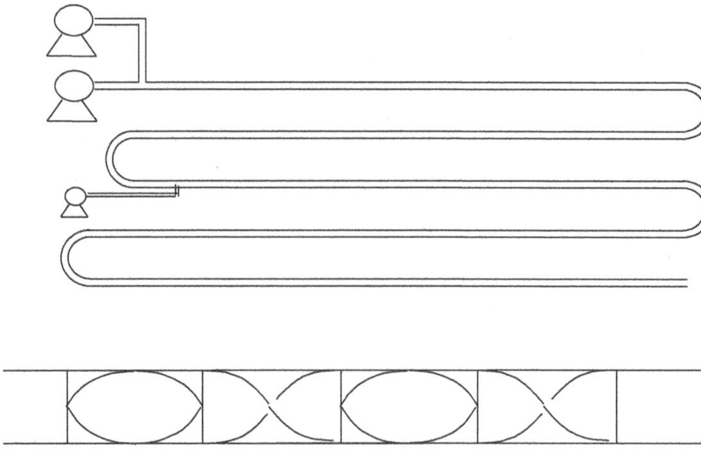

Figure 19.12 Pipeline reactor with static mixer.

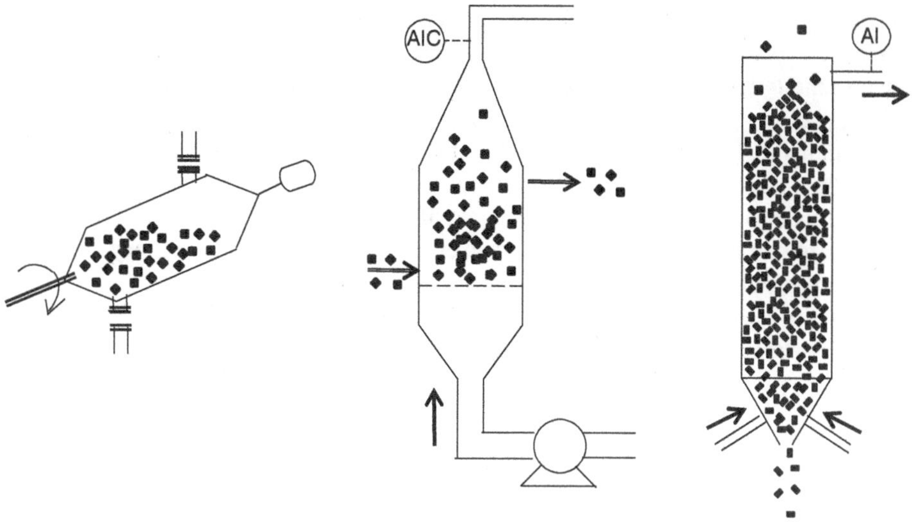

Figure 19.13 Tumbling reactor, fluidized-bed reactor, moving-bed reactor.

3. diffusion within the particle,
4. reaction within the particle, and
5. diffusion of gaseous reaction products back out of the particle.

The diameter of the particles is a major determinant of rate. The surface area of a fixed total mass varies inversely as the diameter of the particles into which it is divided. The external mass transfer coefficient tends to vary inversely as diameter. The internal diffusion rate varies inversely as diameter squared. All effects therefore point to the need to subdivide the solid mass as finely as possible for the benefit of the reaction. Larger particles are favored when lower pressure drop and less entrainment are important.

Fluid and solid reactant combinations are typically brought together in rotating drums or in fluidized beds (Figure 19.13) in order to enhance the surface contact between the two phases. Of course, if the bulk of the mass transfer resistance lies within the particles, the enhanced contact will be ineffective and a packed or slowly moving bed is appropriate.

19.6 Solid-Catalyzed Reactions

Solid or heterogeneous catalysts are used in many industrially important reactions. The fluid medium may be gas or liquid or a mixture of both. The chemical reaction occurs at the solid surface, and, in order to provide as much surface area as possible in a given volume of catalyst, the catalyst particles are porous. The pore diameter is typically in the range 0.001–0.1 micron, providing a surface area of 10 000–100 000 m^2 (20 football fields) per kilogram of catalyst.

Just as in the case of non-catalyst solids (Section 3.10.5), the overall reaction speed is governed by a combination of diffusion and actual chemical reaction. In this case, the principal diffusion mechanism is the movement of reactive molecules up the pores and the movement of product molecules back out of the pores. The ratio of actual reaction rate to the rate that would be achieved if there were no pore diffusion resistance (i.e. the purely chemical rate) is called the *effectiveness factor*, η. Design of a solid-catalyzed system requires some knowledge of the chemical rate and of the effectiveness factor and of the conditions that affect them, e.g., temperature, pressure, and concentration of reactants. This is knowledge that comes from the catalyst manufacturer and from laboratory studies.

Liquid systems tend to have higher diffusional resistance than gas because diffusivity is lower and viscosity is higher.

Other considerations are the selectivity of the catalyst for the desired reaction and also the speed and nature of catalyst deactivation and/or poisoning. Also, the heat of reaction may be such as to make the temperature of the interior of the catalyst significantly different from the bulk temperature. A factor in scale-up is that external mass transfer around the catalyst pellet may also be a limitation: Pilot-plant studies should address this effect.

The reactor type is usually a choice between the fixed bed and fluidized bed.

A fixed bed tends to be used if

1. the particles are large,
2. the catalyst has a long life,
3. plug flow is desired, and
4. heat effects are small.

A fluidized bed tends to be used if

1. the particles are small (e.g. less than 0.2 mm diameter),
2. significant heat must be removed,
3. the catalyst requires frequent regeneration,
4. plug flow is not required, and
5. the catalyst is not prone to attrition.

A variant of the fixed bed reactor is the moving bed, where the pellets are cycled into, through, and out of the reactor so that they can be regenerated offline.

There are two considerations in building a solid-catalyzed system:

(1) choosing the catalyst;
(2) designing the reactor.

For background on the subject of solid catalysts, Heinen (2012) provided a history of 90 years of development. Assistance in choosing a catalyst is provided by Bartholomew & Hecker (1994), Loffler (2001), Maldonado & Rothenberg (2009), and Jenkins (2015). Problems with catalysts are discussed by Worstell et al. (2000), Worstell & Worstell (2004), and Loffler (2002).

Designing the reactor often starts with a pilot plant. Scaling up is discussed by Worstell (2000), Turaga et al. (2006), and Worstell (2007). Other design issues are taken up by Fontes et al. (2003) and Nehlsen et al. (2007). Ahn et al. (2017, 2019) were helpful on the internals of catalytic reactors in which the catalyst is held fixed in one or more beds. One paper presented a high-performance vapor–liquid distributor. The other describes mixing chambers between beds, where mixing of fluids can be reestablished and if necessary the stream can be cooled in order to protect the catalysts from excessive temperatures.

19.7 Bio-Reactions

Chemical reactions deal with inanimate substances. Biochemical reactions deal with living substances, in the form of microbes, sometimes as small as consisting of just a single cell. Microbe types are bacteria, archaea, fungi (mold or yeast), algae, protozoa, and also mammalian cells, all of which are the province of biochemistry.

From the point of view of bio-reaction chemistry, the goal is either

1. to grow a living organism in a controlled manner or
2. to use an organism to convert one chemical species into another.

An example of the first case is the production of yeast, a single-cell microorganism of which the variety *saccharomyces cerevisiae* is the desired product. Yeast exists beneficially in the body of mammals, it is used in food, and it is used to power the production of other substances (such as wine).

Another example of the growth of an organism is the bio-treatment of liquid wastes. A mixture of bacteria and protozoa grows by ingesting impurities from the liquid and thereby purifying it.

The second application of bio-reactions, to convert substances into useful products, is the subject of continuing development. The main products at present are pharmaceuticals. There is hope for a future in which biology can replace petroleum-based processes, with attendant reduction in greenhouse gas (GHG) emission and reduction in energy consumption. These developments are in early stages. One very successful example is the production of 1,3-propane diol from corn-based sugars. Developed by DuPont and Tate & Lyle, it uses a DNA-modified *Escherichia Coli* bacteria to produce the product, with about 72% molar yield, 30–40% less energy than an oil-based route and significantly less (GHG) emissions. Recent studies are presented by Oh et al. (2018).

The prospects for the bio industry are discussed by Demain & Solomon (1981), van Hoek et al. (2003), Jenkins (2010), Davison & Lievense (2016), and Schuster (2016).

Background on bio-chemistry is provided by Monk (2010), Jenkins (2011), Bhatia (2013, 2014), and Le et al. (2016).

19.7.1 Bioreactors

Much of the technology of ordinary chemical reactors carries over to bioreactors. Some aspects that are specific to, or more important for, bio-systems are examined here.

1. At present, most bio-reactions are carried out batch-wise. Batch time is often long (days), so labor requirements are not a factor.
2. A large reactor is often preceded by a small "seed" reactor (or even a series of increasing-size seed reactors) where the microbe concentration grows. The addition of one reactor's contents to the next bigger reactor is an *inoculum*.
3. The "fed batch" mode of batch processing is often used in order to avoid excessive concentrations of the feed materials.
4. The reaction may include the consumption of oxygen (aerobic), or it may require the absence of oxygen (anaerobic).
5. There must be some form of agitation in order to avoid stagnant zones. If oxygen is needed, then the agitation must ensure an adequate mass transfer rate. However, the physical action of agitation must respect the fragility of some microbial types, especially mammalian cells.
6. In most cases, avoidance of contamination is vital. This applies to the reactor itself and also to agitators, pumps, valves, and piping around the vessel. Only the desired organisms must be present, not the variety that float around in the world at large.
7. Control of conditions, e.g., pH, temperature, and oxygen concentration, is frequently important in order to achieve the desired products and in desired yield.
8. If the product is a pharmaceutical, a proven record of the time-wise profile of each batch must be kept for certification purposes.

Many publications provide guidance in the design of bioreactors. Benz (2011) is specifically aimed at bio design for chemical engineers. Williams (2002) presents keys to bioreactor selections.

Adequate mass transfer for aerobic reactors is stressed by Doble (2006), not simply to make sure that the oxygen requirement is met but also to ensure that the carbon dioxide waste product is flushed out to keep its concentration low and avoid its toxic effects.

Benz (2008) treated the scale-up of agitators, and Benz (2016) discussed mass transfer in aerobic fermenters. Bartow & Spark (2004) pointed out the restrictions on agitators when dealing with fragile mammalian cell cultures.

Gregory & Green (2013), Gregory et al. (2014) discussed heat transfer and mass transfer respectively in fermentation reactors. Heat transfer can be inadequate with large vessels if the only heating surface is the vessel walls. Additional heat can be provided by an external heat exchanger, through which some part of the reaction mixture is continually pumped, reheated, and returned to the batch vessel. Excessively hot surfaces can be fatal to microbial populations.

As with all chemical processes, there is an impetus to switch from batch to continuous operation. This transformation has been slow because mainly of the fragility of the reaction components and because of the need for sterility. However, there are opportunities for continuous handling of feed and product *around* a batch reactor. Le et al. (2015) proposed to replace fed-batch mode by instead cycling a stream from inside the reactor through a *perfusion* cell, which draws off the liquid and returns a concentrated stream of the solids to the reactor. By contrast, the fed-batch mode requires drawing off a portion of the whole mixture (in order to keep constant the contents of the reactor). Przybycien & Titchener-Hooker (2015) discussed making continuous some of the operations that go on after the reactor.

Bioreactors, like any reactors, require process control. Particularly, for pharmaceutical production, the control must be tight and the relevant process variables must be recorded for product validation purposes. In 2001, the US Food and Drug Administration launched the Process Analytical Technology (PAT) initiative, with the goal of ensuring product safety and quality. Dream (2005) gave several examples of PAT. Alford (2011) discussed the application of automation for this purpose. Wiss (2015) showed how various aspects of PAT come together in pharmaceutical manufacturing. This concept extends beyond the reactor itself and includes feed preparation, subsequent processing, and equipment maintenance.

19.7.2 Auxiliary Equipment and Operations for Bio-Systems

Miley et al. (2015) discussed the general principles of hygiene that are so necessary for bio systems.

Keeping equipment contamination-free often requires sterilization. Baez & Assaf-Anid (2008) described novel and conventional approaches to this operation. Jenkins (2011) listed and described five approaches. Rodriguez (2013) argued for single-use equipment, given that some stainless steel units are being replaced by cheaper molded or flexible plastics.

Ingredients must be contamination-free. Baird & Williams (2005) presented eleven purification options for water.

Leak-free facilities at and away from the reactor include pumps, valves, and seals. Weeks & Bennett (2006) spoke about all the three items. Smith (2002) and Hiroyasu & Backes (2017) dealt with pumps. Perusek & Orlando (2005) explained how diaphragm valves are leak proof. Benz & Teng (2008) presented a selection platform for agitator seals. Birkmire (2007) focused on the reactor itself and the use of bolted head closures.

The product from a bioreactor is never pure enough in form to be saleable or usable. Following the reactor is a bio-separation train, which isolates the desired product. Methods include filtration, centrifugation, extraction, reverse osmosis, chromatography, and drying. If the product is a living entity, then the intensity of a separation step must respect the allowable range of pH, temperature, and other conditions that affect the product or any recyclable components. The separation sequence should follow the usual rules for chemical separation, as stated by Harrison (2014):

- remove the most plentiful impurities first,
- remove the easiest-to-remove impurities first,
- make the most difficult and expensive separations last,
- select processes that make use of the greatest differences between the properties of the product and those of its impurities, and
- select sequence processes that exploit different separation-driving forces.

Isolation and purification of the product is often the most difficult challenge in developing a complete bio-production system. Harrison, in his paper, focuses on two separation techniques – filtration and chromatography.

References

Ahn E. S., Park S. I., Song I. C., Kim G. B., Chang G. S., Develop reactor internals for optimizing reactor design: high-performance vapor-liquid distributor, *Hyd. Process.* 2017 96 Issue 7, p43–48 Jul.

Ahn, E.S., Park, S.I., Song, I.C. et al. (2019 Apr, 2019). Develop reactor internals for optimizing reactor design – part 2: high-performance mixing chamber. *Hyd. Process.* 98 (4): 49–52.

Alford, J.S. (2011 Oct). Using automation to produce quality pharmaceuticals. *Chem. Eng. Prog.* 107 (10): 19–26.

Baez, H.A. and Assaf-Anid, N.M. (2008 Aug). Novel and conventional approaches to sterilization. *Chem. Eng.* 115 (8): 42–45.

Baird, A. and Williams, R. (2005 May). Making high-purity water. *Chem. Eng.* 112 (5): 36–43.

Bakker, A., Haidari, A.H., and Marshall, E.M. (2001 Dec). Design reactors via CFD. *Chem. Eng. Prog.* 97 (12): 30–39.

Bartholomew, C.H. and Farrauto, R. (2006). *Fundamentals of Industrial Catalytic Processes*, 2nde. New Jersey: Wiley.

Bartholomew, C.H. and Hecker, W.C. (1994 Jun). Catalytic reactor design. *Chem. Eng.* 101: 70–75.

Bartow, M. and Spark, E. (2004 Jan). Bioreactor design for mammalian cell cultures. *Chem. Eng.* 111 (1): 49–54.

Benz, G.T. (2008 Feb). Piloting bioreactors for agitator scale-up. *Chem. Eng. Prog.* 104 (2): 32–34.

Benz, G.T. (2011 Aug). Bioreactor design for chemical engineers. *Chem. Eng. Prog.* 107 (8): 21–26.

Benz, G.T. (2016 Mar). Optimize aerobic fermenter operation. *Chem. Eng. Prog.* 113 (3): 60–62.

Benz, G.T. (2019 Apr). Optimize aspect ratio in industrial fermenters. *Chem. Eng. Prog.* 115 (4): 26–30.

Benz, G.T. and Teng, W.H.L. (2008 Jan). Selecting agitator seals for bioprocessing applications. *Chem. Eng. Prog.* 104 (1): 44–52.

Bhatia, S.K. (2013 Jul). Biology as a basis for biochemical engineering. *Chem. Eng. Prog.* 109 (7): 40–44.

Bhatia, S.K. (2014 Jan). Fermentation fundamentals: brewing bugs for bioengineering. *Chem. Eng. Prog.* 110 (1): 38–42.

Birkmire, J.C. (2007 Apr). Selecting and using bolted head closures. *Chem. Eng. Prog.* 103 (4): 47–52.

Conesa, J.A. (2019 July). *Chemical Reactor Design: Mathematical Modeling and Applications*. Hoboken, NJ: Wiley.

Cusack, R.W. (1999A Oct). A fresh look at reaction engineering. *Chem. Eng.* 106: 134–146.

Cusack, R.W. (1999B Dec). Reaction engineering - part 2: choosing the right reactor. *Chem. Eng.* 106: 80–85.

Cusack, R.W. (2000 Feb). Reaction engineering - part 3: optimize design and operation. *Chem. Eng.* 107: 88–95.

Davison, B.H. and Lievense, J.C. (2016 Jun). Technology challenges and opportunities. *Chem. Eng. Prog.* 112 (6): 35–41.

Demain, A.L. and Solomon, N.A. (1981 Sep). Industrial microbiology. *Sci. Am.* 245 (3): 66–75.

Doble, M. (2006 Aug). Avoid the pitfalls of bioprocess development. *Chem. Eng. Prog.* 102 (8): 34–41.

Dream, B. (2005 Jan). Process analytical technology. *Chem. Eng. Prog.* 101 (1): 19.

Dutta, S. and Gualy, R. (2000 Oct). Build robust reactor models. *Chem. Eng. Prog.* 96: 37–51.

Fair, J.R. (1967 Jul). Designing gas-sparged reactors. *Chem. Eng.* 3 (74): 67–74.

Fontes, E. (2015 Feb). Modeling chemical reactors. *Chem. Eng. Prog.* 111 (2): 46–50.

Fontes, E., Byrne, P., and Hernell, O. (2003 Mar). Put more punch into catalytic reactors. *Chem. Eng. Prog.* 99 (3): 48–53.

Gregory, J. and Green, B. (2013 Nov). Heat transfer for huge-scale fermentation. *Chem. Eng.* 120 (11): 44–46.

Gregory, J., Green, B., Courtemanche, N., and Kehn, R. (2014 Mar). Mass transfer in fermentation scaleup. *Chem. Eng.* 121 (3): 44–47.

Harrison, R.G. (2014 Oct). Bioseparation basics. *Chem. Eng. Prog.* 110 (10): 36–42.

Heinen, R. (2012 Jul). Catalyst developments: the last 90 years. *Hyd. Proc.* 91 (7): 39–45.

Hiroyasu, G. and Backes, S.R. (2017 Sep). Addressing pumping issues in biopharmaceutical operations. *Chem. Eng.* 124 (9): 71–75.

Jenkins, S. (2010 Jul). Bio-based chemicals get real. *Chem. Eng.* 117 (7): 17–21.

Jenkins, S. (2011 Sep). Sterilization approaches. *Chem. Eng.* 118 (9): 32.

Jenkins, S. (2015 Apr). Industrial microrganisms. *Chem. Eng.* 122 (4).

Jenkins, S. (2015 May). Catalysis fundamentals. *Chem. Eng.* 122 (5): 47.

Kwasny, R. (2008 Apr). A checklist for safer chemical batch reactions. *Chem. Eng.* 115 (4): 61–62.

Lapinskas, E. (2010 Nov). Scaling up research to commercial manufacturing. *Chem. Eng. Prog.* 106 (11): 44–55.

Le, H., Chen, C., and Goudar, C.T. (2015 Dec). Continuous process in upstream operations. *Chem. Eng. Prog.* 111 (12): 32–37.

Le, T.S., McCann, M., Azarin, S.M., and Hu, W.-S. (2016 Apr). An introduction to mammalian cell culture. *Chem. Eng. Prog.* 112 (4): 34–40.

Levenspiel, O. (1999). *Chemical Reaction Engineering*, 3rde. New York: Wiley.

Loffler, D.G. (2001 Jul). Avoid pitfalls in evaluating catalyst performance. *Chem. Eng. Prog.* 97: 74–77.

Loffler, D.G. (2002 Dec). Which catalyst. *Chem. Eng. Prog.* 98 (12): 47.

Maldonado, A.G. and Rothenberg, G. (2009 Jun). Predictive modeling in catalyst – from dream to reality. *Chem. Eng. Prog.* 105 (6): 26–32.

Marchall, R. (2006 Mar). Electrochemistry. *Chem. Eng.* 113 (3): 57.

Marshall, R. (2007 Apr). Preventing runaway reactions. *Chem. Eng.* 114 (4): 65.

McConville, F.X. (2007 Oct). The do's and don'ts of scale up. *Chem. Eng. Prog.* 103 (10): 18–19.

Miley, B., Riley, J., and Zelmanovich, Y. (2015 Nov). Hygienic design principles. *Chem. Eng.* 122 (11): 59–65.

Milne, D., Glasser, D., Hildebrandt, D., and Hausberger, B. (2006A Mar). Graphically assess a reactor's characteristics. *Chem. Eng. Prog.* 102 (3): 46–51.

Milne, D., Glasser, D., Hildebrandt, D., and Hausberger, B. (2006B Apr). Reactor selection: plug flow or continuously stirred tank. *Chem. Eng. Prog.* 102 (4): 34–37.

Monk, J. (2010 Nov). Stem cell engineering 101. *Chem. Eng. Prog.* 106 (11): 35–36.

Nehlsen, J., Muckherjee, M., and Porcelli, R.V. (2007 Feb). Apply an integrated approach to catalytic process design. *Chem. Eng. Prog.* 103 (2): 31–38.

Oh, B.-R., Lee, S.-M., Heo, S.-Y. et al. (2018). Efficient production of 1,3-propanediol from crude glycerol by repeated fed-batch fermentation strategy of a lactate and 2,3-butanediol deficient mutant of *Klebsiella pneumoniae*. *Microb. Cell Factories* 17: 1–9.

Perusek, R. and Orlando, M.D. (2005 Jan). Making the cast for diaphragm valves. *Chem. Eng.* 112 (1): 25–30.

Przybycien, T. and Titchener-Hooker, N. (2015 Dec). Continuous process in downstream operations. *Chem. Eng. Prog.* 111 (12): 38–34.

Ranade, V.V. (2002 Aug). At the heart of the process – reactor engineering. *Chem. Eng.* 109 (8): 84–85.

Rodriguez, R. (2013 Nov). Consider single-use equipment for biomanufacturing. *Chem. Eng. Prog.* 109 (11): 54–57.

San Martin, M.F., Barbosa Canovas, G.V., and Swanson, B.G. (2003 Mar). Innovations in food processing. *Chem. Eng. Prog.* 99 (3): 54–60.

Schuster, D. (2016 Jun). Is there a future for biotechnology in the commercial production of chemicals. *Chem. Eng. Prog.* 112 (6): 30.

Shinskey, G. (2004 Mar). Exothermic reactors: the stable, the unstable, the uncontrollable. *Chem. Eng.* 111 (3): 54–59.

Smith, B. (2002 Apr). What makes a pump for high-purity fluids. *Chem. Eng.* 109 (4): 87–89.

Streiff, F.A. and Rogers, J.A. (1994 Jun). Don't overlook static mixer reactors. *Chem. Eng.* 101: 76–82.

Torzewski, K. (2008 Nov). Biodiesel production. *Chem. Eng.* 115 (12): 39.

Turaga, U.T., Engelbert, D.R., and Osbourne, J.T. (2006 Jun). Succeed at catalyst scale-up. *Chem. Eng. Prog.* 102 (6): 29–33.

Van Hoek, P., Aristidou, A., Hahn, J.J., and Patist, A. (2003 Jan). Fermentation goes large-scale. *Chem. Eng. Prog.* 99 (1): 37S.

Venugopal, B. (2002 Jun). Avoiding runaway reactions. *Chem. Eng.* 109 (6): 54–58.

Weeks, D.T. and Bennett, T.M. (2006 Aug). Specifying equipment for high-purity fluid flow. *Chem. Eng.* 113 (8): 27–30.

Williams, J.A. (2002 Mar). Keys to bioreactor selections. *Chem. Eng. Prog.* 98 (3): 34–41.

Wiss, J. (2015). Process analytical technology: tools and applications in pharmaceutical manufacturing. *Chim. Oggi.* 33: 30–33.

Worstell, J.H. (2000 Jun). Succeed at reactor scale-up. *Chem. Eng. Prog.* 96: 55–60.

Worstell, J.H. (2001 Mar). Don't act like a novice about reaction engineering'. *Chem. Eng. Prog.* 97: 68–72.

Worstell, J.H. (2007 Mar). Scaling fixed-bed catalyzed processes. *Chem. Eng. Prog.* 103 (3): 31–35.

Worstell, J.H. and Worstell, J.H. (2004 Jan). Improve fixed-bed reactor performance without capital expenditure. *Chem. Eng. Prog.* 100 (1): 51–57.

Worstell, J.H., Doll, M.J., and Worstell, J.H. (2000 Sep). What's causing your catalyst to decay. *Chem. Eng. Prog.* 96 (9): 59–64.

20

Process Intensification and Integration

To meet its goals, the technology of chemical engineering comprises a wide variety of generic processes and equipment, many of which have been described in Chapters 11–19. However, Chemical Engineering also has a dynamism whereby its slate of processes and equipment is periodically being expanded, being modified, being integrated, and/or simply being rearranged. The incentives for these developments are a selection of the following:

- increase processing capacity,
- improve quality and/or uniformity of product,
- increase yield from raw materials,
- reduce energy requirement,
- reduce emissions, and
- reduce safety hazards.

Suppliers of process equipment regularly make changes that satisfy some of these objectives, but these changes are usually incremental, as are likewise the changes that are made from time to time by the operators of such processes. On the other hand, when a large step change with significant benefits is made, often with innovative thinking "outside the box", then this change qualifies as *Process Intensification*. The term incorporates the making of such technological upgrades anywhere across the entire field of processes. One major type of intensification is *Process Integration*, and that large class of *PI* is specially featured in this chapter.

Process intensification has been going on for as long as chemical-engineering types of processes have been around, but the idea of defining it as a subset of technology goes back to around the 1990s. Some early references are Green et al. (1999), Hallale (2001), Hendershot (2000), Stankiewicz and Moulijn (2000), Tsouris & Porcelli (2003), Van Gerven & Stankiewicz (2009), and Stankiewicz (2020). Kiss & Bildea (2015) presented a good list of the variety and benefits of *PI*. Bielenberg & Bryner (2018) reviewed the current situation. Two recent books on the subject are Reay et al. (2013) and Stankiewicz et al. (2019).

Any chemical-engineering process is a candidate for intensification. Several for which such work has been done or proposed and which are instructive for general knowledge have been sought out, grouped by similarity, and described in the next five sections (20.1 to 20.5).

20.1 Mixing and Comminution

Comminution is the physical dividing into smaller parts of a volume of solid or liquid. Mixing is a broader term: It may act by causing comminution, but it may also act through other mechanisms.

Practical Process Design for Chemical Engineers, First Edition. Keith Marchildon and David Mody.

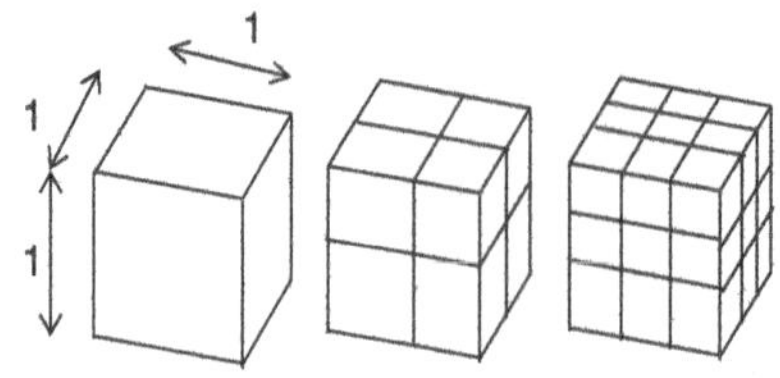

Volumetric subdivision:	1	8	27
Surface area:	6	12	18

Figure 20.1 Subdivison of volumes.

The arithmetic of subdivision is illustrated by the model of a cube of dimensions $1 \times 1 \times 1$ cubic units. Its surface area is $6 \times 1 \times 1$ or 6 surface units. If the cube is divided into eight parts by bisecting it in the three dimensions, then the surface area becomes $8 \times 6 \times (1/2 \times 1/2)$ or 12 units.

If it is trisected into a Rubik's Cube, the surface becomes $27 \times 6 \times (1/3 \times 1/3)$ or18 units (Figure 20.1).

The surface area is key for a simple mass transfer operation like melting or freezing. From Eq. (3.1),

$$\text{Rate} = \text{Rate coefficient} \times \text{zone of action} \times \text{driving force}$$

In this case, the "zone of action" is the particle surface. Hence, for change of phase and for any interphase transfer where there is negligible other resistance in series with the surface resistance, the rate of the process doubles when the solid is divided weight-wise by a factor of 8, and it triples when the division is 27. However, if significant resistance to transfer also resides within the zone in question or within the surrounding milieu, then the benefits of comminution are not as great: see the example in Figure 3.2.

Comminution, which is brought about by mechanical means such as milling, requires energy typically proportional to the new surface area produced. There are several theories to predict the energy as a function of subdivision and of hardness and friability of the material. A comparative review of these methods is found in the paper by Hukki (1961).

Polymers are often cast into pellet form and may require further modification such as control of moisture or of solvent content, for further polymerization. Smaller pellets are generally an advantage and are achieved by adjusting the extrusion and cutting of the molten polymer. Potential for process intensification is moderate because of transfer resistance within the pellets.

High-speed mixing and agitation of liquid–liquid or liquid–vapor mixtures can, in many cases, produce major increases in process capacity. Chapter 16 presented the procedure for agitator selection and speed for three mixing situations:

- blending of liquids,
- suspension of solid particles, and
- dispersion of gas phase in liquid.

These calculations produce the desired *momentum* behavior. With that aspect satisfied, it may be possible, by enhancing agitator design and speed, to increase *heat* and *mass* transfer, which are beneficial in themselves and which may have a direct effect on the rate of any reaction.

References on the effect of mixing on reaction rate and selectivity are Czerlinski et al. (2003) and Shah et al. (2012B).

Chapter 16 also describes in-line static mixing whereby a flow of disparate liquids is continually broken into mixed separate streams (i.e. comminuted), rotated, and then re-combined. The result of a cycle is to disperse the liquids more finely into each other. A succession of these cycles makes the dispersion finer and finer. Reaction between components may be speeded up, the benefit depending on how fast the chemical rate compares with the rate of physical mixing. One application would be to replace a continuously stirred reactor with a pipeline fitted internally with mixers. This would replace a back-mixed unit with a close-to-plug-flow unit. Heating of the pipe would be simple, especially if it were shaped into a coil and contained inside a shell (Myers et al. 2018).

20.2 Enhanced Energy

The vast majority of chemical engineering processes operate at a temperature other than, and usually higher than, ambient. Heat must be provided to raise the temperature, as well as to support evaporative phase change. In a small fraction of cases, the heat is provided electrically or sometimes the heat of reaction is sufficient. However, in the large majority, the source is conduction and convection from a hot heat-transfer fluid (liquid or vapor). This is the "heavy duty" mode of providing energy, and it will likely continue to be so. The burning of the fuel to heat the heat-transfer fluid imposes a contribution to atmospheric greenhouse gases, which can sometimes be reduced by judicious design of a heat transfer network (HEN). For instance, the hot effluent from a process step can cool itself down by transferring its heat to an incoming stream. Provision of effective HENs anywhere in a process qualifies as *Process Intensification*.

In this section, we look at some "niche" sources of energy. In a specific geometrical situation, they may add to or replace the usual mode of heat supply, but this does not generally qualify as an intensification. Of greater interest is that these sources may affect the process in ways that direct thermal heating cannot. Three of these sources are examined here: ultrasound, microwaves, and ultraviolet radiation. They are reviewed in Stankiewicz (2007).

20.2.1 Ultrasound

Sound (and ultrasound) is a mechanical phenomenon, as opposed to electromagnetic phenomena such as microwaves and ultraviolet waves. Ultrasound applications in chemistry comprise waves of typical frequency 20–50 kHz, although much higher frequencies have been reported. The lower limit sets it just above human hearing. The actual frequency is generally not important. The corresponding wavelengths are set by dividing the speed of sound (in water it is $1481\ \mathrm{m\,s^{-1}}$) by the frequency: at 20 kHz 0.74 m and at 50 kHz 0.30 m. The technique is applicable to liquids and to liquid–solid mixtures. A pure liquid is susceptible only to ultrasounds of impractical power, but any admixture of liquid or solid reduces required power greatly.

Sound waves consist of a sequence of high and low pressure, generated by the pulsing of an electrically stimulated crystal – the piezoelectrical effect. With a high enough amplitude, the mechanism is for the low-pressure swings of the sound cycle to generate mini cavities in the liquid, which then collapse and produce large localized thermal energy input and agitation into the mixture. Several cycles may be needed to build a cavity before it collapses. The benefits can be

- bringing reactants into more intimate contact, emulsification,
- breaking up agglomerates,
- speeding up phase change, e.g., dissolution of solids, mass transfer of a component,

- continually renewing the surface of a catalyst or solid reactant,
- momentarily creating free radicals, and
- enhancing heat transfer.

Some experiences with ultrasound are reported:

- by Wierschem et al. (2017A), a 50% increase in rate for one form of ethyl butyrate production,
- by Sarnaik & Gogate (2019), a 23% increase in the rate of sunflower hydrogenation,
- by Suslick (1989), a reduction in required temperature for cracking of crude oil, from 500 °C to room temperature, and
- by Gogate et al. (2002), beneficial use of ultrasound in combination with other agents or with electromagnetic waves.

20.2.2 Microwaves

A microwave is an electromagnetic wave (generated by a magnetron) in the frequency range 3–30 MHz. It interacts with three types of material:

(1) transparent, through which it simply passes unchanged. Nonmetals and nonpolar substances like many polymers are in this class,
(2) reflective, mainly metals, of which the mobile electrons at the surface cause the waves to bounce off in different directions, and
(3) absorbent – polar substances such as water.

The third type is most interesting, because microwaves excite molecular motion, causing the creation of thermal energy and other effects on mass transfer and chemical reaction.

As in domestic use, microwaves are often used simply for thermal heating. A particular geometry may lend itself to this type of heating rather than to hot-fluid or direct electrical heating. Microwaves are extensively used in the drying of food, where they are competitive with and/or have advantages over other methods of drying. Wang et al. (2015) have studied the energy efficiency of microwave heating and have concluded that it never exceeds 80%. Other estimates are closer to 50%, which could be compared with other systems. It is not low enough to keep this type of heating from significant use.

The energy of this relatively low-frequency radiation is not high enough to break most chemical bonds, but it can augment other methods and produce some of the following benefits:

- reaction rate acceleration,
- milder reaction conditions,
- higher chemical yield,
- lower energy usage, and
- different reaction selectivities.

The components in a mixture of immiscible liquids may have different susceptibilities to microwaves and will heat up differently. The difference between temperatures can accelerate any extraction of a compound from one fluid to the other.

A solid heterogeneous metal catalysis can attain higher than ambient temperature and become more effective.

Some references are

- Wang et al. (2015) factors affecting energy efficiency of microwave heating,
- Kappe et al. (2012) other-than-thermal effects in microwave reaction chemistry, and
- Diaz-Ortiz et al. (2019) computer study of microwaves on reaction chemistry.

20.2.3 Ultraviolet Radiation

It was seen that the primary effect and usage of microwaves are in their ability to convert into thermal energy. With regard to reaction chemistry, microwaves act by adding heat; there may be other, more direct, effects, but they are not well recognized and play little role in technology.

The facts are different for ultraviolet waves, as the calculation below shows.

Figure 20.2 shows part of the electromagnetic spectrum, including the region of the microwaves that have just been discussed. The frequency and the wavelengths are on the two scales and they are related to each other by the relation

$$\text{Frequency} \times \text{wavelength} = \text{speed of light, approx. } 3 \times 10^8 \text{ m s}^{-1} \tag{20.1}$$

This relation may be checked by comparing any pair of corresponding numbers in Figure 20.2.

An electromagnetic wave has an energy which is called the *photon*, which is pure energy and has no rest mass. The energy of a photon is given by

$$\begin{aligned} \text{Energy (J)} &= \text{Planck's constant (J s)} \times \text{speed of light (m s}^{-1})/\text{wavelength (m)} \\ &= 6.626 \times 10^{-34} \times 3 \times 10^8/\text{wavelength} \\ &= 2.2 \times 10^{-26}/\text{wavelength} \end{aligned} \tag{20.2}$$

The low-frequency boundary of the ultraviolet zone has a wavelength of 4×10^{-7} m, therefore from Eq. (20.2) its photons have energy 0.55×10^{-19} J. The high frequency boundary at the top of the zone has a wavelength of 1×10^{-8}, giving the photons an energy of 22×10^{-19} J.

A bond in a chemical compound has an energy associated with it, namely the energy that would be required to pull the bond apart. A typical energy is for the carbon–carbon bond, 348 kJ mol^{-1}. The energy per bond itself is obtained by dividing by Avogadro's number, $6.023 \times 10^{+23}$, giving 57×10^{-23} kJ or 5.7×10^{-19} J.

The energies may be compared. 5.7 lies between the lower and upper bounds of 0.55 and 22. It is evident that a low-frequency UV wave (and any waves with still lower frequencies, like

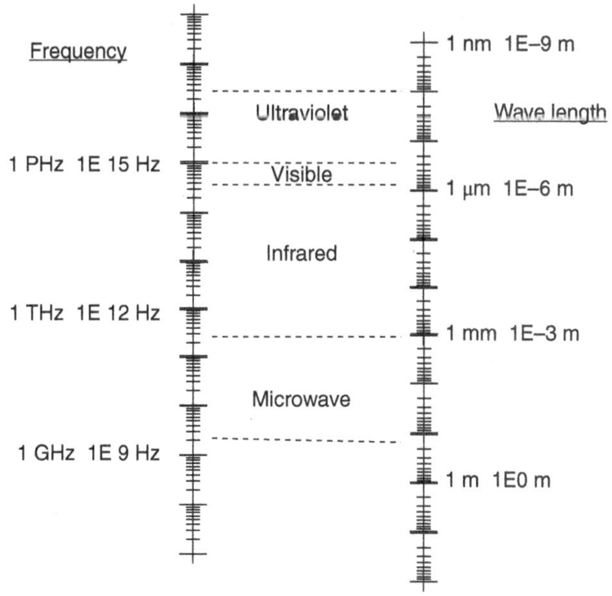

Figure 20.2 Electromagnetic waves.

visual light and microwaves) is unable to disrupt the carbon–carbon connection. However, at the high-frequency end, it is a possibility, and this action is what constitutes the field of *Photochemistry*.

Not all chemical bonds are as strong as the C—C bond, so waves with somewhat lesser energy may interact with them. There is a wide variety of reactions that can be photocatalyzed, including

- addition to an unsaturated molecule,
- addition of unsaturated molecules to each other,
- splitting off of molecular fragments,
- oxidation,
- hydration,
- *cis*-trans isomerization,
- rearrangement, and
- intermolecular electron transfer.

There are some reactions that cannot be done by thermal stimulation alone.

A review of the field is given by Glusac (2016).

In combination with transition-metal catalysis, the range of photocatalysis can extend into the visible range – see Schultz & Yoon (2014).

Photocatalysis is used in the purification of water – see Laroussi et al. (2002).

Bremus-Kobberling et al. (2012) used a laser as a source of light and compared the speed of a reaction with catalysis by a conventional xenon lamp. Time was reduced from hours to minutes, and better selectivity was obtained.

A difficulty with photochemistry is conveying the energy-containing wave to its target, which prompts research into suitable forms of reactors.

20.3 Solvent Development

Solvents are a ubiquitous liquid component in chemical processing. In chemical reactions, they may act by bringing the reactive components into a single phase, where contact is at the molecular level. In the process of extraction, they provide a welcoming phase into which the extracted material is transferred.

The least expensive and safest solvent is water. Being polar itself, it is effective with polar solutes. Unfortunately, it does not dissolve a large majority of organic compounds. Other common solvents such as methanol, benzene and toluene are generally used.

The traditional solvents, including water, are volatile enough at process temperatures that pressure is required. Subsequent to the process step, the solvent must be separated from its solute, generally requiring an energy-consuming operation and a concomitant loss of a small fraction to the environment.

A Process Intensification objective would be to eliminate some of the disadvantages of current solvents. Much of the work on this goal is centered around the concept of *Ionic Liquids*. We examine this topic, as well as related ones of *Deep Extraction Solvents* and *Switchable Solvents*.

20.3.1 Ionic Liquids

Ionic liquids are organic salts and, more specifically, organic salts with a melting point below 100 °C. Figure 20.3 shows examples of some popular cations and anions of such salts. There are a multiplicity of cations and anions: these variations include the many functional groups that can constitute part of the cations and which affect the nature of the particular liquid.

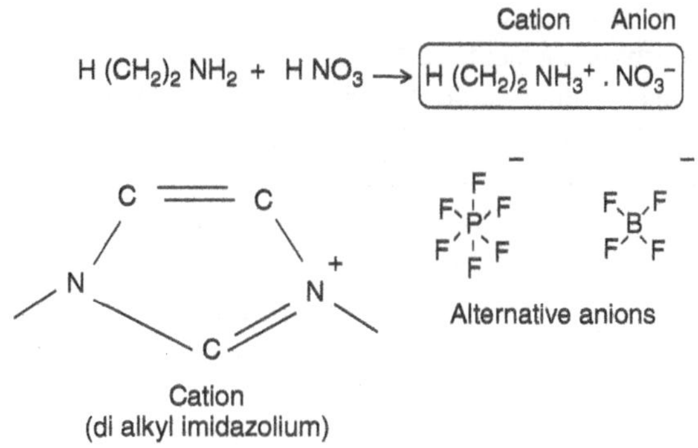

Figure 20.3 Typical cations and anions for ionic liquids.

Many of these materials turn out to be good solvents, for both polar and nonpolar substances. A major advantage is their very low volatility, thus eliminating the major problem of current solvents mentioned above. Brennecke & Maginn (2001) provide one of the earliest general discussions of ionic liquids in chemical processes. They list seven possible applications:

- replacing common solvents for reactions, with no loss in rates and selectivities,
- gas separations (including CO_2 capture),
- liquid separations,
- solvents for cleaning,
- electrolytic liquids in batteries and fuel cells,
- lubricants, and
- heat transfer fluids.

For a brief history of ionic liquids, consult Welton (2018). For a well-balanced consideration of the benefits of ionic liquids, see Jessop (2018) who cautioned that the complete life cycle of the solvent needs to be studied. Such a study includes the cost and environmental effects of manufacturing the ionic liquid, as well as the process, parallel to the main process, of separating, recovering, and reusing the solvent.

20.3.2 Deep Eutectic Solvents (DESs)

These solvents are often considered as part of the ionic liquid family, but they have some important differences and applications. They are formed by the interaction of a quaternary ammonium salt (see Choline chloride in Figure 20.4) and either a metal hydrate or a hydrogen-bond donor (see Urea in Figure 20.4). The ratio is widely adjustable.

The term "eutectic" is very apt, as shown by the dramatic lowering of the following melting points:

Choline chloride melting point	302 °C
Urea melting point	133 °C
1 : 2 mole ratio ChCl/urea melting point	12 °C

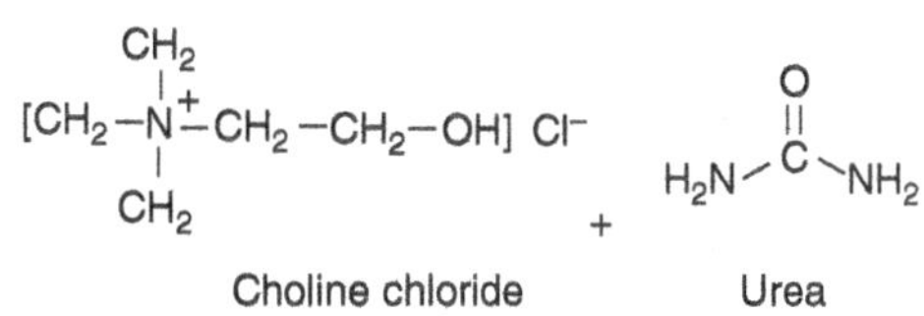

Figure 20.4 A deep eutectic solvent.

Some DESs can stay liquid down to still lesser temperatures. The whole concept of this type of solvent is the fact that they stay liquid at these temperatures.

Like fellow ionic liquids, the DESs have very low vapor pressures and are therefore non-flammable. They have the disadvantage of high viscosity. Relative to other ionic liquids, the DESs have the advantage of components of low cost and low toxicity: the popular choline chloride is often part of feeds for animals. As with ionic liquids in general, the DESs can comprise a large variety of components and, unlike other liquids, the ratio can vary because the components are not fixed into ionic bonds with each other.

Smith et al. (2014) provided authoritative discussions of DESs. With regard to applications, their paper focused mainly on metal finishing, but also spoke of several successful examples of use in chemical synthesis, in biotransformations, and in the purification and manufacturing of biodiesel.

20.3.3 Carbon Dioxide-Switchable Solvents

A disadvantage with a low-volatility ionic solvent is its inability – at the end of a process step – to be separated by vaporization from the process stream. It would be advantageous to effect some change in its nature to make it less-of-a solvent at this point. That is, to make it "switch". Various switch agencies are possible – temperature, pressure, addition of strong acid or base – but they are either ineffective or permanently change the solvent. The ideal reagent turns out to be carbon dioxide. For certain organic liquids, sparging with CO_2 converts the material into a polar form. Subsequently, if the CO_2 is removed (again by sparging, but with an inert gas), the liquid reverts to a nonpolar form. This technology of the switching of a single solvent benefits from the fact that carbon dioxide is cheap, nontoxic, noncorrosive, and abundant.

Many ionic-liquid solvents that are susceptible to this treatment have been identified over the years. Cunningham & Jessop (2019) tell us that most switchable solvents are neutral nitrogen-containing compounds, some of which are shown in Figure 20.5.

The switching is usually done in the presence of water or another proton-supplying species, as shown in Figure 20.6. For liquids that contain N—H bonds, the water may be dispensed with. Weak organic acids may also be switchable.

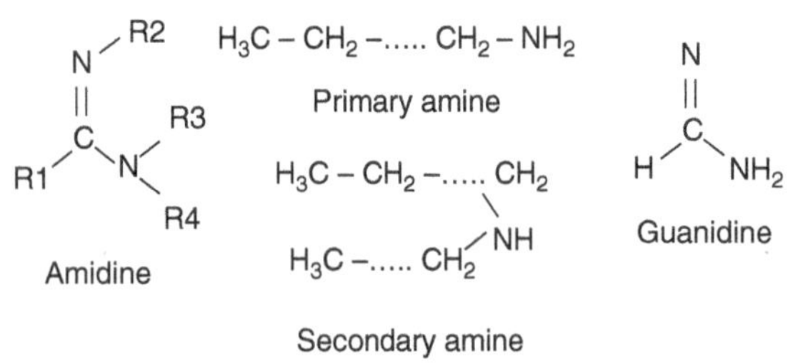

Figure 20.5 Some CO_2-switchable organic liquids.

$$B + CO_2 + H_2O \rightleftarrows [BH^+][HCO_3^-] \quad (1)$$

$$R, R\,NH + CO_2 \rightleftarrows [R, R\,NH_2^+][R_2\,NCO_2^-] \quad (2)$$

$$A^- + CO_2 + H_2O \rightleftarrows HA + HCO_3^- \quad (3)$$

Figure 20.6 Effect of carbon dioxide on ionization.

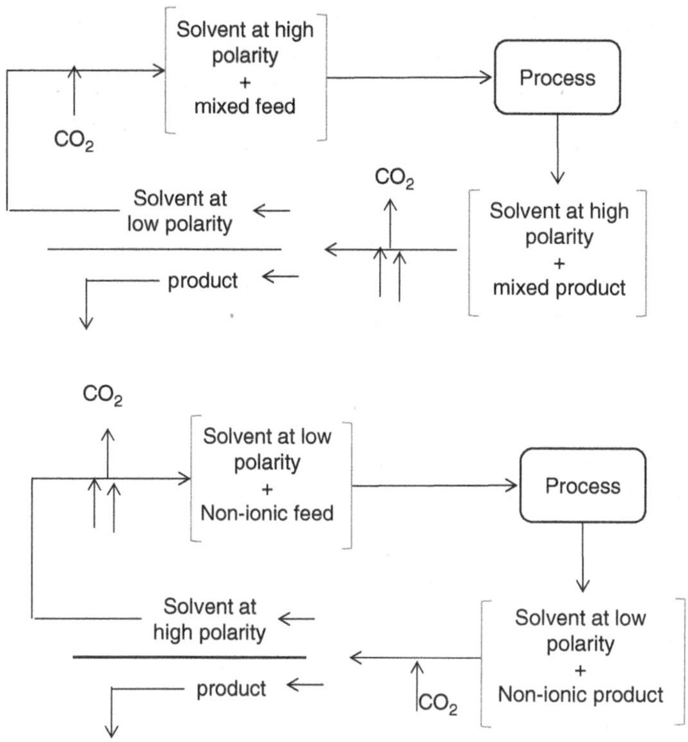

Figure 20.7 Two paths for switchable solvent.

The suitability of a candidate depends on whether the addition or deletion of CO_2 makes a significant change in polarity. Jessop (2013) identified three different categories of switchable solvents:

- switchable-polarity solvents (SPSs), which were described above,
- switchable-hydrophilicity solvents (SHSs), where the switch is between miscibility and immiscibility in water, and
- switchable water (SW), where the aqueous solution of a polyamine with low ionic strength has the ability to dissolve organic compounds, but with its ionic strength increased by addition of CO_2 it separates from the organics.

Figure 20.7 illustrates how a switchable solvent could work with a process, acting as a normal solvent in the reaction (or liquid–liquid extraction), then separating itself, and recycling back to the start of the process.

20.4 Novel Reactors

The essential qualities of a reactor are:

- uniformity of product, at least to the extent that is required in downstream usage or in further processing,
- management of temperature, whether heat must be added or must be removed, and
- mass transfer of components, to bring them into proximity to each other or to a catalyst for reaction.

For solid-catalyzed reactions, Moulijn et al. (2003) recommended the use of a structured design instead of random packing. Random packing often results in uneven fluid flow and inactivity at vessel walls. Instead, the catalyst should be shaped into three-dimensional ("monoliths") that promote good contact with reactants and which promote plug, radially uniform flow. The Sulzer Chemtech company market their Katapak-SP structure: elements for vertical flow that have alternative vertical layers of catalyst separated by a layer of mesh or packing that ensures good contact with the catalyst. These improvements are logical steps in the drive for process intensification.

For continuous flow, the alternate extremes are a reactor that is fully back-mixed and a reactor that operates in plug-flow. Since the latter alternative considerably economizes on reaction volume, there is always some consideration given to converting a back-mixed reactor into a plug-flow configuration. Consider the following case:

- a back-mixed reactor of hold-up time 10 minutes (M/F),
- a through-flow of 5000 kg h^{-1} or 83.3 kg min^{-1} (F),
- entering concentration of reactant: 1.0 units per kilogram ([AI]), and
- *in situ* and exit concentration of reactant: 0.05 ([A]).

The steady-state mass balance on A is

$$\mathrm{d}[A]/\mathrm{d(time)} = 0 = F\,[\mathrm{AI}] - F\,[A] - \mathrm{k}\,M\,[\mathrm{A}] = 83.3 \times 1.0 - 83.3 \times 0.05 - \mathrm{k}\,(10 \times 83.3) \times 0.05$$

from which the rate constant, k, is inferred as 1.9 min^{-1}.

If the same analysis is applied to a plug-flow version, using k as 1.9, the steady-state mass balance is

$$F\,[A] - F\,([A] + \Delta[A]) - \mathrm{k}\,\Delta M\,[A] = 0$$

$$\int \mathrm{d}[A]/[A]1.0 \text{ to } 0.05 = -\mathrm{k}\,M/F$$

from which M is calculated as 132 kg. This compares with 833 kg for the mixed reactor.

While this 84% reduction in hold-up mass is a great objective, the implementation is not as impressive. Considering the density of fluid to be 1000 kg m^{-3}, the mixed reactor occupies 0.833 m^3. Considering the plug flow reactor, choosing a 0.05 m internal diameter pipe (5 cm), a length of 68 m is required. The likeliest plug flow configuration is for 10 lengths of 7-meter-long pipe sections, each section doubling back on the previous one and the whole reactor encased in a heated shell – about 0.25 m in diameter. The whole assembly occupies about 0.5 m^3, just a 40% reduction from the mixed reactor.

However, the plug-flow system has other benefits:

- less residence time for undesirable side reactions to occur,
- no un-mixed stagnant zones,
- powered by a pump rather than a more complicated agitator, and
- possibly adding other components at different points along the flow.

For smaller scale operations, some other plug-flow configurations may be available. Consider the above case but at 50 kg h^{-1}, say, for a pilot-plant or for manufacture of a fine chemical. All conditions of hold-up mass, entering and leaving concentrations, kinetic constant being the same, a plug-flow reactor of 17 m of 1 cm internal diameter pipe would be suitable. Noble (2001) described a reactor similar to a plate-and-frame heat exchanger proposed by the *Heatric* heat-exchanger company. One-half of the pipe cross section is chemically milled onto the surface of plates, and then the plates are stacked and joined to provide a path for fluid to flow along them and from one plate to the next. Figure 20.8 shows the concept. The stack is encased in a heat-transfer jacket.

Figure 20.8 Milli-channel reactor element.

20.5 Combined Operations

Prologue

On July 7, 2021, the principal author of this text passed away. Keith Marchildon was a great family person, a respected engineer, and a role model for all. In Engineering and life, he was always one to help others be better. His success and experience in bridging research and development with engineering and the physical realities of a plant brought him the respect of everyone he touched. Few people can bridge the gap between theory, differential equations, programming, and the realities of a physical plant and have a humorous personality that people would always want to hear and learn from. I had the fortunate timing to be a newly graduated engineer he tutored, a colleague he taught professional development courses with and wrote chapters together, a university lecturer with whom together we helped students prepare for the world outside the university, the person who helped to publish his last contribution to the engineering world, and finally an engineer who thrived under his guidance. If I have taken one thing to heart that he told me, it was, "Never think you can't learn something from someone who knows less about a subject than you". Those lasting words from a man so knowledgeable rang true many times in my career dealing with engineers, designers, operators, business people, and students. I hope you will forgive that this book ends with his last written words, "Combined Operations," which seems appropriate given Keith's commitment to family and church, community service, his success in engineering, and what I call his "saving the butts of engineers everywhere." I think he will be happy you have read this book and learned something from someone who knows less about a subject than you.

– David Mody, 2024

References

Abufares, A.A. and Douglas, P.L. (1995). Mathematical modeling and simulation of an MTBE catalytic distillation process using SPEEDUP and ASPENPLUS. *Trans. I. Chem. Eng.* 73A: 3–12.

Agreda, V.H., Partin, L.R., and Heise, W.H. (1990). High-purity methyl acetate via reactive distillation. *Chem. Eng. Prog.* 86 (2): 40–46.

Bielenberg, J. and Bryner, M. (2018 Mar). Realize the potential of process intensification. *Chem. Eng. Prog.* 114 (3): 41–45.

Brearley A. M., Lang J. J., Marchildon E. K., Continuous polymerization process for polyamides, US Patent 5,674,974, 1997 October 7.

Bremus-Kobberling, E., Gillner, A., and Brase, S. (2012). Photochemistry with laser radiation in condensed phase using miniaturized photoreactors. *Beilstein J. Org. Chem.* 8: 1213–1218.

Brennecke, J.F. and Maginn, E.J. (2001 Nov). Ionic liquids: innovative fluids for chemical processing. *AICHE J.* 47 (11): 2384–2389.

Brown, R.C. (2019 Jun). Intensification of fast pyrolysis. *Chem. Eng. Prog.* 115 (6): 23.

Cicci, A., Sed, G., Jessop, P.G., and Bravi, M. (2018). Circular extraction: an innovative use of switchable solvents for the biomass biorefinery. *Green Chem.* 20: 3908–3911.

Cleary, W. and Doherty, M.F. (1985). Separation of closely boiling mixtures by reactive distillation. 2. Experiments. *Ind. Eng. Chem. Process. Des. Dev.* 24: 1071–1074.

Cunningham, M.F. and Jessop, P.G. (2019 Aug). Carbon dioxide-switchable polymers: where are the future opportunities. *Macromolecules* 52 (18): 6801–6816.

Czerlinski, G., Levin, R., and Ypma, T. (2003 Jul). The effect of the mixing process on reaction kinetics. *Int. J. Chem. Kinet.* 35 (10): 484–496.

Diaz-Ortiz, A., Peieto, P., and de la Hoz, A. (2019). Computational calculations can be used for the determination of parameters that have an important influence in MAOS. *Chem. Rec.* 19: 85–97.

Fan, L.-S., Zeng, L., and Luo, S. (2015 May). Chemical looping rises to the challenge. *Chem. Eng. Prog.* 111 (5): 3–38.

Franssen, I.S., Irimia, D., Stefanidis, G.D., and Stankiewicz, A.I. (2014 Oct). Practical challenges in the energy-based control of molecular transformations in chemical reactors. *AICHE J.* 60 (10): 3392–3405.

Garcia, G.C. and Kiss, A.A. (2017 Jan). A review on process intensification in HiGee distillation. *J. Chem. Technol. Biotechnol.* 92 (6): 1136–1156.

Gautam, P., Singh, P.P., Mishra, A. et al. (2013). Simulation of reactive distillation column. *Int. J. Chem. Tech. Res.* 5 (2): 1024–1029.

Glusac, K. (2016). What has light ever done for chemistry? *Nat. Chem.* 8 (7): 734–735.

Gogate, P.R., Mujumdar, S., and Pandit, A.B. (2002). A sonophotochemical reactor for the removal of formic acid from wastewater. *Ind. Eng. Chem. Res.* 41 (14): 3370–3378.

Green, A., Johnson, B., and John, A. (1999). Process intensification magnifies profits. *Chem. Eng.* 106 (13): 66–73.

Hallale, N. (2001 Jul). Burning bright trends in process integration. *Chem. Eng. Prog.* 97 (7): 30–41.
Hanik, P. (2018 Dec). Functional contradictions: insights into process improvement. *Chem. Eng. Prog.* 114 (12): 30–36.
Hendershot, D.C. (2000 Jan). Process minimization: making plants safer. *Chem. Eng. Prog.* 96 (1): 35–40.
Hendershot, R.J., Lebrecht, T.D., and Easterbrook, N.C. (2010 Jul). Use oxygen to improve combustion and oxidation. *Chem. Eng. Prog.* 106 (7): 57–61.
Hukki, R.T. (1961). Proposal for a solomonic settlement between the theories of von Rittinger, Kick and Bond. *Trans. AIME* 220: 403–408.
Huss, R.S., Chen, F., Malone, M.F., and Doherty, M.F. (2003). Reactive distillation for methyl acetate production. *Comput. Chem. Eng.* 27 (12): 1855–1866. https://doi.org/10.1016/S0098-1354(03)00156-X.
Isla, M.A. and Irazoqui, H.A. (1996). Modeling, analysis, and simulation of a methyl tert-butyl ether reactive distillation column. *IEC Res.* 35 (8): 2696–2708.
Jessop P. G., Switchable solvents, 10th Green Chemistry Conference 2013, Barcelona Spain.
Jessop, P.G. (2018). Fundamental properties and practical applications of ionic liquids: concluding remarks. *Faraday Discuss.* 206: 587–601.
Johnson, K.E. (2007 Spring). What's an ionic liquid. *Electrochem. Soc. Interface* 38–41.
Jonsson, M. and Johnson, B. (2009 Dec). Millichannel reactors: a practical middle ground for production. *Chem. Eng.* 116 (13): 44–50.
Kappe, C.O., Pieber, B., and Dallinger, D. (2012 Dec). Microwave effects in organic synthesis: myth or reality. *Angew. Chem. Int. Ed.* 52 (4): 1088–1094.
Kenig, E. and Seferlis, P. (2009 Jan). Modeling reactive absorption. *Chem. Eng. Prog.* 105 (1): 65–73.
Kiss, A.A. and Bildea, C.S. (2015 Dec). Revive your columns with cyclic distillation. *Chem. Eng. Prog.* 111 (12): 21–27.
Laroussi, M., Dobbs, F.C., Wei, Z. et al. (2002 Aug). Decontamination of water by excimer UV radiation. *IEEE Trans. Plasma Sci.* 30 (4): 1501–1503.
Moulijn, J.A., Stankewicz, A., and Kapteijn, F. (2003). The potential of structured reactors in process intensification. *Chem. Sustain. Dev.* 11: 3–9.
Mullick, S. and Dhole, V. (2007 Dec). Consider integrated plant design and engineering. *Hyd. Proc.* 81–85.
Myers, K.J., Janz, E.E., Cathie, N., and Jones, M.A. (2018 Mar). Employ static mixers for process intensification. *Chem. Eng. Prog.* 114 (3): 55–63.
Noble, T. (2001 Jul). Microreactors – smaller can be better. *Chem. Eng. Prog.* 97 (7): 10–11.
Pulati, N., Lupinsky, A., Miller, B., and Painter, P. (2015). Extraction of bitumen from oil sands using deep eutectic ionic liquid analogues. *Energy Fuel* 29 (8): 4927–4935.
Reay, D., Ramshaw, C., and Harvey, A. (2013 June). *Process Intensification*, Seconde. Oxford: Butterworth-Heinemann.
Rossiter, A.P. (2004 Jan). Succeed at process integration. *Chem. Eng. Prog.* 100 (1): 58–62.
Sarnaik, K.D. and Gogate, P.R. (2019). Intensified catalytic transfer hydrogenation of sunflower oil using ultrasound. *Ind. Chem. Eng. Res.* 58: 22942–22950.
Schaefer, G.W. (2010 Spring). Switchable solvents: a combination of reaction and separations. *The Tower, Georgia Institute of Technology* 2: 71–77.
Schultz, D.M. and Yoon, T.P. (2014 Feb). Solar synthesis: prospects in visible light photocatalyis. *Science* 343 (6174): 1239176.
Schuur, B., Nijland, M., Blahusiak, M., and Juan, A. (2018). CO_2 – switchable solvents as entrainer in fluid separations. *ACS Sustain. Chem. Eng.* 6 (8): 10429–10435.

Shah, M., Kiss, A.A., Zondervan, E., and de Haan, A.B. (2012A). Pilot – scale experimental validation of unsaturated polyester synthesis by reactive distillation. *Chem. Eng. J.* 213: 175–185.

Shah, S.I.A., Kostiuk, L.W., and Kresta, S.M. (2012B Apr). The effects of mixing, reaction rates, and stoichiometry on yield for mixing sensitive reactions – part 1: model development. *Int. J. Chem. Eng.* 2012: 750162.

Smith, E.L., Abbott, A.P., and Ryder, K.S. (2014). Deep eutectic solvents (DESs) and their applications. *Chem. Rev.* 114 (21): 11060–11082.

Stankiewicz, A. (2007). On the application of alternative energy forms and transfer mechanisms in microprocessing systems. *Ind. Eng. Chem. Res.* 46 (12): 4232–4235.

Stankiewicz, A.I. and Moulijn, J.A. (2000 Jan). Process intensification: transforming chemical engineering. *Chem. Eng. Prog.* 96 (1): 22–34.

Stankiewicz, A., Van Gerven, T., and Stefanidis, G. (2019 August). *The Fundamentals of Process Intensification*. Hoboken, NJ: Wiley.

Stankiewicz A. *Chem. Eng. Prog.*, The Principles and Domains of Process Intensification Vol. 116, Iss. 3, (Mar 2020): 23-28.

Suslick, K.S. (1989 Feb). The Chemical Effects of Ultrasound. *Sci. Am.* 260: 80–86.

Terrill, D.L., Sylvestre, L.F., and Doherty, M.F. (1985). Separation of closely boiling mixtures by reactive distillation. 1. Theory. *Ind. Eng. Chem. Process. Des. Dev.* 24: 1062–1071.

Thompson, L.H. and Doraiswamy, L.K. (1999). Sonochemistry: science and engineering. *Ind. Eng. Chem. Res.* 38 (4): 1215–1249.

Tsouris, C. and Porcelli, J.V. (2003 October). Process intensification – has its time finally come? *Chem. Eng. Prog.* 50–55.

US Patent 4,435,595 to EASTMAN Chemical Company, 1984.

Van Gerven, T. and Stankiewicz, A. (2009). Structure, energy, synergy, time – the fundamentals of process intensification. *Ind. Eng. Chem. Res.* 48 (5): 2465–2474.

Vanderveen, J.R., Patiny, L., Chalifoux, C.B. et al. (2015). A virtual screening approach to identifying the greenest compound for a task: application to switch-hydrophilicity solvents. *Green Chem.* 17 (12): 5182–5188.

Wang, W., Zhao, C., Sun, J. et al. (2015 July). Quantitive measurement of energy utilization efficiency and study of influence factors in typical microwave heating process. *Energy* 1 (87): 678–685.

Welton, T. (2018 June). Ionic liquids: a brief history. *Biophys. Rev.* 10 (3): 691–706.

Wierschem, M., Walz, O., Mitsos, A. et al. (2017A). Enzyme kinetics for the transesterification of ethyl butyrate with enzyme beads, coated packing and ultrasound assistance. *Chem. Eng. Process. Process Intensif.* 111: 25–34.

Wierschem, M., Skiborowski, M., Gorak, A. et al. (2017B). Techno-economic evaluation of an ultrasound-assisted enzymatic reactive distillation process. *Comput. Chem. Eng.* 111: 25–34.

Index

Note: *Italic* page numbers refer to *figures* and **Bold** page numbers reference to **tables**.

t